AF698223
WITHDRAWN

SYMPOSIA IN THIS SERIES

[1] Virology in Agriculture
May 10–12, 1976
Published, 1977

[2] Biosystematics in Agriculture
May 8–11, 1977
Published, 1978

[3] Animal Reproduction
May 14–17, 1978
Published, 1979

[4] Human Nutrition Research
May 8–9, 1979
Published, 1981

[5] Biological Control in Crop Production
May 18–21, 1980
Published, 1981

[6] Strategies of Plant Reproduction
May 17–20, 1981
Published, 1983

[7] Genetic Engineering: Applications to Agriculture
May 16–19, 1982
Published, 1983

[8] Agricultural Chemicals of the Future
May 16–19, 1983
Published, 1984

[9] Frontiers of Membrane Research
May 20–24, 1984
Published, 1985

[10] Biotechnology for Solving Agricultural Problems
May 5–9, 1985
Published, 1986

FORTHCOMING SYMPOSIUM

[11] Instrumentation in Agricultural Research
May 4–8, 1986

Beltsville Symposia in Agricultural Research

[10] Biotechnology for Solving Agricultural Problems

Patricia C. Augustine, Harry D. Danforth,
Murray R. Bakst, Editors

Invited papers presented at a symposium held May 5–9, 1985, at the Beltsville Agricultural Research Center (BARC), Beltsville, Maryland

Organized by the BARC SYMPOSIUM X COMMITTEE
Judith B. St. John, Chairman

Sponsored by
THE BELTSVILLE AGRICULTURAL RESEARCH CENTER
Agricultural Research Service
United States Department of Agriculture

1986 **MARTINUS NIJHOFF PUBLISHERS**
a member of the KLUWER ACADEMIC PUBLISHERS GROUP
DORDRECHT / BOSTON / LANCASTER

S
494.5
.B563
B56
1986

Distributors

for the United States and Canada: Kluwer Academic Publishers, 190 Old Derby Street, Hingham, MA 02043, USA
for the UK and Ireland: Kluwer Academic Publishers, MTP Press Limited, Falcon House, Queen Square, Lancaster LA1 1RN, UK
for all other countries: Kluwer Academic Publishers Group, Distribution Center, P.O. Box 322, 3300 AH Dordrecht, The Netherlands

ISBN 90-247-3311-1 (this volume)

Copyright

© 1986 by Martinus Nijhoff Publishers, Dordrecht.

All rights reserved. No part of this publication may be reproduced, stored in a retrieval system, or transmitted in any form or by any means, mechanical, photocopying, recording, or otherwise, without the prior written permission of the publishers,
Martinus Nijhoff Publishers, P.O. Box 163, 3300 AD Dordrecht,
The Netherlands.

PRINTED IN THE NETHERLANDS

FOREWORD

The Annual Beltsville Symposium provides a forum for interaction among scientists involved in research that has vital impact on agriculture and on the agricultural sciences. The 10th Symposium in the series, Biotechnology for Solving Agricultural Problems, focuses on the use of a revolutionary new set of tools, biotechnology, and attempts to define the set in terms of its applications in agriculture.

Biotechnology has already contributed to the genetic improvement of agricultural products. Procedures that were impossible to test or to implement in the past because of technological limitations are now routinely used by many scientists. Four areas that have benefitted from advances in biotechnology are covered in the symposium proceedings. These areas include genetic manipulation, nutrition, health and disease, and natural resource management.

The 31 invited speakers have identified programs of basic and applied research on plants, animals, and insects that fall within these broad areas. Their research strategies included such techniques as germline modification, gene mapping, monoclonal antibody production, and gene transposition. These strategies have tapped new well springs of information and technologies ranging from the regulation of gene expression (and with it, the regulation of development, growth, disease resistance, and nutrient metabolism) to degradation of pesticides and toxic wastes. The applications of biotechnology to agricultural research have opened virgin vistas with enormous potential.

The new biotechnological techniques and those that will evolve with their use will contribute markedly to the capacity of the agricultural sciences to advance the well-being of the human race.

WALDEMAR KLASSEN, Director
Beltsville Area

CONTRIBUTORS AND THEIR AFFILIATIONS

Paul Armstrong
330 Bellerica Road
Chelmsford, MA 01824
U.S.A.

Dean Befus
Department of Pathology
McMaster University
Hamilton, Ontario, L8N 3Z5
CANADA

Winston J. Brill
Cetus Corporation
Middleton, WI 53562
U.S.A.

Bruce C. Carlton
ECOGEN, Inc.
Princeton, NJ 08540
U.S.A.

Anada M. Chakrabarty
Department of Microbiology and Immunology
University of Illinois
Chicago, IL 60612
U.S.A.

Victor Corces
Department of Biology
John Hopkins University
Baltimore, MD 21218
U.S.A.

James B. Courtright
Department of Biology
Marquette University
Milwaukee, Wisconsin 53233
U.S.A.

Robert J. Cousins
University of Florida
Gainesville, Florida 32611
U.S.A.

Harry D. Danforth
USDA Agricultural Research Service
Animal Parasitology Institute
Protozoan Diseases Laboratory
Beltsville, MD 20705
U.S.A.

Flora DePablo
National Institutes of Health
National Institute of Arthritis, Diabetes, and Digestive Kidney Diseases
Bethesda, MD 20014
U.S.A.

David A. Evans
DNA Plant Technology Corp.
Cinnaminson, NH 08077
U.S.A.

Rudolf Fries
Department of Biology
Yale University
New Haven, CT 06520
U.S.A.

H. Ray Gamble
USDA Agricultural Research Service
Animal Parasitology Institute
Helminthic Diseases Laboratory
Beltsville, MD 20706
U.S.A.

Jon Gordon
Department of Ob-Gyn and Reproductive Science
Mt. Sinai Hospital
New York, NY 10029
U.S.A.

Terrence Graham
Biological Control
Monsanto Company
St. Louis, MO 63166
U.S.A.

Jonathan Gressel
Plant Genetics Department
Weigmann Institute of Science
Rehovot,
ISRAEL

Robert J. Griesbach
USDA Agricultural Research Service
Horticultural Science Institute
Florist and Nursery Crop Laboratory
Beltsville, MD 20705
U.S.A.

Dean H. Hamer
National Institutes of Health
National Cancer Institute
Bethesda, MD 20014
U.S.A.

Lawrence A. Johnson
USDA Agricultural Research Service
Animal Science Institute
Reproduction Laboratory
Beltsville, MD 20705
U.S.A.

Jeffrey S. Karns
USDA Agricultural Research Service
Agricultural Environment Quality Institute
Pesticide Degradation Laboratory
Beltsville, MD 20705
U.S.A.

Terry B. Kinney
Administrator
USDA Agricultural Research Service
Washington, DC 20250
U.S.A.

Noel T. Keen
Department of Plant Pathology
University of California
Riverside, CA 92521
U.S.A

Christopher J. Lamb
The Salk Institute for Biological Sciences
San Diego, CA 92138
U.S.A.

Joan K. Lunney
USDA Agricultural Research Service
Animal Parasitology Institute
Helminthic Disease Laboratory
Beltsville, MD 20705
U.S.A.

Clement L. Markert
Department of Biology
Yale University
New Haven, CT 06520
U.S.A.

Bernard Moss
National Institutes of Health
National Institute of Allergy and Infectious Diseases
Bethesda, MD 20205
U.S.A.

Karel Newman
Salsbury Laboratories, Inc.
Charles City, IA 50616
U.S.A.

J. B. Neilands
Biochemistry Department
University of California
Berkeley, CA 94720
U.S.A.

Frank Press
President
National Academy of Sciences
Washington, DC 20418
U.S.A.

Gerard E. Riedel
Genetics Institute
Boston, MA 02115
U.S.A.

Eladio Vinuela
Centro Biologia Molecular
Universidad Autonoma Canto Blanco
Madrid 34,
SPAIN

Lila O. Vodkin
USDA Agricultural Research Service
Plant Genetics and Germplasm Institute
Seed Laboratory
Beltsville, MD 20705
U.S.A.

SYMPOSIUM ORGANIZATION

Beltsville Agricultural Research Center (BARC)
Waldemar Klassen, Director, Beltsville Area

BARC SCIENCE SEMINAR COMMITTEE
John G. Moseman, Chairman
Patricia C. Augustine
Gordon C. Carpenter
Alan Lefcourt
Barbara A. Leonhardt
Otho E. Michaelis, IV
David A. Nickle
Steven L. Sinden
John Vandenberg
William S. VanDerWoude

BARC SYMPOSIUM X SUBCOMMITTEE
Judith B. St. John, Chairman
Ralston Rhodes, Co-Chairman

PROGRAM COMMITTEE
James D. Anderson
Rufus L. Chaney
Edward T. Engman
H. Ray Gamble
Dora K. Hayes
R. Heutill
Jeffrey S. Karns
Philip C. Kearney
L. David Kuykendall
Joan K. Lunney
Karl N. Norris
Robert A. Owens
Lowell D. Owens
M. J. Paape
Caird E. Rexroad
Michael D. Ruff
Claude Veillon
Lila O. Vodkin

REGISTRATION COMMITTEE
James A. Saunders

FINANCE COMMITTEE
Elizabeth Anderson
William P. Wergin

LOCAL ARRANGEMENTS
Ralston R. Rhodes
Elizabeth Anderson
Nicki O'Neill
William S. VanDerWoude

LOGO
Robert B. Ewing
Vincent Hughes

PUBLICITY
Steven M. Berberich

POSTER SESSION
Judith A. Abbott
Freddi A. Hamerschlag

EDITORIAL
Patricia C. Augustine, Chairman
Harry D. Danforth
Murray R. Bakst

CONTENTS

Biotechnology in perspective

Good morning, ladies and gentlemen, and welcome to "Biotechnology for Solving Agricultural Problems," the tenth in the series of Beltsville Agricultural Research Center Symposia.

The Agricultural Research Service is pleased to sponsor this symposium on the rapidly changing world of biotechnology. Over the next 4 days, 29 international experts in molecular biology and genetic engineering will provide us with a current look at the promise of biotechnology for improving agriculture, human nutrition, and the environment.

To say that this area of research is dynamic is an understatement. In fact, research in the biotechnologies is so rapid that articles written in scientific journals are often eclipsed by new developments before the articles are off the press. Biotechnology promises to yield an infinite number of improvements in just about every enterprise from health care to waste management. Most predictions point to agriculture as the industry that will reap the greatest benefits.

For the purpose of my remarks this morning, I suggest that we define biotechnology simply as those biological means used to develop processes and products employing organisms or their components. Those biological means include:

1. Bioreactors and bioreactor support systems (bioreactions).
2. Immobilized cells, organelles, or cell components (enzymes).
3. Plant and animal tissue and organ culturing.
4. Recombinant DNA (genetic engineering, gene transfer).
5. And, hybridoma techniques.

These new techniques – microculture, cell fusion, regeneration of plants from single cells, and embryo recovery and transfer – are creating vast new opportunities all across the agricultural sciences. The tools of biotechnology are being applied to viruses and bacteria, to insects and weeds, to plants and animals, in all stages of their life cycles from replication through aging.

ARS progress in biotechnology goes back decades, long before the concept was so named. A milestone achievement came from classic research in photoperiodism. This pioneering effort, begun by Wrightman Garner and Harry Allard in 1918, culminated when Harry Borthwick and Sterling Hendricks showed that flowering and seed formation are

Remarks by Dr. Terry B. Kinney, Jr., Administrator, Agricultural Research Service, U.S. Department of Agriculture, before the BARC Symposium X "Biotechnology for Solving Agricultural Problems," Beltsville, Maryland, May 6, 1985.

controlled by a chemical – now called phytochrome. And phytochrome can be manipulated as desired to switch plant growth on or off.

The ground-breaking work of these four scientists pointed the way toward exploiting other natural substances to control plant growth and development. Today, this field of research – conducted at the level of cells and molecules – is called bioregulation.

Another ARS milestone in biotechnology was deciphering the molecular structure of a ribonucleic acid, or tRNA, by a team of agency and Cornell University scientists. That achievement won the research team's leader, former ARS biochemist Robert W. Holley, a share of the 1968 Nobel Prize for medicine or physiology.

The team's work enabled other scientists to determine the structure of other tRNAs. A few years later, the method was modified to identify the sequence of nucleotides in various bacterial, plant, and human viruses. Modified further, the Holley team's approach is playing a role in determining DNA sequences in today's chromosomal research.

Other examples of ARS achievements in biotechnology include:

* A vaccine against foot-and-mouth disease, developed through recombinant DNA technology, with collaboration by scientists from Genentech.

* New rice plants, developed through tissue culture, with more and better quality protein.

* Gene transfer, through recombinant DNA technology, from one plant through a bacterium to another plant.

* A vaccine against Marek's disease developed with cell culture techniques. Since 1971, it is estimated that this vaccine has saved the U.S. poultry industry from losses of 2 billion dollars.

* A genetically engineered antigen which helps protect chickens against one parasite that causes coccidiosis. This disease, which afflicts chickens and turkeys, is estimated to cost the poultry industry 300 million dollars a year in lost production and medication.

ARS research on cell membranes is internationally known. A top membrane research priority is developing crop varieties that "harvest" more sunlight, resulting in healthier, more efficient plants and larger yields.

Other current ARS research involving biotechnology includes:

* Devising a way to microinject genetic material into plants whose tough cell walls currently limit the practice to animals and human cells.

* Genetically engineering a vaccine against vesicular stomatitis, a viral disease of livestock that also affects humans.

* Transferring organelles and their DNA between plant species so that breeders can raise crop yields or impart resistance to herbicides, pests, or diseases.

The ARS 1985 budget includes 26 million dollars for research in biotechnology. Some 200 ARS scientists are focusing on biotechnology in 165 projects at laboratories throughout the United States.

ARS will expand biotechnology research at two major locations – here at Beltsville and at the Western Research Center in Albany, California.

Last year, a new Plant Gene Expression Center was opened at the Albany facility. Charged with accelerating genetic engineering of crops, the new center will receive Federal funding of 4 million dollars annually, with that amount matched by industry and various public institutions.

Here at Beltsville, the funding support has reached nearly 8 million dollars per year to intensify biotechnology endeavors.

We also increased the number of postdoctoral scientists in our elite Research Associates program from 26 last year to 50 in 1985. We can hire these scientists without the usual paperwork. They will concentrate mainly on biotechnology projects in the animal and plant sciences during tours of employment with the agency lasting from 1 to 2 years.

Major efforts are being placed on finding ways to culture single cells from a wide number of agronomic plants and to regenerate them into whole organisms with the genetic message intact and properly expressed. In animals, embryonic single cells – the fertilized egg – may well be the recipients for gene transplants. Modern techniques of embryo recovery, splitting, and transfer are playing critical roles in genetic engineering efforts.

Other new and promising techniques now make it possible to determine much more quickly the precise chemical makeup of genes and their protein products, condensing into days what might once have required years to accomplish.

Genes can also be located in their chromosomal packages relative to each other much more easily than in the past. More difficult, however, is the identification of the mechanisms regulating gene expression – what turns any given gene's activity on or off. Finding the answers will require years of experimentation.

It is not gene transfer capability that will constrain biotechnological solutions to problems. Rather, it is the lack of base data relating to gene identification for selection of genes to transfer, gene expression, plant differentiation control, and a host of other functions.

A number of research areas in agriculture are being considered as candidates for the new biotechnologies – research areas that were not approachable before recent technological breakthroughs. For example, new techniques now allow much more detailed study of the defense systems in plants and animals than was possible in the past. New knowledge of the immune system in animals promises improved approaches for disease and parasite control and faster, more accurate diagnosis.

We envision more effective ways for increasing the plant's or animal's own genetic resistance to pests and other stresses. Conversely, improved techniques for sabotaging the defense systems of insects, weeds, and other pests also seem possible.

A growing understanding of how the chemical messengers we know as hormones operate as intermediaries in growth and development processes also shows promise. This is an area that has been studied extensively and one that has produced much progress in the past.

Another area of challenge is to develop entirely new food products using plant protein. This is a challenge to food scientists – to produce an acceptable product that does not attempt to mimic animal protein. The result could give us entirely new foods unlike anything we consume today.

Let me suggest to you a few examples of major innovations that I believe will be available within the next 20 years.

* Crops that will be less susceptible to diseases caused by viruses, bacteria, fungi and insects.

* More efficient crops that will better absorb and use fertilizer.

* Crops that are genetically modified to fix their own nitrogen.

* An increase in plants' photosynthetic efficiency by manipulating their energy conversion systems.

* A greater resistance to stresses brought on by drought, salinity, chill, and frost.

Genetic regulation of plant growth to achieve such goals as:

* Higher ratios of edible to nonedible parts, longer seed filling times, corrected structural weaknesses, and higher yields of economically important plant constituents.

* Improved food quality, such as improving the amino acid balance and nutritive value of small grains for animal and human consumption.

* Regulation of plant growth to allow harvest of fruits and vegetables of uniform ripeness. This will help maintain and deliver desired quality produce to consumers through complex systems of transport, processing and marketing.

Applied and fundamental research in agriculture are so intimately intertwined in the biotechnologies that many practical spin-offs may emerge. At the same time, fundamental inquires will continue to pursue the basic knowledge necessary to make wide-ranging applications.

Thus far, my remarks have concerned the relatively short-term application of the new biotechnologies over the next 15 to 20 years.

Now let us consider the impact of the new biotechnologies over a longer timeframe – 50 to 100 years. Some futurists – both outside and within ARS – have described a totally revised system of agriculture in the mid-21st Century.

In the interest of conserving our natural resources – soil, water and fossil fuels – needed for long-term sustained agricultural production, they foresee more conservative production or components of foods by biotechnological or combined systems as opposed to producing plant organs.

They visualize reconstitution of the components into consumer-acceptable traditional foods.

They foresee these technologies deriving food components from chemical feed stocks made from nonedible parts of plants such as celluloses and lignins. These would then be converted, for example, into sugar syrups.

These syrups might then be moved by pipeline or tank car to biotechnological food production facilities close to population centers in any part of the country.

This is the agricultural equivalent of reaching the moon. As a research agency, ARS has the responsibility to develop all the future food production options that we can. I expect that the economics of agricultural production – and the availability of water, land and energy – will shape the future application of these biotechnologies.

Having provided a brief perspective on some of the exciting challenges and opportunities presented by the biotechnologies – where we are now, and where I think we are going – let me conclude these remarks with three personal observations.

First, I remind you that biotechnologies are not a panacea. They will not displace traditional plant breeding anytime in the near term – perhaps never. Biotechnologies are tools . . . but they are not our only tools.

Specifically, the biotechnologies will provide the means for creating new and pure sources or combinations of genetic variation having specific desired traits. Certainly the use of biotechnology will allow germplasm enhancement and plant breeding to proceed at a faster rate with a more predictable outcome. But, we still have to use classical breeding techniques to develop new varieties.

Second, I suggest that recommendations for new biotechnology centers – both at the Federal and State levels may not represent the best use of our limited funds. Rather than investing in more brick and mortar, I believe in putting the money directly into research. I think it is more important to give the money to the scientists to pursue the research, rather than build yet another tower.

Therefore, I think that our research system needs to have a low tolerance for the administrator – or the dean – or the vice president for research – who sees the new biotechnologies primarily as opportunities to develop more real estate. I don't know what to call this point. It's not intellectual honesty, and it's not academic honesty. Maybe it's administrative honesty.

Third, I want to reflect my concern that America's higher education system be stimulated to supply the large numbers of talented young scientists trained in the biotechnologies who will be in demand for the balance of the century. The Federal system, the State system, and the private sector are presently competing avidly for the available talent – and this competition is very aggressive. Those of us in the public sector need to be equally assiduous in recruiting and retaining first rate scientists.

Some people assume that pumping enough new dollars into the agricultural research system will automatically result in more graduate students and, ultimately, in more scientists of exactly the sort needed. I feel strongly, on the other hand, that higher education is a science unto itself and not just a by-product of research.

In any case, it behooves all of us working in agricultural research to give increasing attention to the educational system. Our universities need to turn out more young scientists – especially with graduate training in the biotechnologies – if the United States is to have a pool of scientific talent adequate to meet the demands of the research marketplace.

I would be very pleased if some clearcut recommendations are made by the symposium concerning our work in biotechnology. The National Program Staff will look for opportunities to translate the report and recommendations of the symposium into a plan of action for using existing agency resources.

You have my very best wishes for a successful symposium.

Part 1

An overview of biotechnology

Recombinant DNA – an introduction

DEAN H. HAMER
National Institutes of Health

Recombinant DNA is not something that has been invented in the last 10 years by scientists. Recombination has probably been around as long as DNA has been around, and occurs naturally in vivo at a fairly high frequency. However, there are limitations to the recombination that occurs in vivo. First, it usually, although not always, occurs only between molecules that are strongly homologous to one another. Therefore recombination in vivo only offers limited scope for diversified genetic information.

A second very important constraint on recombination in vivo is that it occurs only when two DNA molecules get inside the same cell. In general, although again not always, that only occurs if the cells come from the same species, i.e. from two organisms that are capable of mating with one another. In contrast, the new recombinant DNA technology allows scientists to recombine DNA from completely different and diverse species. The reason for this is that the joining reaction is done not inside of cells as in vivo but rather inside of test tubes.

In general, in a recombinant DNA experiment, there are three essential components. The first is called the vector DNA molecule; it is the molecule that will act as the vehicle or the carrier for the genes that are to be cloned. The second component is the so-called foreign DNA. This is the DNA that is to be cloned and manipulated. Finally there is the host cell that acts as the in vivo recipient of the recombinant DNA molecules.

In brief, the foreign DNA is cut into approximately gene-size pieces. The usual way to do it is by use of a restriction endonuclease which is an enzyme that cleaves DNA molecules at specific short nucleotide sequences. Likewise, the vector is cleaved with a restriction endonuclease that specifically opens up the DNA molecule at a non-essential site. These two DNA molecules then are mixed together and are treated with an enzyme, DNA ligase, which has the capability of joining the DNA molecules at their terminae. The essential point is that DNA ligase does not recognize the sequence of DNA molecules, but only their ends, and therefore it has no problem with joining together molecules that are from diverse species or diverse sources. Once the recombinant molecules are constructed in vitro, they must be introduced back into living cells in order to be propagated. For that purpose, they are introduced into the host cells from which the vector was originally derived.

There are several critical features of these three components. First, what are the important features of the vector? There are really two important features. One, the vector molecule has to be capable of replicating itself. One way in which replication can be accomplished is by the use of extra-chromosomal elements, pieces of DNA which normally exist independent of the chromosome. The two best known examples of extra chromosomal elements are plasmids, small circular DNA molecules which are found in many species of bacteria and which typically carry traits such as drug resistance, and viruses. Virtually all viruses exist at one point independently of the chromosome and have the information for replicating themselves. Both plasmids and viruses contain origins of DNA replication and in some cases, carry transacting factors, for example polymerases, that are required for replication.

Another type of vector that is useful is one that instead of replicating independently of the chromosome, actually integrates into the chromosome. These vectors thus act essentially as parasites, using the host information to replicate. Examples of this are transposable elements, which were originally discovered in plants, and certain viruses which at some point in their life cycle, are capable of integrating into the chromosome.

The other essential feature of the vector is that it must have some trait that can be scored and selected for, so that cells that have received the recombinant DNA molecules can be distinguished from those that have not. There are many examples of this. One of the best known is antibiotic resistance. Cells can simply be plated on a medium containing the antibiotic and thus be selected out. Another example is the ability to form viral plaques, another readily recognizable feature.

With regard to the host cell, there is really only one restriction; the host must be able to take up DNA in one form or another. There are some organisms that naturally take up DNA as part of their life cycle, but this is not true for the majority. In order to use these latter organisms, the scientist must use different techniques to "trick" the cells into taking up DNA. One technique is to incubate the cells with various divalent cations, such as calcium or lithium. This essentially punches holes into the membrane so that DNA can enter the cells. Additionally, there are more complicated techniques such as protoplast fusion with polyethyleneglycol and electrophoresis of DNA molecules into cells by putting them into a directed electric field.

There are no restrictions on the third component in recombinant

DNA experiments, the foreign DNA. Essentially any piece of DNA from any species and of any size can be used.

The next question is what are the applications of DNA recombination? The most powerful application is the isolation and purification of individual genes. If you think about purifying a gene, it is really a rather daunting thought. Most organisms contain thousands to millions of genes in each cell, and from a biochemical or biophysical standpoint, all of these genes look very much the same. They have the same composition and differ only in the order of the nucleotides. Purification of genes by conventional biochemical techniques would, in fact, be impossible. With gene cloning, however, an investigator may start with a mixture of several different DNA fragments but end up with a single cell containing a single recombinant DNA molecule or many identical copies of that molecule. From this single cell, a clone of cells can be grown up to any desired size and every cell will contain exactly the same DNA molecule. In this way, the molecule is purified.

The problem that arises at this point is that although each clone contains a single DNA molecule, if you started out with a million different genes you are going to end up with a million different clones. In a typical experiment where DNA has been cloned into a bacteria phage (lambda) then plaqued out, thousands or millions of plaques could be formed that contained an entire genome's equivalent worth of DNA. The problem, of course, is how to decide which of these plaques contains the particular gene that we are interested in.

There are numerous techniques to solve this problem. One of the most powerful is DNA hybridization. The procedure involves the transfer of clones from a culture plate onto a nitrocellulose filter, saving a master copy of the plate. The filter is treated in such a way that the bacteria and plaques are lysed and the DNA is immobilized on the filter. This filter is then incubated with radioactively-labeled DNA that contains the gene or sequence of interest. Where the labeled DNA is complimentary to the DNA on the plaque it will hybridize or hydrogen bond and, following a washing protocol, only those plaques that contain the gene of interest will contain radioactivity. When the filter is exposed to X-ray film, the radioactively-labeled plaques can be identified (Fig. 1).

Of course, you might ask "Where did you get the radioactively-labeled DNA?" Again, there are several approaches. One is that if you know anything about the structure of the protein and any of its sequence, you can predict from the genetic code where the DNA should be. The prediction may be slightly ambiguous due to the ambiguity of the code, but by choosing appropriate parts of the protein sequence, you can have a limited number of DNA sequences. These DNA sequences

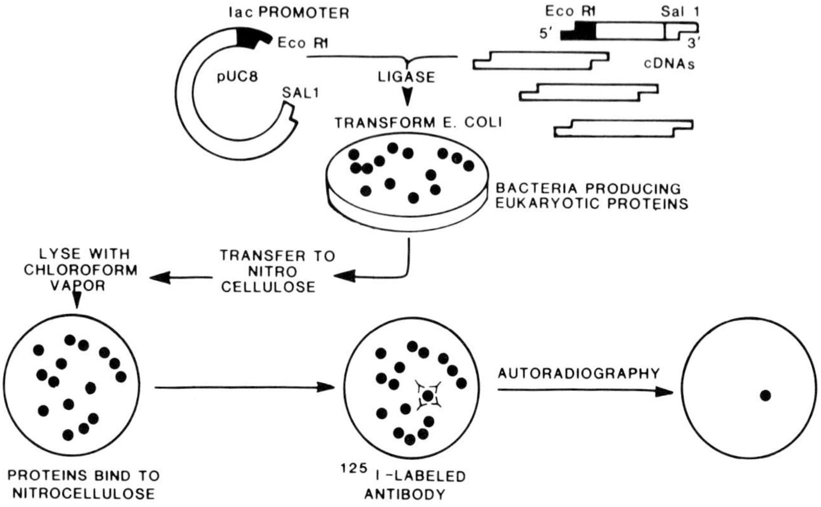

Figure 1. Identification of genes by DNA hybridization techniques using a radioactively-labeled DNA probe.

can actually be synthesized by standard organic chemical techniques. This used to be a very difficult problem but there are now commercially available machines that allow the synthesis of 40 or 50 oligo nucleotides with relatively little difficulty. Such oligo nucleotides can be used as probes.

Another approach which is very powerful is the so-called plus/minus method. Here you need to know a tissue or cell type wherein the gene of interest is expressed strongly and another cell type where it is not expressed. Total RNA is then prepared from the two different cell types and labeled with P^{32}. The plus and minus populations are then hybridized to the filter, and the spots which hybridize with one population and not with the other are identified. Basically as long as you know something about the protein or where it is made, you can find the clone of interest.

Still another approach involves the use of a vector such as *Escherichia coli*, which allows the expression of inserted foreign pieces of DNA. The vector contains a promoter that will start transcription. It also contains sites for ribosome binding and for termination of transcription. The population of DNA molecules is inserted into this vector and subsequently makes plaques. To determine which of those plaques are making the protein of interest, antibodies against the protein are labeled directly or indirectly, and applied to the filter. Again, only the plaque or colony of interest will show the label. Using this approach, the

the protein need not be purified or sequenced. Only an antibody against the protein is necessary for the isolation of the gene of interest.

Finally, there is a one-third approach to isolating specific clones by making use of their function. In some experiments this is very simple. For example, to clone an *E. coli* gene for an enzyme, a library of *E. coli* clones would be made and inserted into host cells that contained a mutation in this gene. Cells that were able to grow in the absence of the substrate would be selected. The approach works well with homologous experiments, *E. coli* genes into *E. coli*, or perhaps other bacterial genes into *E. coli*. However, it does not work well with heterologous experiments; the reason being that the control sequences that start transcription, for example, are very different among organisms, so the heterologous gene may not work in *E. coli*.

There is an alternative approach, which actually exhibits one of the most powerful applications of this technology – namely the cloning of oncogenes (genes that are involved in cancer and tumor formation, in this case in human beings). Until relatively recently, it was known that there were genes that caused tumor formation, but it was not known what they encoded, what proteins were involved, exactly how the gene worked, or what the sequence was.

Nevertheless, investigators were able to clone these genes. DNA from a tumor cell, in this case from a bladder carcinoma line, was cleaved into a million or so different genes. The DNA was ligated to a marker gene, in this case a bacterial antibiotic resistance gene, so that every one of the genes contained a bacterial marker. This whole mixture was transferred into mammalian cells. The cells were then incubated and foci or transformed cells were isolated. Perhaps one or two out of these millions of different genes got into the mammalian cells and caused a neoplastic event.

The transformed cells were then isolated, the DNA was extracted from them, and reinserted into mammalian cells as a second round of purification or enrichment. Finally DNA was extracted from the transformed cells arising from this round and put back into *E. coli*. The only gene in these transformed cells that should contain a bacterial marker gene is the one that actually caused the focus formation to occur. This gene can be taken out of the mammalian cells, put back into a phage vector, and plaqued onto *E. coli*. Those few plaques that form contain the gene that was responsible for transforming these cells neoplastically. The cloned gene can be put back into mammalian cells. The increase in the frequency of retransfer from one out of a million to one out of one indicated that the oncogene had been cloned.

Collectively, these approaches demonstrate a very important

application of recombinant DNA, namely the ability to clone and purify individual genes. Once the gene is cloned, it then provides a critical substrate first for studying its structure, second for studying its expression, and lastly for understanding the physiology of its gene products. This technology is not limited to animals. It is applicable to plants and has been used to clone a whole variety of different plant genes.

A second application of recombinant DNA technology is the use of recombinant DNA to detect mutations in DNA. We frequently encounter situations where we know from genetics that the expression of a trait is caused by one gene, two genes, or perhaps more. We may also be able to infer from the genetic behavior whether the gene is chromosomal or extrachromosomal, and whether it is unstable or stable. However, detection of changes and the mutations at the molecular level requires recombinant DNA technology.

For example, a specific mutation in a globin gene causes sickle cell anemia to occur – the so-called sickle cell trait. This particular mutation is a change of a single base. There is a restriction at the nuclease site which encompasses this particular change. It is a site for a particular enzyme called Dde1, one of almost 100 different enzymes that recognize different sites in DNA. The sickle cell lacks this site. To recognize the mutation that gives rise to sickle cell anemia, total DNA is isolated from a normal person and from a sickle cell person. The DNA is cleaved with the enzyme, Dde1 and the fragments are stretched out according to length by gel electrophoresis. The gel will have a million different fragments; it would be impossible to visually determine differences between normal and sickle cell DNA. Therefore, the DNA on the gel is hybridized to a radioactively-labeled cloned probe that is specific for the globin gene. Only the fragments which contain the globin gene will become radioactive and show up on an autoradiogram. With a normal individual, two different bands will be detected because there is a Dde1 restriction site in the gene. A sickle cell individual will give only a single band because there is no Dde1 site.

Again this technology can be applied equally well to plants, and may be used to detect unstable elements and mutator elements in plants. The technology has recently developed even further and, in fact, it is possible by using oligonucleotides to detect single base changes in the DNA even if they do not alter any restriction site. One simply hybridizes with an appropriate oligo and then washes under conditions which can distinguish between a single mismatch.

A third application of recombinant DNA is the use of this technology to overproduce useful or interesting gene products. Often in biology the gene products of the most interest are the ones that are the hardest; for example, receptors – genes whose products are present in small amounts are present in small amounts and act as signals inside the cell. Recombinant DNA technology can be used to isolate such proteins and make them in large quantity. For example, a human hormone called somatostatin can be produced in large quantities in bacterial cells. The basic idea is to take advantage of the fact that the genetic code is a universal code. Therefore, coding sequences can be translated in any organism and express the same protein. The difficulty, of course, is that although the genetic code is universal, the epigenetic code or the code that determines how genes are expressed is not universal and differs from organism to organism. Therefore, the coding sequence must be presented to the organisms in such a way that it is considered a naturally occurring gene and will be translated with high efficiency.

To achieve this presentation, the somatostatin coding sequence was inserted into a specialized vector which contains several critical features. In this case, the vector is an intact bacterial gene containing its own promoter and termination sites; this particular gene encodes the product betalactamase. The coding sequence for somatostatin was inserted so it would be in frame with the lactamase gene. As a result, the translation, now inside *E. coli* cells, would yield a hybrid protein, mostly containing betalactamase, but at its carboxy terminus, containing a very short peptide corresponding to somatostatin. Short peptides like this would be unstable if synthesized alone and would be very rapidly degraded. When this peptide is incorporated into a large, normally folded protein, it is stable.

The hybrid molecule is extracted from the bacteria and treated with cyanogenbromide which cleaves at methionine residues. A major constraint of this method is that you cannot have methionine residues inside your protein sequence. In this case the only methionine residue in the protein occurs between the betalactamase and the somatostatin molecules. With the cleavage, large quantities of the hormone of interest, somatostatin, are generated in a pure form.

Again, technology has advanced appreciably. It is now possible to use this technology to produce virtually any protein of interest. Investigators have developed techniques that are more efficient than cyanogen-bromide cleavage for separating hybrid molecules. Specific linkers of protein sequences that are susceptible to specific proteases have been engineered. For example only collagen is susceptible to collagenase, so it is a simple matter then to make a hybrid molecule that is linked with collagen, treat it with collagenase, and isolate the protein of interest. This technology could be used to engineer virtually any protein of interest.

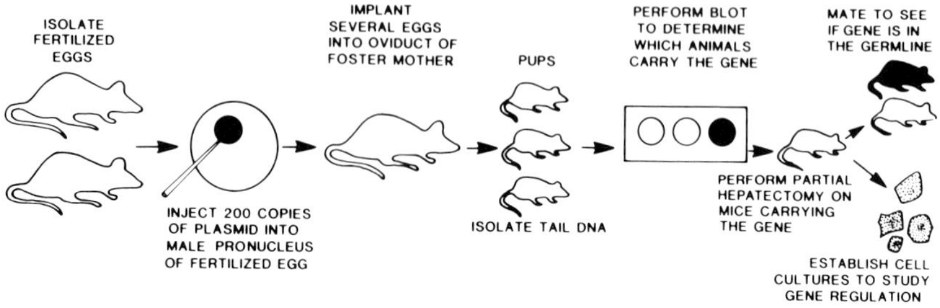

Figure 2. Use of recombinant DNA technology to change the genotype and phenotype of an intact organism.

The final and perhaps one of the most dramatic applications of recombinant DNA technology is the engineering of organisms with new genotypes and therefore new phenotypes. This, of course, is a technology that is of great interest both in the study of basic molecular biology, for example, in the study of development, and for practical purposes such as custom-tailoring organisms to your own desires.

Such an experiment has been done in mammalian cells. A fertilized egg was extracted from a mouse at the one-cell stage and microinjected with DNA solution using a small needle and a micromanipulator. When this was done, about 50% of the eggs that were injected actually received DNA molecules. The fertilized and injected egg was then inserted into a sham-pregnant female mouse. The mouse was allowed to give birth, and the pups were assayed for the presence of the foreign gene. In fact, in this experiment, one of three animals incorporated the foreign gene into the germline. When the animal was backcrossed, the pups were found to be heterozygous with regard to the gene. The heterozygotes could then be backcrossed and ultimately produce homozygous strains.

The question then is does the DNA molecule actually work inside the mice. In an experiment involving a gene for rate growth hormone, the gene was put under the control of a heterologous promoter sequence. In this case the promoter was from a metalthymine gene (a gene expressed very strongly in the liver and kidney). The mice that received the growth hormone gene grew rapidly and attained about twice the size of siblings that did not contain the gene. This is a spectacular example of how recombinant DNA technology can be used to change not only the genotype, but also the phenotype of an intact organism (Fig. 2).

Plant biotechnology also fits into this picture. Several different systems for transferring genes into plants are now under investigation Certainly, the best known is the TI system. There are certain bacteria which grow on the root nodules of specific legumes. These bacteria contain plasmids which are capable of integrating into the chromosomal DNA. When that occurs, crown gall, which is essentially a tumor of the plant, develops.

Many groups now have been working on the molecular biology of these bacterial plasmids because they seem to offer a natural way for inserting genes into plant cells. The plasmids are large and contain many different genes. It has been possible by deletion mutation analysis to localize specific regions of the DNA that are involved in processes such as integration into the chromosome. So, by putting foreign genes next to this region, one can construct plasmids where foreign DNA genes are capable of being transmitted into plant DNA.

Plant cells and eventually entire plants that contain an integrated gene can be generated. This technology that is very powerful and allows one to go from cloned plant genes back to entire organisms.

Other technologies are also being developed. One of them is the use of viral vectors, particularly the use of a virus known as cauliflower mosaic virus. This would be much like the use of phage in *E. coli* or the use of animal DNA tumor viruses like SV40 in mammalian cells. There are some difficulties with viral vectors in that there are no large non-essential regions of the genome and that there are packaging restraints on the virus. These problems may be solved using the same techology that was used in the mammalian cells – that is, by making host cells that express all the information that is required.

A third technology that will certainly be developed is the use of transposable elements. These are elements which are capable of moving in and out of chromosomes. If foreign DNA is put in the middle of a transposable element with its inserting parts intact, the element should be able to move the foreign DNA into different positions on the chromosome. This technology has already proven to be tremendously powerful in *Drosophila* with the use of P elements, and has been used to generate flies that contain foreign information.

There are two major limitations in the field of plant biotechnology that will need to be addressed in the near future. The first is the absence of a so-called transient assay system. This is a system whereby DNA molecules can be inserted into cells, and the gene expression examined within a few hours or a few days. Why is that important? It certainly is no way to generate new plant species. The answer is that this technology is essential for determining how DNA molecules work; it will reveal basic information such as which nucleotides are important for transcription or translation and which amino acid residues are required for a given function.

As an example, studies on the metallaothionine gene of mammalian cells were conducted to determine which sequences in front of the gene are important for transcription and how they bind to regulatory factors. A number of mutations were produced in the gene. The genes were put back into cells, and using a transient assay, their transcription and ability to bind or compete for regulatory factors were assessed. From that type of analysis we found that there is a repeated oligonucleotide sequence which is the main determinant for transcription for bonding to regulatory factor. This same experiment if conducted with TI plasmids would take a much longer time.

The second very important technology that is needed is the development of systems for homologous recombination in plants. This technology is necessary for the manipulation of genes on the chromosome. The ability to manipulate genes is absolutely essential for understanding what genes are doing physiologically in vivo and also very important for engineering plants that have desired characteristics. Homologous recombinant DNA systems would allow the removal of the endogenous gene coding for a protein and its replacement with another version. The techniques currently used do not allow this to occur; they either integrate the cloned DNA into a foreign location or keep the foreign DNA as an extrachromosomal element.

An experiment with yeast shows what can be accomplished with a homologous recombination system. The function of a particular yeast gene, Cup 1, that encodes a copper-binding metallothionine gene in yeast, was unknown. The gene was cloned and then manipulated in vitro by standard recombinant DNA techniques so that all of the coding sequences for metallothionine were replaced with a selectable marker such as antibiotic resistance. Basically a double recombination event occurred so that the metallothionine gene is replaced by the selectable marker. Strains can then be generated which do not contain the metallothionine gene, and function of the metallothionine can be determined by comparison of metallothionine positive and negative strains.

From these experiments, we were able to learn two very interesting things about metallothionine. First, although it is involved in protecting cells against metal poisoning, it is not involved in normal metal metabolism, or in transport, storage, or activation of metallo-enzymes. Secondly, this protein autoregulates itself. In other words, the protein actually turns off its own transcription in the absence of heavy metals – an unexpected finding. Experiments such as this show the usefulness of homologous recombination systems in studying the basic biology of the cell.

Gene mapping in domestic animals

RUDOLF FRIES and FRANK H. RUDDLE
Yale University, Department of Biology, Kline Biology Tower, New Haven CT 06511, U.S.A.

Abstract Gene maps are constructed by the synthesis of data obtained by different methods which include family analyses, somatic cell hybridization, direct mapping of DNA segments by Southern blot analysis and in situ hybridization to fixed metaphase chromosomes. Gene mapping has already contributed significantly to a better understanding of the mammalian genome, in particular the human genome but the gene maps of economically important domestic species are not well characterized. The application of somatic cell genetics and recombinant DNA methodologies now allows rapid progress to be made in the construction of detailed gene maps for domestic animals. Such gene maps will serve as tools for selection in applied breeding and for the analysis of polygenic traits.

Introduction

During the last few decades, animal breeders have been very successful in breeding animals that are superior in the production of milk, eggs, meat and wool. The strategies for optimized breeding of livestock have been based on the concepts of quantitative genetics which assume that there are several genes, possibly on different chromosomes, each of which has a certain effect on a particular quantitative trait. However, we do not know how many genes are actually involved in the expression of a given quantitative trait nor how much each gene contributes to the trait and where it is located on the chromosomes. Such knowledge will become crucial for better success in the breeding of animals with a greater resistance to diseases in unfavorable environments or greater fertility and concomitant increased production of milk, eggs, meat or wool. Modern methods of gene mapping, so far mostly applied in mice and man, offer new ways towards a better understanding of the genetic determination of animal performance.

Geneticists have been very successful in mapping the human chromosomes. Hundreds of genes have been assigned to particular chromosomes (Genetic Maps, 1984) and knowledge gained by human gene mapping has been applied in prenatal diagnosis (Ruddle, 1981). As a result of gene mapping the chromosomal regions containing the defects responsible for Huntington's disease and Duchenne muscular dystrophy have been determined and the isolation of the genes which cause these diseases will become possible in the near future (White et al., 1985). Human gene mapping has also produced important insights into the etiology of cancer by showing that genes which cause cancer are sometimes found near genes that enhance their activity (Croce and Klein, 1985).

In this article we would like to give an overview of the methods and strategies involved in mapping genes and describe the present status of the gene maps of domestic species. Further, we will focus on a potential application of gene maps: the identification and isolation of genes for complex traits.

Methods for mapping genes

There are three different approaches towards the mapping of genes: those that are based on the study of the transmission of genes in families (family studies); those that make use of the possiblity of gene transfer between mammalian somatic cells from different species (somatic cell hybridization, chromosome mediated gene transfer); and those that involve the direct assignment of genes by in situ hybridization of DNA of a specific type to fixed chromosomes.

Family studies. Up to the late 1960's family studies were the only means by which the location of genes on the same chromosome (linkage) and the assignation of genetic loci to particular chromosomes could be carried out. To determine whether two genetic loci are on the same chromosome the meiotic recombination frequency is measured between the loci of interest in informative families, i.e. families in which one parent is heterozygous at the two loci. If the recombination frequency (r) is smaller than 0.50 and if the assumption of random segregation at the loci is correct, one can conclude that the two loci are on the same chromosome. The statistical test applied to determine whether an observed segregation fraction is significantly different from 0.50 is usually the lod score test (Wald, 1947; Morton, 1955). The lod (logarithm of the odds) score is calculated as shown in Fig. 1. The lod scores from different pedigrees can be added and are usually presented in standard lod score tables for the values of r = 0.05, 0.10, 0.20, 0.30, and 0.40. The recombination fraction giving the highest lod value is taken as the maximum likelihood estimate. Lod scores of +3.0 or more, meaning that the likelihood of linkage is at least 1000 times greater than the likelihood of nonlinkage, are considered as sufficient evidence for linkage, while lod scores of −2.0 are sufficient for the rejection of linkage in human studies.

Family studies can also be used for the assignation of a gene to a specific chromosome. In such analyses a chromosomal variant present in heterozygous form is used as one locus in the linkage test. Linkage

$$z = \log \frac{\text{Likelihood of a pedigree, assuming } r < 0.5}{\text{Likelihood of a pedigree, assuming } r = 0.5}$$

or

$$z = \log 2^{(a+b-1)} [r^a(1-r)^b + r^b(1-r)^a]$$

a = Number of recombinant progeny
b = Number of parental progeny
r = Recombination frequency

Figure 1. Calculation of the lod score (z).

detected between a chromosomal variant on the long arm of chromosome 1 with a blood group locus made possible the first assignation of a gene to an autosome in man (Donahue et al., 1968). Fries, Stranzinger, and Vögeli (1983) took a similar approach in an attempt to determine the chromosomal position of several genes in swine. They studied the linkage relationships of natural, as well as radiation induced, marker chromosomes with genetic loci. The outcome of this study was exclusion of linkage upto a level of 30% recombination for several combinations of marker chromosomes and loci for blood groups, enzymes and serum proteins, and the provisional assignment of the G-blood group locus to chromosome 15. The evidence for this assignment is presented in Table 1. There was a positive lod score above the significance limit of 3.0 for the combination of the G-blood group locus and a natural marker chromosome found in hybrids between domestic pigs and European wild pigs (rob (15; 17)). The positive lod scores calculated for the

Table 1 Lod scores for the G-blood group locus and several porcine marker chromosomes

Marker combination	Recombination frequency (r)					
	0.05	0.10	0.20	0.30	0.40	N[a]
G-rob(15; 17)[b]	0.48	2.39	3.36	2.97	1.82	43
G-rcp(2p+; 15q–)[c]	0.74	1.55	1.87	1.60	0.96	21
15C+[d]	–2.44	2.21	4.84	4.38	2.31	103
17C+[e]	–3.09	–1.79	–0.70	–0.25	–0.05	13

[a]Number of informative progeny; [b]centromeric ("Robertsonian") fusion of chromosomes 15 and 17; [c]reciprocal translocation of chromosomes 2 and 15; [d]enlargement of the centromeric region of chromosome 15; [e]enlargement of the centromeric region of chromosome 17. (From Fries et al., 1983 and Fries et al., 1984.)

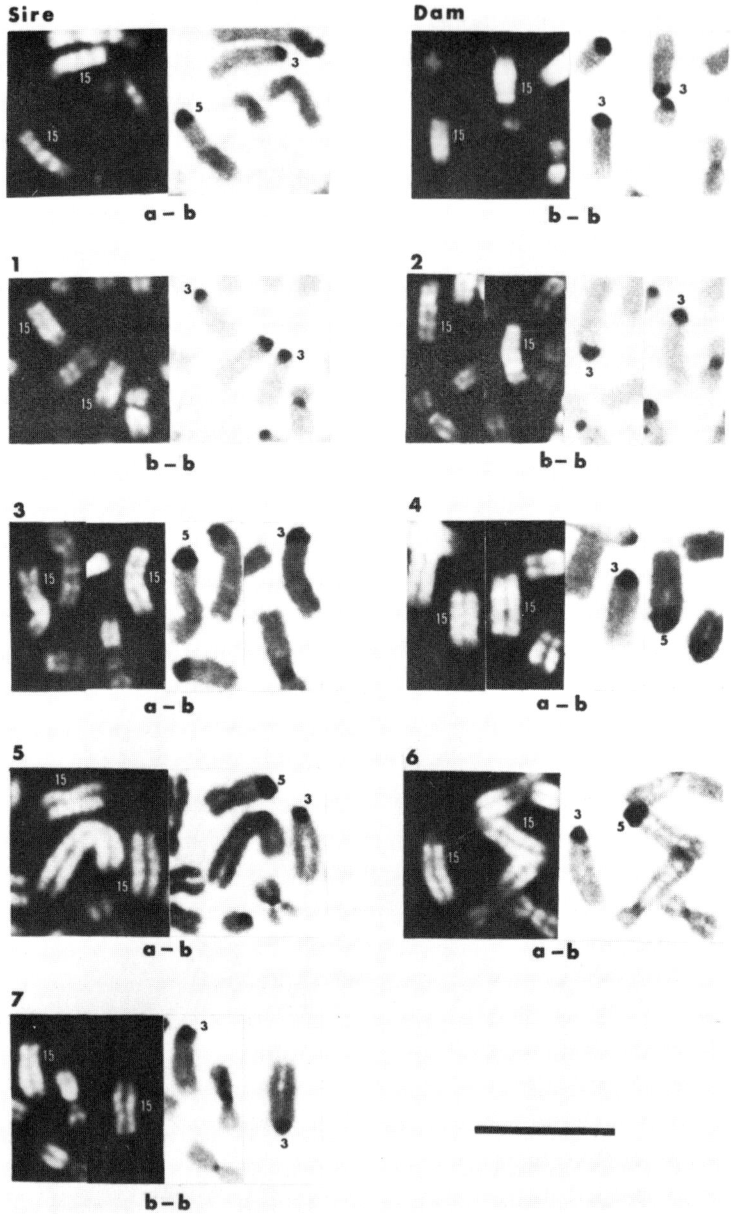

Figure 2. Partial metaphase spreads with chromosomes 15, centromere scores and G-blood types in one porcine family without recombinants. Sequential Q- (left) and C- (right) banding was carried out according to Caspersson et al. (1969) and Sumner (1972) respectively. The centromere scores are verbally described as very small (1), small (2), intermediate (3), large (4) and very large (5). (From Fries et al., 1984, used with permission.)

combination of the G-blood group locus and a radiation induced marker chromosome 15 (rcp (2p+; 15q−)) (Fries and Stranzinger, 1982), as well as the negative lod scores from the combination of a chromosome 17 marker (17 C+) suggested that genes encoding the G-blood antigens reside on chromosome 15. In a subsequent study (Fries et al. 1984) the provisional assignment of the G-blood group locus to chromosome 15 was confirmed based on positive lod scores calculated from segregation data of G-blood group alleles and a centromere variant of chromosome 15 (15 C+). Figure 2 depicts the inheritance of the so called C-band variants of chromosome 15 and the G-blood group alleles in an informative family of pigs.

The major drawback of gene mapping with family studies is that a locus can only be assigned to a particular chromosome when the following requirements are met: (1) An already assigned polymorphic marker locus or a polymorphic cytogenetic locus is not further from the locus to be mapped than a genetic distance characterized by 20% recombination. Linkage can only be detected with a reasonable number of animals when the recombination frequency is not much greater than 20%. (2) There must be at least two alleles at the locus to be mapped and they must both segregate with alleles at the marker locus in informative families. It is obvious that these requirements are met for only a few genes, mainly because of the lack of a sufficient number of polymorphic marker loci. The human genome is estimated to span 3,000 recombination units or centimorgans (cM) (Renwick, 1969). One can assume that this number is approximately the same for most mammalian genomes (at least there is no contradictory evidence thus far). Therefore, at least 75 marker loci evenly spaced at 40 cM intervals are required to cover the genome and to place any new gene locus within 20 cM of a marker locus. However, before evenly spaced loci can be obtained a much larger number of marker loci must first be placed on the map (White et al., 1985).

Gene mapping by somatic cell genetics. Gene mapping by somatic cell genetics is based on parasexual events which facilitate the transfer of genetic material between cells (gene transfer) and the partial loss of the transferred genetic material from hybrid cells. The most common approach for the delivery of genes in a random fashion into recipient cells for the purpose of gene mapping is cell hybridization. Somatic cells of one species undergo spontaneous fusion with those from another species at a low frequency when they are cultured together. This frequency can be enhanced considerably by exposure of the cells to inactivated Sendai virus (Yerganian and Nell, 1966) and to chemical agents such as polyethylene glycol (PEG; Pontecorvo, 1975). The use of

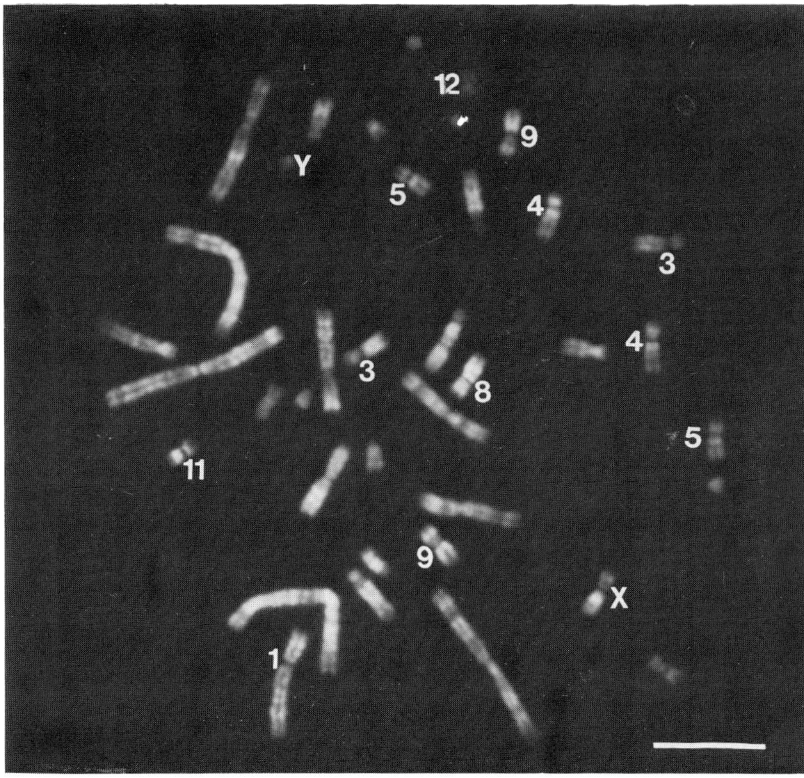

Figure 3. Metaphase spread of a hamster–swine hybrid cell. The numbers indicate the pig chromosomes. Q-banding was carried out according to Caspersson et al. (1969). Bar represents 10 μm.

mutant cell lines that lack one of the enzymes HPRT (hypoxanthine-guanine phosphoribosyltransferase, an enzyme in the purine salvage pathway) or TK (thymidine kinase, an enzyme in the pyrimidine salvage pathway) as one fusion partner and cells that are TK^+ or $HPRT^+$ as the other, with subsequent culture in medium that contains HAT (hypoxanthine, aminopterin, thymidine) allows the efficient selection of hybrid cells from the $HPRT^-$ or TK^- parental cells (Littlefield, 1964). Some cell lines which are $HPRT^-$ or TK^-; are resistant to Ouabain (Baker et al., 1974). The combination of $HPRT^-$ or TK^-, Ouabain resistant cells with $HPRT^+$ or TK^+, Ouabain sensitive cells and growth in medium that contains HAT allows selection against all parental cells which did not undergo fusion. If mouse and hamster cell lines are combined with human or cells from other species, such as swine or cattle, there is a progressive and preferential loss of the chromosomes of the non-mouse or non-hamster species, as first observed

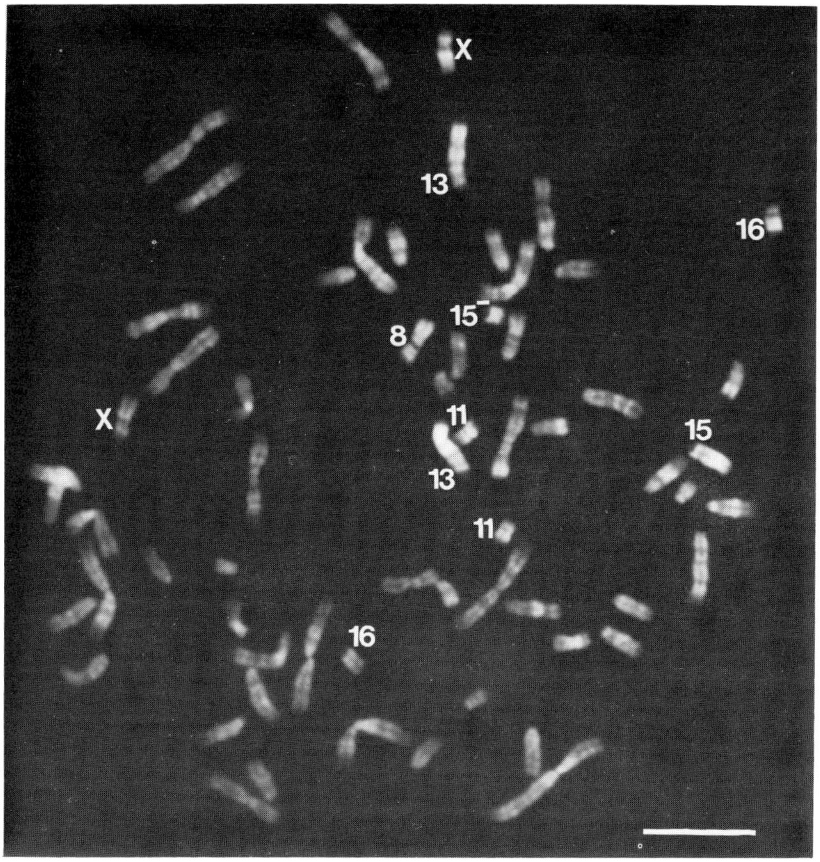

Figure 4. Metaphase spread of a mouse–swine hybrid cell. The numbers indicated the pig chromosomes. Combined Q- and H33258-staining was carried out according to Yoshida, Ikeuchi, and Sasaki (1975). Bar represents 10 μm.

by Weiss and Green (1967) in human–mouse hybrids. Clones of hybrid cells have only some of the chromosomes of the original parental cell. Figures 3 and 4 show metaphase spreads of hamster–swine and mouse–swine hybrid cells.

It is possible to draw conclusions about synteny (location of genes on the same chromosome) or to assign gene loci to particular chromosomes after the investigation of the chromosomal complement and the determination of the presence or absence of gene products in several hybrid clones. When a gene probe is available, the presence of a gene can be directly determined by restriction fragment analyses (Ruddle, 1981). A collection of hybrid lines providing these kinds of information is called a hybrid panel. Kamarck et al. (1984) reviewed the problems involved in optimizing a hybrid panel. The minimal number of hybrid lines with multiple chromosomes in unique configurations is specified by $C = 2^m$, where C is the haploid chromosome number and m the minimum number of panel members needed to assign a gene to any of the chromosomes. In man, with 24 different chromosomes, and in swine, with 20 different chromosomes, the minimum number of panel members would be only 5. However, the panels which are currently used for mapping human genes usually consist of 20 to 30 members, because it is practically impossible to obtain hybrid cell lines with the specific chromosome complements needed for minimizing the number of hybrid lines of a mapping panel. Moreover, a panel with the minimum number of hybrid lines would not provide any internal controls or experimental redundancy.

Once a gene is assigned to a chromosome, the next step is to localize the gene to a region within the chromosome. One way to achieve such a regional assignment is in situ hybridization, as we will see below. Another way to regionally map a gene is by using subchromosomal mapping panels. A subchromosomal mapping panel can be constructed with somatic cell hybrids that have deletions or translocations which involve the chromosome to which the gene has been mapped. It was for this reason that we chose porcine cell lines known to carry well characterized, heterozygous translocations as one of the parental lines in fusions with hamster and mouse cell lines. We can expect that, after extensive segregation and subcloning of the hybrids, cell lines will be obtained which will have either a normal copy of certain chromosomes or only parts of them. A series of hybrid lines with different parts of a chromosome can then be used to delimit the position of the gene or, in other words, to determine the "shortest region of overlap" (SRO).

The mapping resolution which can be achieved with a suitable subchromosomal mapping panel is 5–10 cM or 5×10^6 to 10×10^6 base pairs at the most, when high resolution chromosome banding techniques are applied (Ruddle, 1981). However, at the DNA level, recombinant DNA procedures that include so called "chromosome walking" approaches can provide mapping data from the level of the single nucleotide base pair up to 1×10^5 base pairs (100 kilo base pairs, kb). The gap between the level of resolution by panel mapping or in situ hybridization and the level achieved by recombinant DNA methodologies can be bridged by an approach based on chromosome mediated gene transfer. This technique allows the transfer of intact segments of donor chromosomes, in the size range of 10–10,000 kb, into suitable murine cells (Klobutcher and Ruddle, 1981). Dominant selection markers can be introduced into the donor genome by retrovirus expression vectors beforehand (Miller et al. 1983; Brennand, Konecki, and Caskey, 1983). Segments of the donor genome carrying the inserted selectable markers can then be isolated in rodent cells after chromosome mediated gene transfer. By this method one should be able to clone donor segments in the range of 10–10,000 kb. A collection of not more than 1,000 rodent clones would contain most of the donor genome and might be designated as a chromosome segment library (Ruddle, 1984). The donor DNA of clones, shown to carry donor genes of interest, could then be recovered by microbial cloning and selection by filter hybridization with donor specific repetitive DNA. A maximum of only about 300 cosmid inserts may be sufficient to saturate an entire segment that can span from 10 up to 10,000 kb.

In situ hybridization. In situ hybridization is a straightforward mapping technique involving the hybridization of radiolabelled DNA probes to fixed metaphase chromosomes and the subsequent visualization of the signal as silver grains after autoradiography. Hybridization in situ was originally developed by Gall and Pardue (1969) and John, Birnstiel and Jones (1969). This technique first allowed only the localization of highly reiterated or amplified genes. Improvement in the hybridization procedure, including the use of dextran sulphate in the reaction mixture, made possible the routine mapping of single copy genes in man by in situ hybridization with probes labelled to high specific activities by nick translation with ^{3}H- or ^{125}I-labelled nucleotides (Harper and Saunders, 1981; Rabin et al., 1985a). An essential step in gene mapping by in situ hybridization is the unambiguous identification of the chromosomes. This can be achieved by prephotographing Giemsa (G)-banded chromosomes after trypsin treatment (Seabright, 1971) or Quinacrine (Q)-banded chromosomes after quinacrine mustard staining (Caspersson et al., 1969). Q-bands can also be visualized by quinacrine mustard staining after hybridization, (Morton et al., 1984). The resolution of mapping by in situ hybridization, like the maximum resolution achieved by subchromosomal panel mapping, depends on the band resolution of the chromosomes and is, therefore, in the range of 5–10 centimorgans (Ruddle, 1981).

Rabin et al. (1985b) assigned the major histocompatibility complex (SLA) of the domestic pig to chromosome 7 with an SLA class I specific recombinant DNA probe. Autoradiographic silver grains were scored in a total of 84 metaphase spreads, prepared from two different cell lines. Of the 300 silver grains associated with chromosomes, 97 were found to be concentrated within the region q12 → p12 of chromosome 7 (Figs. 5 and 6). Geffrotin et al. (1984) have previously used a ^{35}S labeled

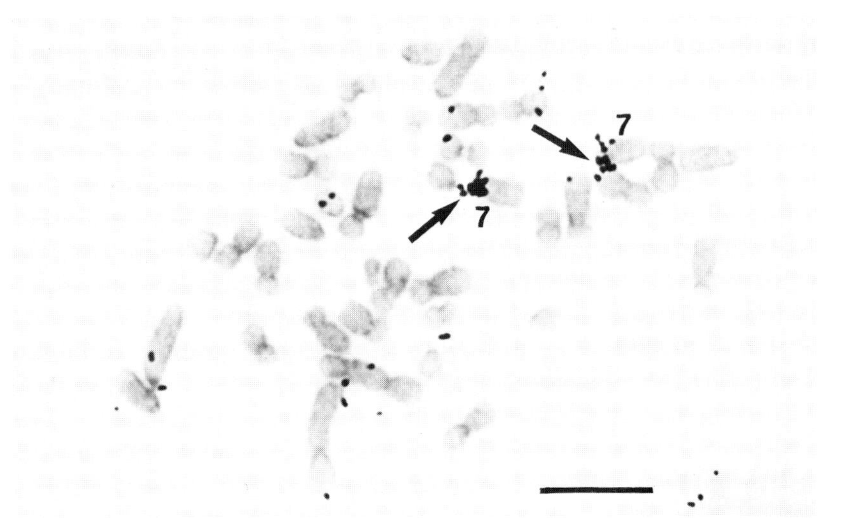

Figure 5. Porcine metaphase spread after hybridization with a ^{125}I-labeled DNA clone, specific for a porcine major histocompatibility complex (SLA) class I gene, and autoradiography. A weak G-banding pattern induced by the hybridization procedure is sufficient for the identification of most of the chromosomes including 7 (see numbers). The arrows indicate the localization of the SLA class I genes on chromosome 7. (From Rabin et al., 1985b, used with permission.) Bar represents 10 μm.

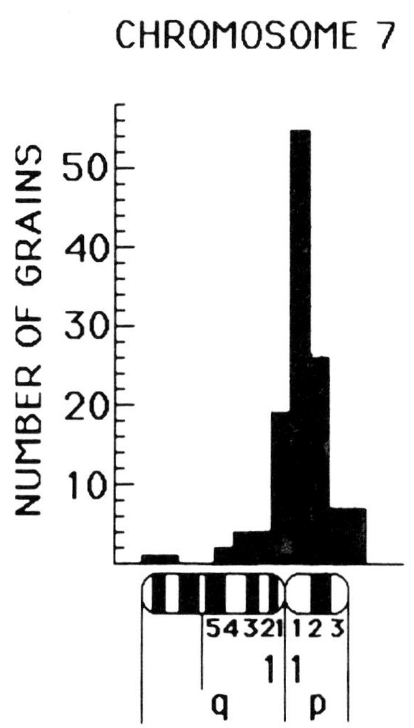

Figure 6. Histogram illustrating the distribution of silver grains found over chromosomes 7 from 84 porcine metaphase spreads after hybridization with a ^{125}I-labeled DNA probe of a porcine major histocompatibility (SLA) class I gene. The short chromosome arm is designated by "p" and the long arm by "q". (From Rabin et al., 1985b, used with permission.)

Table 2 Status of the gene maps of some domestic species

Species	Number of linkage or synteny groups	Number of chromosomal assignments	Haploid chromosome number
Cow (*Bos taurus*)	25	5	30
Sheep (*Ovis aries*)	11	0	27
Goat (*Capra hircus*)	0	0	30
Pig (*Sus scrofa*)	13	12	19
Chicken (*Gallus gallus*)	10	8	39

human major histocompatibility complex cDNA probe for hybridization in situ with pig metaphase chromosomes. They also found significant labelling on chromosome 7.

Status of the gene maps

The gene maps of the economically most important domestic species: cattle, sheep, swine, goat and chicken, are not far advanced when compared to the human or mouse gene maps. Updated information about linkage and synteny groups and gene assignments in some of these species can be found in the Genetic Maps (1984). However, most of the gene maps of domestic animals consist of only a few linkage or synteny groups which in many cases are not assigned to specific chromosomes (Table 2). In this article we present the gene map of the pig as an example of a gene map of a domestic species (Table 3, Fig. 7).

A major drawback of gene mapping in domestic species is the lack of sufficient standardization of the karyotypes. At the First International Conference for the Standardization of Banded Karyotypes of Domestic Animals (Reading Conference, 1976) agreement was reached only on how karyotypes should be arranged, and only a verbal description of the main G-bands was published. A standardized, numerical description of the bands, similar to the description of human banded chromosomes (Paris Conference, 1971) is not yet available; however, standardization committees for the different species are at work and are expected to publish standardized band idiograms in the near future. The cattle and goat karyotypes pose special problems; all 58 autosomes in cattle and all chromosomes except the Y-chromosome in the goat are teleocentric. In each of the two species the G-banding patterns of some chromosomes of similar length look so similar that the chromosomes are not easy to identify. Fluorescent banding techniques seem to provide better banding resolution and therefore facilitate considerably the identification of the chromosomes (Di Berardino and Jannuzzi, 1982). Gene mapping of the chicken is restricted by the fact that this species has 39 pairs of

Table 3 Chromosomal assignments in swine

Chromosome	Gene locus	References
3	MDH1 (Malate dehydrogenase, soluble)	Förster and Hecht, 1984
4	LDHA (Lactate dehydrogenase A)	Förster and Hecht, 1984
5	LDHB (Lactate dehydrogenase B)	Förster and Hecht, 1984
7	SLA (Swine leukocyte antigen)	Andresen and Baker, 1964
	C (C-blood group)	Hruban et al., 1976
	J (J-blood group)	Gellin et al., 1981
	NP (Nucleoside phosphorylase)	Dolf, 1984
	MPI (Mannosephosphate isomerase)	Förster and Hecht, 1984
	PKM2 (Pyruvate kinase-3)	Geffrotin et al., 1984
		Rabin et al., 1985b
8	NOR (Nucleolar organizer region)	Mayr, Schweizer and Geber, 1984
9	SOD1 (Superoxide dismutase, soluble)	Leong, Lin and Ruth, 1983a
		Förster and Hecht, 1984
10	NOR (Nucleolar organizer region)	Christensen, 1980
	PGM1 (Phosphoglucomutase-1)	Förster and Hecht, 1984
	M (M-blood group)	Mayr et al., 1984
	PEPB (Peptidase B)	Förster and Hecht, 1984
15	G (G-blood group)	Andresen and Jensen, 1977
	GPI (Glucosephosphate isomerase)	Rasmusen, 1981
	HAL (Halothane sensivity)	Juneja et al., 1983
	S (A–O inhibition)	Tikhonov et al., 1983
	H (H-blood group)	Fries et al., 1984
	PO2 (Postalbumin-2)	
	PGD (6-Phosphogluconate dehydrogenase)	
16	A-23 (Serum protein A-23)	Knyazev and Tikhonov, 1984
X	TRAIII (Paralytic tremor AIII)	Lax, 1971
	SPL (Splay leg condition)	Harding et al., 1973
	HPRT (Hypoxanthine-guanine phosphoribosyltransferase)	Förster, Stranzinger, and Hellkuhl, 1980
	G6PD (Glucose-6-phosphate dehydrogenase)	Gellin et al., 1980
	PGK (Phosphoglycerate kinase)	Leong, Lin, and Ruth, 1983b
	GLA (Alpha-galactosidase A)	

chromosomes, of which only 10 pairs are large enough to permit identification. The remaining 29 pairs cannot be identified by any of the available banding techniques.

Strategies for mapping genes

A first step in obtaining more detailed gene maps of domestic species should consist in studying the synteny of genes known to be closely linked in man and mouse. Table 4 gives a summary of synteny groups

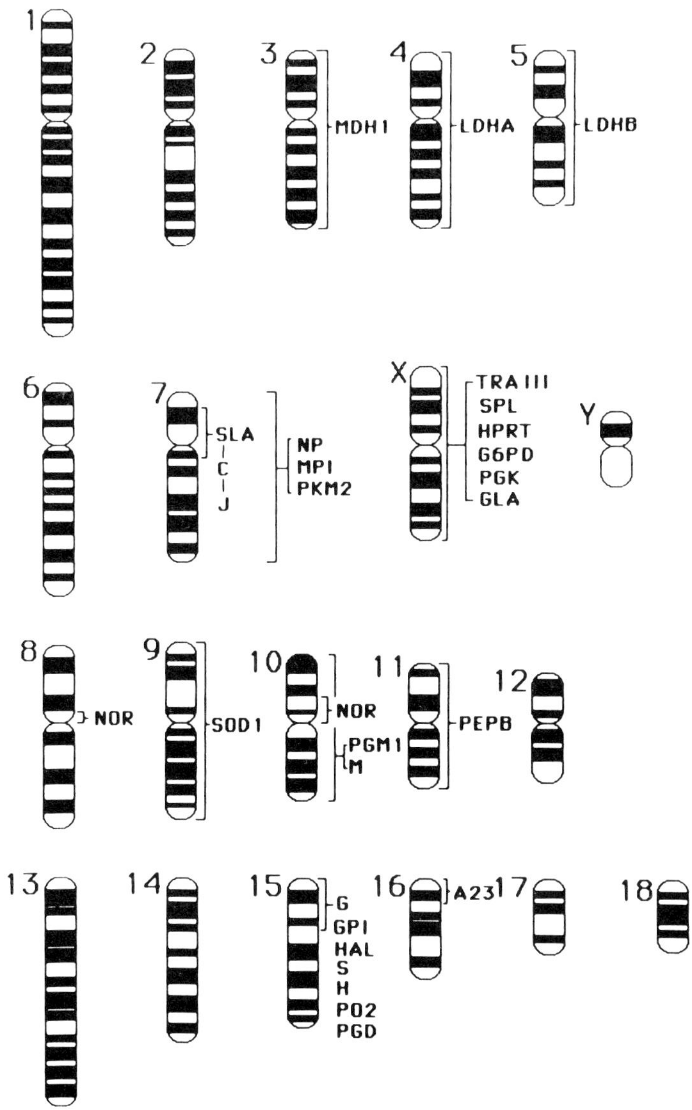

Figure 7. The gene map of the pig. For explanation of gene symbols and references see Table 3.

which have already been shown to be conserved in man, mouse, pig, sheep and cow. The comparative mapping approach allows the identification of homologous chromosome regions in different species and, most importantly, it can provide preliminary information about the location of gene loci. Comparative mapping is particularly helpful in evolutionarily closely related species such as goat, cow and sheep.

As described above, gene assignments to chromosomes and linkage groups by family studies is applicable in only a few cases. It is therefore

Table 4 Some conserved synteny groups

Gene locus[c]	Chromosome[a] or syntenic group[b]				
	Man	Mouse	Pig	Sheep	Cow
PGD	1	4	15	U1	U1
ENO1	1	4		U1	U1
PGM1	1	4	10	U1	U6
TF	3	9			Un
CP	3				Un
C4	6	17		U6	
HLA	6	17	7	U6	
GLO1	6	17			
ME1	6	9		B	U2
PGM3	6	9		B	U2
SOD2	6	17		U8	U2
TPI	12	6		U2	U3
LDHB	12	6	5	U2	U3
PEPB	12	10	11	U2	U3
MPI	15	9	7	U9	U4
PKM2	15	9	7	U4	U5
SOD1	21	16	9	U10	U10
IFRC	21	16			U10
PAIS	21			U10	
PRGS	21	16		U10	
PGK	X	X	X	X	X
GLA	X	X	X	X	X
HPRT	X	X	X	X	X
G6PD	X	X	X	X	X

[a]Designated by a number; [b]designated by a letter plus number or letter; [c]for nomenclature see Genetic Maps, 1984.

important that somatic cell hybrid panels be established for all species. It can be expected that the bulk of mapping data will be collected by panel mapping. Confirmation and regional assignments will be accomplished by in situ hybridization in cases where gene probes are available. However, gene assignments by somatic methods (panel mapping, in situ hybridization) are part of the so called "physical" gene map and do not provide information about genetic distances. The translation of the physical distance, in terms of base pairs, into the genetic distance, expressed as centimorgans, is not straightforward, because the frequency of crossing over seems to vary throughout the genome. As the physical maps fill up as a result of assignments by somatic methods, family studies will become important again in the determination of genetic distances.

Genes that have no known phenotype in cultured cells or for which there are no cloned DNA probes available cannot be mapped by somatic cell hybridization or by in situ hybridization. The map position of such genes, however, can be determined by studying linkage with genetic markers which have already been mapped by somatic methods. Genetic markers, based on polymorphisms detected at the level of the DNA by restriction enzymes may be present in an abundant number in domestic animals. Botstein et al. (1980) first described this new type of genetic marker, called restriction fragment polymorphism (RFLP). Over 200 loci of RFLP were reported in man by 1983 (Skolnick, Willard, and Menlove, 1984). It has already been possible to map indirectly the defects responsible for muscular dystrophy (Murrary et al., 1982) and Huntington's disease (Gusella et al., 1983) based on associations found between loci of RFLPs and the inherited diseases. Searching for RFLPs will most likely become an integral part of gene mapping in domestic species. Randomly cloned DNA fragments, free of repetitive sequences, as well as cloned unique DNA sequences that encode known genes, can be used as probes in Southern blot analyses of DNA samples from members of large families to determine loci of RFLPs. Sources for random DNA sequences are both cDNA and genomic libraries. Newly detected RFLP loci can be assigned chromosomally by physical methods (somatic cell hybridization, in situ hybridization) and then be used as marker loci for the study of the map positions of genes which have no known phenotype in cell cultures.

Finally it should be emphasized that any strategy for mapping genes in domestic species should adopt the general rule of human gene mapping, namely that gene assignments must be confirmed by several groups independently before complete confidence can be placed on their accuracy.

Application of gene mapping

We do not believe that there are many direct applications to practical animal breeding of the knowledge gained by gene mapping. As the gene map fills up, one might come across further linkage groups like the one found between the locus for Halothane sensitivity (HAL), an indicator for the porcine stress syndrome (PSS), and blood group and enzyme loci (Vögeli et al., 1984); and Marek's disease and the major histocompatibility locus in chicken (Briles et al., 1983). These linkage groups have turned out to be very useful tools for selecting against PSS and Marek's disease, respectively.

However, we do believe that gene mapping will play an important role in the definition and isolation of genes that determine or modify complex traits such as lactation, fertility, growth and disease resistance.

It can be assumed that not more than ten genes affect these characteristics to a useful extent (Lande, 1981). In spite of this limitation, most of the investigations to find associations between biochemical markers or blood group loci and productive traits have been unsuccessful. However, once a large enough number of polymorphic loci have been placed throughout the genome, and RFLPs can be expected to facilitate this undertaking considerably, it should become possible to narrow down the possible locations of some of these unknown genes. A few large families, typed for as many as possible RFLPs and other marker loci, and consisting of crosses and backcrosses of breeds with large phenotypic differences in productive traits or resistance against diseases, will provide the relevant information.

Once linkage between an RFLP locus, which has been localized chromosomally, and a gene that contributes to a complex trait has been established, the next step consists in searching within a chromosome segment library for a clone that carries the RFLP locus. A genomic library constructed from this clone could then serve as a source for unique DNA fragments to define more RFLP loci, all within not more than 10,000 kb from the gene of interest. From further linkage studies it should then be possible to localize the gene between four RFLP loci, two on either side. In a next step one would again screen a chromosome segment library for a segment that contains the proximal, but lacks the distal, flanking markers. A cosmid library of less than one hundred cosmid inserts would then be sufficient to cover the entire region that contains the two proximal markers. Depending on the nature of the trait of interest different approaches may be applied to detect the cosmids that contribute to the expression of the trait. In the case of a quantitative trait, such as milk yield, one could compare the frequency of reiteration of candidate sequences in animals with high and low milk yields, assuming that quantitative differences are reflected by the frequency of reiteration of the gene. Candidate sequences might also be tested by microinjection into embryos and subsequent study of the effect of the candidate sequence on the performance of the transgenic animal. Another way of determining the sequences that contribute to the trait of interest could consist in screening cDNA libraries from various tissues that are functionally related to the trait and tissues that are not related.

It is obvious that the isolation of genes contributing to animal performances is a long and tedious process. However, it is very likely that various steps involved in this procedure might have interesting applications that cannot be foreseen.

Conclusions

Gene mapping techniques are now sufficiently advanced to permit rapid progress in the establishment of detailed gene maps of domestic animals. The search for polymorphic DNA markers is expected to become an integral part of gene mapping. Restriction fragment mapping and in situ hybridization can be used to place DNA markers at random sites throughout the genome. Family studies should enable us to find linkage between unknown genes for complex traits and the markers. The polymorphic DNA markers may then serve to predict an animal's performance. A series of DNA markers flanking one or several genes for complex, polygenic traits might eventually lead to the isolation and cloning of the genes for the complex traits.

Acknowledgments This work was supported by Fellowships from the Swiss National Science Foundation to Rudolf Fries. We would like to thank Suzy Pafka for excellent technical assistance.

References

Andresen, E and L.N. Baker. 1964. *The C-blood group system in pigs and the detection and estimation of linkage between the C and J systems.* Genetics 49:379–386.

Andresen, E and P. Jensen. 1977. *Close linkage established between HAL locus for halothane sensitivity and PHI (phosphohexose isomerase) locus in pigs of the Danish Landrace breed.* Nord. Vet. Med. 29:502–504.

Baker, R.M., D.M. Brunette, R. Mankovitz, L.H. Thompson, G.F. Whitmore, L. Siminovitch and J.E. Till. 1974. *Ouabain-resistant mutants of mouse and hamster cells in culture.* Cell 1:9–21.

Botstein, D., R.L. White, M. Skolnick and R.W. Davis. 1980. *Construction of a genetic linkage map using restriction fragment length polymorphisms.* Am. J. Hum. Genet. 32:314–331.

Brennand, J., D.S. Konecki and C.T. Caskey. 1983. *Expression of human and chinese hamster hypoxanthine-guanine phosphoribosyltransferase cDNA recombinants in cultured Lesh-Nyhan and chinese hamster fibroblasts.* J. Biol. Chem. 16:9593–9596.

Briles, W.E., R.W. Briles, R.E. Taffs and H.A. Stone. 1983. *Resistance to a malignant lymphoma in chicken is mapped to a subregion of major histocompatibility (B) complex.* Science 219: 977–979.

Caspersson, T., L. Zech, E.J. Modest, G.E. Foley, U. Wagh and E. Simonsson. 1969. *Chemical differentiation with fluorescence alkylating agents in vicia faba metaphase chromosomes.* Expl. Cell Res. 58:128–140.

Christensen, K. 1980. *Evidence of polymorphism of the nuclear organizer region (N-Band) in pig chromosomes.* Proceedings of the 4th European Colloquium on Cytogenetics of Domestic Animals. 464–468. Uppsala.

Croce, C.M. and G. Klein. 1985. *Chromosome translocations and human cancer.* Scientific American 252:54–60.

Di Berardino, D. and L. Iannuzi. 1982. *Detailed description of R-banded bovine chromosomes.* J. Hered. 73:434–438.

Dolf, G. 1984. *Genkartierung beim Schwein mit Hilfe von somatischen Zellhybriden.* Thesis no. 7644. Eidgenössische Technische Hochschule Zürich.

Donahue, R.P., W.B. Bias, J.H. Renwick and V.A. McKusick. 1968. *Probable assignment of the Duffy blood group locus to chromosome I in man.* Proc. Nat. Acad. Sci. USA 61:949–955.

Förster, M., G. Stranzinger and B. Hellkuhl. 1980. *X-chromosomal gene assignment of swine and cattle.* Naturwissenschaften 67:48.

Förster, M. and W. Hecht. 1984. *Some provisional gene assignments in pig.* Proceedings of the 6th European Colloquium on Cytogenetics of Domestic Animals. 351–355. Zürich.

Fries, R., B.A. Rasmusen, V.L. Jarrell and R.R. Maurer. 1984. *Mapping of the gene for G blood antigens to chromosome 15 in swine.* Anim. Blood Grps. Biochem. Genet. 15:251–258.

Fries, R. and G. Stranzinger. 1982. *Chromosomal mutations in pigs derived from X-irradiated semen.* Cytogenet. Cell Genet. 34:55–66.

Fries, R., G. Stranzinger and P. Vögeli. 1983. *Provisional assignment of the G-blood group locus to chromosome 15 in swine.* J. Hered. 74:426–430.

Gall, J.G. and M.L. Pardue. 1969. *Formation and detection of RNA–DNA hybrid molecules in cytological preparations.* Proc. Nat. Acad. Sci. USA 63:378–383.

Geffrotin, C., C.P. Popescu, E.P. Cribiu, J. Bosher, C. Renard, P. Chardon and M. Vaiman. 1984. *Assignment of MHC in swine to chromosome 7 by in situ hybridization and serological typing.* Ann. Génét. 27:213–219.

Gellin, J., F. Benne, M.C. Hors-Cayla and M. Gillois. 1980. *Gene mapping in the pig (sus-scrofa L.). Study of two syntenic groups G6PD, PGK, HPRT and PKM2, MPI.* Ann. Génét. 23:15–21.

Gellin, J., G. Echard, F. Benne and M. Gillois. 1981. *Pig gene mapping: PKM2–MPI–NP synteny.* Cytogenet. Cell Genet. 30:59–62.

Genetic Maps. 1984. S.J. O'Brien, ed., National Institute of Health, Bethesda, Maryland. Volume 3.

Gusella, J.F., N.S. Wexler, P.M. Conneally, S.L. Naylor, M.A. Anderson, R.E. Tanzi, P.C. Watkins, K. Ottina, M.R. Wallace, A.Y. Sakaguchi, A.B. Young, I. Shoulson, E. Bonilla and J.B. Martin. 1983. *A polymorphic DNA marker genetically linked to Huntington's disease.* Nature (London) 306:234–238.

Harding, J.D.J., J.T. Done, J.F. Harbourne and F.R. Gilbert. 1973. *Congenital tremor type AIII in pigs: an hereditary sex linked cerebrospinal hypomyelinogenesis.* Vet. Rec. 92:527–529.

Harper, M.E. and G.F. Saunders. 1981. *Localization of single copy DNA sequences on G-banded human chromosomes by in situ hybridization.* Chromosoma 83:431–439.

Hruban, V., H. Simon, J. Hradecky and F. Jilek. 1976. *Linkage of the pig main histocompatibility complex and the J blood group system.* Tissue Antigens 7:267–271.

John, H.A., M.L. Birnstiel and K.W. Jones. 1969. *RNA–DNA hybrids at the cytological level.* Nature (London) 223:582–587.

Juneja, R.K., B. Gahne, I. Edfors-Lilja and E. Andresen. 1983. *Genetic variation at a pig serum protein locus, Po-2 and its assignment to the Phi, Hal, S, H, Pgd linkage group.* Anim. Blood Grps. Biochem. Genet. 14:27–36.

Kamarck, M.E., P.E. Barker, R.L. Miller and F.H. Ruddle. 1984. *Somatic hybrid panels.* Expl. Cell Res. 152:1–14.

Klobutcher, L.A. and F.H. Ruddle. 1981. *Chromosome mediated gene transfer.* Ann. Rev. Biochem. 50:533–554.

Knyazev, S.P. and Tikhonov, V.N. 1984. *Gene mapping of the pig chromosome no. 16.* Proceedings of the 6th European Colloquium on Cytogenetics of Domestic Animals. 395–399. Zürich.

Lande, R. 1981. *The minimum number of genes contributing to quantitative variation between and within populations.* Genetics 99:541–553.

Lax, T. 1971. *Hereditary splayleg in pigs.* J. Hered. 62:250–252.

Leong, M.M.L., C.C. Lin and R.F. Ruth. 1983a. *Assignment of superoxide dismutase (SOD-1) gene to chromosome no. 9 of domestic pig.* Can. J. Genet. Cytol. 25:233–238.

Leong, M.M.L., C.C. Lin and R.F. Ruth. 1983b. *The localization of genes for HPRT, G6PD and α-GAL onto the X-chromosome of domestic pig (sus scrofa domesticus).* Can. J. Genet. Cytol. 25:239–245.

Littlefield, J.W. 1964. *Selection of hybrids from matings of fibroblasts in vitro and their presumed recombinants.* Nature (London) 256:495–497.

Mayr, B., D. Schweizer and G. Geber. 1984. *NOR activity, heterochromatin differentiation, and the Robertsonian polymorphism in sus scrofa L.* J. Hered. 75:79–80.

Miller, A.D., D.J. Jolly, T. Friedmann and I.M. Verma. 1983. *A transmissible retrovirus expressing human hypoxanthine phosphoribosyltransferase (HPRT): Gene transfer into cells obtained from humans deficient in HPRT.* Proc. Nat. Acad. Sci. USA 80:4709–4713.

Morton, C.C., I.R. Kirsch, R. Taub, S.H. Orkin and J.A. Brown. 1984. *Localization of the β-globin gene by chromosomal in situ hybridization.* Am. J. Hum. Genet. 36:576–585.

Morton, N.E. 1955. *Sequential tests for the detection of linkage.* Am. J. Hum. Genet. 7:277–318.

Murray, J.M., K.E. Davies, P.S. Harper, L. Meredith, C.R. Mueller and R. Williamson. 1982. *Linkage relationship of a cloned DNA sequence on the short arm of the X-chromosome to Duchenne muscular dystrophy.* Nature (London) 300:69–71.

Paris Conference (1971). 1972. *Standardization in human cytogenetics.* Cytogenet. Cell Genet. 11:313–362.

Pontecorvo, G. 1975. *Production of mammalian somatic cell hybrids by means of polyethylene glycol (PEG) treatment.* Somatic Cell Genet. 1:397–400.

Rabin, M., C.P. Hart, A. Ferguson-Smith, W. McGinnis, M. Levine and F.H. Ruddle. 1985a. *Two homeobox loci mapped in evolutionarily related mouse and human chromosomes.* Nature (London):175–178.

Rabin, M., R. Fries, D. Singer and F.H. Ruddle. 1985b. *Assignment of the porcine major histocompatibility complex to chromosome 7 by in situ hybridization.* Cytogenet. Cell Genet. (in press).

Rasmusen, B.A. 1981. *Linkage of genes for PHI, halothane sensitivity, A–O inhibition, H red blood cell antigens, and 6-PGD variants in pigs.* Anim. Blood Grps. Biochem. Genet. 12: 207–209.

Reading Conference (1976). 1980. *Proceedings of the first international conference for the standardisation of banded karyotypes of domestic animals.* Hereditas 92:145–162.

Renwick, J.H. 1969. *Progress in mapping human autosomes.* Br. Med. Bull. 25:65–73.

Ruddle, F.H. 1981. *A new era in mammalian gene mapping: somatic cell genetics and recombinant DNA methodologies.* Nature (London) 294:115–120.

Ruddle, F.H. 1984. *The William Allan Memorial Award Address: Reverse genetics and beyond.* Am. J. Hum. Genet. 36:944–953.

Seabright, M. 1971. *A rapid banding technique for human chromosomes.* Lancet 2:971.

Skolnick, M.H., H.F. Willard and L.A. Menlove. 1984. *Report of the committee on human gene mapping by recombinant DNA techniques.* Cytogenet. Cell Genet. 37:210–273.

Sumner, A.T. 1972. *A simple technique for demonstrating centromeric heterochromatin.* Expl. Cell Res. 75:304–306.

Tikhonov, V.N., I.G. Gorelov, S.V. Nikitin, V.E. Bobovich and N.M. Astakhova. 1983. *Mapping of the locus for the H-blood group system on chromosome 15 of domestic pig.* Doklady Akademii Nauk S.S.S.R. 272:486–489.

Vögeli, P., G. Stranzinger, H. Schneebeli, C. Hagger, N. Künzi and C. Gerwig. 1984. *Relationships between the H and A–O blood types, phosphohexose isomerase and 6-phosphogluconate dehydrogenase red cell enzyme systems and halothane sensitivity, and economic traits in a superior and an inferior selection line of Swiss landrace pigs.* J. Anim. Sci. 59: 1440–1450.

Wald, A. 1947. *Sequential analysis.* Dover Pubns., Inc., New York.

Weiss, M.C. and H. Green. 1967. *Human–mouse hybrid cell lines containing partial complements of human chromosomes and functioning human genes.* Proc. Nat. Acad. Sci. USA 58:1104–1111.

White, R., M. Leppert, D.T. Bishop, D. Barker, J. Berkowitz, C. Brown, P. Callahan, T. Holm and L. Jerominski. 1985. *Construction of linkage maps with DNA markers for human chromosomes.* Nature (London) 313:101–105.

Yervanian, G. and M.B. Nell. 1966. *Hybridization of dwarf hamster cells by UV-inactivated Sendai virus.* Proc. Nat. Acad. Sci. USA 55:1066–1073.

Yoshida, M.C., T. Ikeuchi and M. Sasaki. 1975. *Differential staining of parental chromosomes in interspecific cell hybrids with a combined quinacrine and 33258 Hoechst technique.* Proc. Jap. Acad. 51:185–187.

Application of hybridoma technology to problems in the agricultural sciences

H.R. GAMBLE
Helminthic Diseases Laboratory, Animal Parasitology Institute, Agricultural Research Service, U.S. Department of Agriculture, Building 1040, BARC-East, Beltsville, Maryland 20705, U.S.A.

Abstract The development of methods for the production of monoclonal antibodies has provided a powerful tool for virtually all areas of the biological sciences. Monoclonal antibodies allow the dissection of complex mixtures mixtures of antigenic molecules; by applying screening procedures to a library of monoclonals, selection for properties such as viral neutralization or enzyme inactivation is possible, thereby identifying proteins with specific biological activities. Similarly, monoclonal antibodies can be used to map the surface of a protein, define cell surface antigens, and localize molecules within cells or tissues. The property of unique specificity associated with clonally-derived antibodies has served to revolutionize the field of diagnostics where the use of polyclonal antisera once dominated. The use of monoclonal antibodies for antigen identification has greatly benefited the study of the immune response to infectious agents and the development of vaccines. With the advent of better methods for the generation of hybridomas such as electrofusion and oncogene transfection, modifications allowing the exploitation of non-immunogenic molecules, and the development of techniques for producing interspecific hybrids, the contribution of this technology as a tool for research in the agricultural sciences will undoubtedly increase.

Introduction

In the decade since the development of methods for the production of antibody-secreting hybrid cell lines (Kohler and Milstein, 1975) this technology has had major impact in virtually all areas of the biological sciences. The reason for the tremendous application of hybridoma technology lies in the properties of the monoclonal antibody secreting hybrid cells derived from the parent cell lines, namely the antibody secretion of the lymphocyte and the immortality of the myeloma.

Stimulation of the immune system by an antigen or mixture of antigens results in the proliferation of populations of B-lymphocytes which produce antibodies with specificities for an array of antigenic determinants. The formation of hybrids by fusing these lymphocytes with myeloma cells results in populations of immortalized antibody secreting cells. The subsequent cloning of these hybrid cells results in the establishment of unique cell lines, each derived from a single B-cell and myeloma parent, capable of growing continuously in culture and secreting an antibody with a single molecular configuration (heavy and light chain) and variable region antigen specificity.

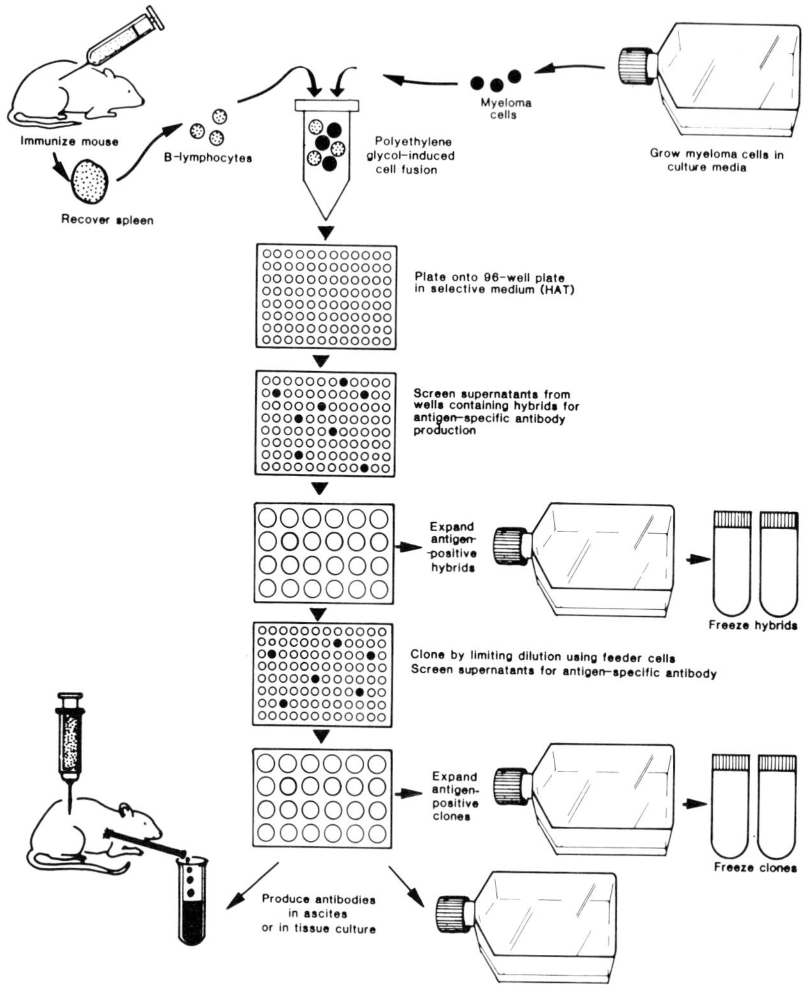

Figure 1. General methods for the production of monoclonal antibodies.

Current technology

The methods employed in the production of antibody-secreting hybrid cell lines are relatively simple, having their basis in somatic cell hybridization (see Goldsby, Srikumaran and Guidry, 1984, for review). Conventional procedures for the production of monoclonal antibodies are detailed in Fig. 1. Appropriate immunization protocols are often determined empirically, however, higher affinity antibodies are usually generated with an increasing number of immunizing doses. It is often useful to check serum for the appropriate specificity prior to fusion. An intravenous boost 3–4 days prior to fusion is generally recommended.

Spleen cells and myeloma cells are normally fused in the presence of polyethylene glycol, however, many modifications and alternative methods exist (Oi and Herzenberg, 1980; Galfre and Milstein, 1981). Following fusion hybrid cells are selected from unfused myeloma cells by growth in selective media (hypoxanthine, aminopterin, thymidine; Littlefield, 1964). The myeloma cells, with a deficiency for a purine salvage pathway enzyme (hypoxanthine phosphoribosyl transferase), are selected against by growth in inhibitors of de novo purine synthesis (aminopterin, amethopterin). Hybrids containing the HPRT gene donated from the spleen cell partner can grow with exogenous hypoxanthine; thymidine is also added since aminopterin inhibits the formation of thymidylate.

Once hybrids have been generated, it is of extreme importance to apply selective screening procedures; subsequent expansion, cloning, cryopreserving, freezing, and antibody production procedures are laborious so it is counterproductive to retain hybrids of no value. A variety of screening procedures are available for selecting hybrid cell lines of interest (Gamble, 1984a). Screening may be tailored to select not only for antigen specificity, but also for antibodies with specific properties such as complement fixations, binding to Staph protein A, or hemagglutination. Selected hybrids should be cloned (McKearn, 1980) as soon as possible, and cryopreserved (Kennett, 1980). Recloning occasionally is necessary to eliminate the subpopulations of hybrids which have stopped secreting.

Preliminary biochemical characterization of monoclonal antibodies may provide useful information. Such characterization often includes determination of antibody isotype by gel diffusion (Letchworth and Appleton, 1984), metabolic labeling (Haas and Kennett, 1980) or isotype specific ELISA (Gamble, 1984a), and the determination of antibody specificity by immunoblotting (Lampson and Fisher, 1985) or immunoprecipitation (Lampson, 1980).

New developments in monoclonal antibody technology

A number of modifications in hybridoma technology have both facilitated the production of hybrids and the recovery of their antibody products.

In vitro immunization

Novel methods have been developed for the immunization of donor B lymphocytes in vitro (Borrebaeck, 1984; McHugh, 1984). In vitro immunization is particularly useful in situations where only small amounts of antigen are available, when antigens are toxic to animals, or when hybrids using human cells are being developed. For in vitro immunization, lymphocytes, grown in medium supplemented with thymocyte growth factors, are exposed to low concentrations of antigen (VanNess, Laemmli , and Pettijohn, 1984). Antigen exposure may be primary or secondary following an initial in vivo stimulation. Blast transformation in response to antigen is then observed as clusters of dividing cells; blast transformed lymphocytes are then used as fusion partners with appropriate myeloma cells.

Fusion technology

Hybridoma technology in the study of immunoglobulins and the immune response to antigens in animals of agricultural and veterinary interest has been hindered by the lack of identified myeloma fusion partners for lymphocytes from these species. However, recent studies have demonstrated the potential of interspecific hybridomas, produced by fusing mouse myeloma cells with bovine lymphocytes (Srikumaran, Guidry and Goldsby, 1983). These studies have shown that although the frequency of stable hybrids is low, due to chromosome loss, rigorous selection allows for the stabilization of some cell lines. Initial studies have used hybridoma secreted monoclonal bovine immunoglobulins to biochemically and immunochemically characterize these immunoglobulin molecules, however, recent studies have generated mouse/bovine hybrids with specificity for a particular pathogen (Raybould et al., 1985). Application of interspecific fusion techniques to other livestock species is possible and should greatly facilitate the study of the immune response in these species to specific pathogens.

Other new methods have been developed for the fusion or immortalization of lymphocytes. One example is the technology of electrofusion (Zimmerman and Vienkin, 1984). Electrical fusion of cells relies on a reversible electrical breakdown of cell membranes in response to high intensity, short duration electrical pulses. Cells to be fused are aligned and allowed to establish membrane contact in an alternating current electrical field. Fusion is then accomplished by applying pulses of direct current of sufficient strength to cause a reversible breakdown in the integrity of the cell membranes. Production of both mouse and human hybridomas by electrofusion has been described. The application of electrical fusion is particularly appropriate if preselection of lymphocytes used for fusion is possible (Zimmerman and Vienken, 1984).

An extremely promising technology for producing immortalized antibody-secreting, as well as other types of cell lines, is transfection. Transfection involves the incorporation of oncogenic DNA into the chromosomes of normal cells, in this case, B lymphocytes, by various methods including calcium phosphate precipitation or electrical pulse mediated cell breakdown as described for electrofusion. Genomic DNA, extracted from myeloma cells, has been incorporated into mouse and pig lymphocytes, resulting in monoclonal antibody secreting cell lines (Jonak et al., 1984; Davis et al., 1985). The application of transfection technology in the veterinary sciences will facilitate the production of antibody producing cell lines for large animal species. With the identification of specific oncogenic DNA sequences and incorporation of these sequences into shuttle vectors the possibilities exist for the immortalization of a wide variety of cell lines.

Antibody production

The growth and production of monoclonal antibodies has been influenced by the development of new technologies. One major advance has been the development of non-secreting myeloma cell lines (Kearney et al., 1979). These non-secreting variants eliminate the dilution of antigen specific antibodies secreted by hybrid cells.

The development of serum-free media for growth of hybridomas has greatly facilitated the subsequent purification of monoclonal antibodies. Studies have shown that the addition of insulin and transferin to a serum-free hybridoma medium results in continuous growth of many antibody secreting cell lines (Chang, Steplewski and Koprowski, 1980).

Because yields of monoclonal antibodies in tissue culture medium are low (10–20 μg/ml), many investigators use antibodies generated in mouse ascites. Recent technology using hybridoma cells encapsulated in polymer microbeads (Nilsson et al., 1983) has greatly increased yields and allows for continuous cultures to be maintained.

Anti-idiotype antibodies

Monoclonal antibodies recognize a specific epitope consisting of linear or conformational regions of a molecule. The variable (binding) region of the monoclonal antibody molecule, the idiotype, then consists of a region complementary to the antigen epitope. The production of antibodies directed against another antibody molecule will result in some antibodies specific for the idiotypic region; these antibodies are called anti-idiotype antibodies (Marx, 1985). Anti-idiotype antibodies (either monoclonal or polyclonal) will have a binding region (idiotype) which is an image of the original antigen epitope and may be utilized in place of antigen in some cases for the identification or induction of antibody. Recently anti-idiotype antibodies have been used in diagnostic tests (Potocjnak et al., 1982) and as an immunogen to induce protective immunity against viral (Uytdehaag and Osterhaus, 1985) and parasitic infections (Sacks, Esser and Sher, 1982).

Application of monoclonal antibody technology in agricultural research

Applications of monoclonal antibody technology in the agricultural sciences may be loosely categorized as either (1) analytical, either biochemical or immunological, (2) diagnostic, or (3) therapeutic or prophylactic. Within these categories several main areas of research within the agricultural sciences may be identified.

Analytical applications of monoclonal antibodies

Monoclonal antibodies have been successfully used in the structural and functional analysis of a variety of agriculturally important proteins. A protein may be considered a series of antigenic regions or epitopes to which specific antibodies bind. Monoclonal antibodies, therefore, may be used to define the number of epitopes on a molecule by determining the number of distinct sites at which non-competitive antibody binding occurs. These methods have been used to identify distinct neutralizing epitopes on the HN glycoprotein of Newcastle disease virus (Iorio and Bratt, 1983), bovine leukemia virus glycoprotein GP51 (Bruck et al., 1982) and the capsid polypeptide VP1 of foot and mouth disease virus A12 (Robertson, Morgan and Moore, 1984), as well as antigenic regions on the phytochrome molecule (Silberman et al., 1985) and tobacco mosaic virus (van Reganmortel, 1984).

Monoclonal antibodies can be used to correlate biologic functions with specific regions on proteins and identify and map these antigenic regions of molecules following enzymatic or chemical degradation. For example, Bruck et al. (1984) demonstrated that monoclonal antibodies specific for neutralizing epitopes of bovine leukemia virus precipitated a proteolytic fragment of molecular weight 15 K from the virus envelop glycoprotein and showed that glycosylation was necessary for antibody binding. Similarly, Robertson et al. (1984) localized two neutralizing epitopes of foot and mouth disease virus within two specific amino acid sequences using competitive binding experiments with labeled monoclonal antibodies and competing virus and synthetic polypeptide fragments.

Monoclonal antibody binding can also shed light on the conformational structure of antigens. Antibody binding can define both continuous and discontinuous epitopes; continuous epitopes retain antibody binding properties with or without conformational integrity. Conversely, discontinuous or conformational epitopes do not bind antibody following denaturation (Van Regenmortel, 1984). Similarly, antibodies may define cryptic epitopes, those antigenic sequences not exposed or available for antibodies in native conformation. During immune processing such regions may be exposed and appear to the host immune systems as foreign (antigenic). Such regions will only be recognized analytically by binding antibodies to denatured antigen.

Monoclonal antibodies have proven extremely useful for identification, quantitation, localization, and isolation of antigenic molecules of interest in the agricultural sciences. In the plant sciences, monoclonal antibodies have been used as immunocytochemical reagents to localize auxin carriers (Jacobs, 1984), phytochrome (Pratt, 1984), and cytokinins (Brandon, 1984).

In the animal sciences, monoclonal antibodies are proving to be extremely valuable as reagents for the identification and monitoring of the complex cellular interactions that regulate the immune response of livestock species. Recent studies have generated monoclonal antibodies that define specific subsets (T_h, T_s, T_c cells) of bovine (Davis, Perryman and McGuire, 1984), equine (Newman, Beegle and Antczak, 1984) and porcine (Peskovitz, Lunney and Sachs, 1984, 1985) leukocyte populations. Other studies have produced monoclonal antibodies that recognize specific classes and subclasses of bovine (Srikumaran et al. 1982) and porcine (Paul, Van Deusen and Mengling, 1985) immunoglobulins. The availability of these reagents will allow the dissection of the immune response to infectious agents in livestock species.

Monoclonal antibodies are currently being used as reagents to identify a subset of leukocyte antigens, gene products of the major histocompatability complex, as a method for studying the genetic regulation of immunity to disease in livestock. Using three independent swine lymphocyte antigen haplotypes in inbred miniature swine, Lunney (1984) have recently correlated these genetic differences with in vivo antibody and in vitro cellular responses to defined antigens. In the future correlation of differences in major histocompatibility complex loci in swine and other species with immune responses to purified antigens from infectious agents should provide information useful in the development of genetically resistant breeds of livestock.

Monoclonal antibodies are also being utilized as biochemical reagents for the affinity purification of biologically important molecules. In comparison to other methods such as chromatographic separations, monoclonal antibodies provide for a high degree of resolution (Chase, 1984), and are particularly useful for the recovery of recombinant proteins from fermentation products. The use of monoclonal antibodies as affinity reagents for the isolation of bovine somatotropin produced in *E. coli* has recently been demonstrated (Krivi and Rowold, 1984).

Diagnostic applications

Monoclonal antibodies are revolutionizing the fields of diagnosis and classification. Because of the property of specificity, monoclonal antibodies are often able to select between very closely related epitopes and hence closely related molecules or organisms. In addition to the properties of specificity, hybridoma produced reagents provide a source of homogeneous antibodies which can be used to standardize testing in all participating laboratories. The use of monoclonal antibodies in enzyme immunoassays and radioimmunoassays result in tests with a high degree of sensitivity.

Plant pathogens including viruses, fungi, bacteria, mycoplasma and nematodes are responsible for billions of dollars in economic losses in agricultural commodities annually. To prevent the spread of disease in plants or seed stock national and international certification and quarantine programs have been established. Antibody-based diagnostic tests have been applied to the detection of plant pathogens including potato viruses, ilarviruses, as well as bacterial, fungal and nematode diseases (Lankow et al., 1984); however, these serologic tests which use polyclonal sera have suffered limitations of specificity, reproducibility, and availability of reagents. Considerable emphasis has been placed on the production of monoclonal antibodies to be used in serology for plant disease diagnosis (Hsu, Jordan and Lawson, 1984). Recent studies have made available monoclonal antibodies with specificity for a variety of plant viruses including potyviruses (Hill, Hill and Durand, 1984) ilarviruses (Halk et al., 1984), alfalfa mosaic virus (Halk et al., 1984), the bacterial agents of citrus canker (Alverez, Benedict and Mizumoto, 1984) and potato ring rot (Magee, Beck and Ristow, 1984), and the fungi causing dwarf bunt disease in wheat (Banowetz, Trione and Krygier, 1984). Further use of monoclonal antibodies should greatly benefit the field of plant disease diagnostics.

In livestock, diagnosis of infectious disease requires the demonstration of the presence of the causative agent or in some cases the presence of a specific immune response to the causative agent. Monoclonal antibodies are rapidly replacing polyclonal antibodies for use in detecting infectious agents in host tissues. Where infectious agents are not detectable due to sequestering of antigens or presence in very low concentrations, detection may rely on a host antibody response to the organism. In this case monoclonal antibodies have been useful in identifying and isolating, by affinity chromatography, specific antigens for detecting antibody responses in binding assays (Gamble, 1984b).

In viral diagnosis, monoclonal antibodies have recently been used to identify the presence of bovine enteric coronavirus (Crouch, Raybould and Acres, 1984) and antibody responses to infection with blue tongue virus (Anderson, 1984) and bovine leukemia virus (Mammerickx et al., 1984). For bacterial disease, monoclonal antibodies have been used to detect the presence of pili (Mills and Tietze, 1984) and enterotoxins (Thompson et al., 1984) produced by enterotoxigenic strains of *E. coli* causing bovine and porcine neonatal diarrhea and for lipopolysaccharide antigens differentiating strains of *Brucella* (Schurig, Hammerberg and Finkler, 1984; Quinn, Campbell and Phillips, 1984).

In the field of parasitology, monoclonal antibodies have been used for the differential diagnosis of coccidia species infecting poultry (Augustine and Danforth, 1984), and for the detection of trichinosis in swine (Gamble and Graham, 1984).

In addition to detecting pathogenic organisms in live animals, monoclonal antibodies are being used to identify and quantitate the presence of pathogenic contaminants (Mattingly, 1984) and chemical residues (Bishop and White, 1984) in food products. For example, an ELISA test has been developed using monoclonal antibodies which can detect all strains of *Salmonella* occurring in meat samples in half the time of conventional detection methods (Mattingly, 1984).

Therapeutic and prophylactic applications

Early in the history of monoclonal antibody technology there was great hope that science had found a "magic bullet" that would cure cancer. While such has not been the case, monoclonal antibodies have proven extremely useful in tumor imaging and diagnosis of other human diseases.

The prevention and treatment of animal disease has also benefited greatly from monoclonal antibody technology. The best examples of this technology have been the identification and isolation of antigens which can be used to vaccinate against infectious agents. Strategies for the use of monoclonal antibodies to identify protective antigens differ with respect to the type of infectious agent being studied. For example, delineation of immunogenic epitopes associated with viral diseases may be accomplished by demonstrating the role of monoclonal antibody binding in viral neutralization in vitro or by passive transfer or antibodies in vivo. Epitopes inducing specific neutralizing antibodies have thus been delineated for foot and mouth disease capsid polypeptides (Morgan et al., 1984; Robertson et al., 1984), Marek's disease virus glycoproteins (Ikuta et al., 1984) and polypeptides of bovine herpesvirus 1 (Collins et al., 1984). Control of *E. coli* diarrhea in neonatal calves and swine has relied largely on maternal vaccination. Recent studies have shown that oral administration of monoclonal antibodies specific for the K99 pilus antigen of enteorotoxogenic *E. coli* passively protects animals in laboratory and field conditions (Morter, 1984; Sadowski, Wilson and Sherman, 1984). This targeted therapeutic use of monoclonal antibodies might obviate the need for universal vaccination of cows or sows.

Significant losses in livestock and poultry are caused by intracellular protozoan parasites including *Theileria* (East Coast fever in cattle), *Babesia* in cattle, and coccidia (*Eimeria* sp. in poultry). Monoclonal antibodies have been developed for *Theileria* (Dobbelaere et al., 1984; Musoke et al., 1984) and *Eimeria* (Danforth, 1983) which have antiparasite activity in in vitro infectivity neutralization assays. These experiments suggest that antigens identified by these monoclonal antibodies may be effective for active immunization of livestock. Such has been the case for antigens of *Babesia bovis;* an antigen with molecular weight of 44 K isolated by affinity chromatography using monoclonal antibodies has been shown to induce significant immunity in cattle (Wright et al., 1983).

Nematode infections in livestock account for hundreds of millions of dollars annually in production losses and medication costs, however, little progress has been made in the search for immunological controls for these parasites. Because of the complexity of antigens and life cycles of these parasites, evaluation of specific antigens has proven difficult. In addition these parasites often act to alter or suppress the host immune response. Monoclonal antibody technology has provided a tool with which parasitologists may dissect the immune response by identifying single antigens and determining their role in immunity. Recent studies have shown that single helminth antigens, isolated using monoclonal antibodies, are capable of stimulating protective immunity against challenge infection with the parasitic nematode of swine *Trichinella spiralis* (Gamble, 1985; Silberstein and Despommier, 1984).

Conclusions

It is clear that the availability of monoclonal antibody technology has contributed significantly to solving problems in the agricultural sciences, as it has in other disciplines. With an increased need for improved agricultural technologies worldwide, the impact of this technology will undoubtedly grow and continue to contribute to the improvement of crop and animal quality and productivity.

References

Alvarez, A., A. Benedict and C. Mizumoto. 1984. *Multiple bacterial etiologies in a citrus epidemic.* P. 311 in N.J. Stern and H.R. Gamble, eds., *Hybridoma Technology in Agricultural and Veterinary Research.* Rowman and Allanheld, Totowa, New Jersey.

Anderson, J. 1984. *Use of monoclonal antibody in a blocking ELISA to detect group specific antibodies to bluetongue virus.* J. Immunol. Meth. 74:139–149.

Augustine, P.C. and H.D. Danforth. 1984. *Use of monoclonal antibodies to locate Eimeria sporozoites (Protozoa) in intestinal sections.* Proc. Helminthol. Soc. Wash. 51:361–362.

Banowetz, G.M., E.J. Trione and B.B. Krygier. 1984. *Immunological comparison of teliospores of two wheat bunt fungi, Tilletia species, using monoclonal antibodies and antisera.* Mycologia 76:51–62.

Bishop, J.R. and C.H. White. 1984. *Antibiotic residue detection in milk – a review.* J. Food Protect. 47:647–652.

Borrebaech, C.A.K. 1984. *Dependence on T-cell-replacing factor and immunogenic dose for the production of monoclonal antibodies using the in vitro immunization technique.* Mol. Immunol. 21:841–845.

Brandon, D.L. 1984. *Monoclonal antibodies for phytohormone research.* Pp. 246–250 in N.J. Stern and H.R. Gamble, eds., *Hybridoma Technology in Agricultural and Veterinary Research.* Rowman and Allanheld, Totowa, New Jersey.

Bruck, C., S. Mathol, D. Portetelle, J.D. Franssen, P. Herion and A. Burny. 1982. *Monoclonal antibodies define eight independent antigenic sites on the bovine leukemia virus envelope glycoprotein gp51.* Virology 122:342–352.

Bruck, C., N. Rensonnet, D. Portetelle, Y. Cleuter, M. Mammerickx, A. Burny, R. Manoun, B. Guillemain, M.J. Vander Maaten and J. Ghysdael. 1984. *Biologically active epitopes of bovine leukemia virus.* Virology 136:20–31.

Change, T.H., Z. Steplewski and H. Koprowski. 1980. *Production of monoclonal antibodies in serum free media.* J. Immunol. Meth. 39:369–375.

Chase, H.A. 1984. *Affinity separations utilizing immobilized monoclonal antibodies – a new tool for the biochemical engineer.* Chem. Eng. Sci. 39:1099–1125.

Collins, J.K., A.C. Butcher, C.A. Riegel, V. McGrane, C.D. Blair, Y.A. Teramoto and S. Winston. 1984. *Neutralizing determinants defined by monoclonal antibodies on polypeptides specified by bovine herpesvirus 1.* J. Virol. 52:403–409.

Crouch, C.F., T.J.G. Raybould and S.D. Acres. 1984. *Monoclonal antibody capture enzyme-linked immunosorbent assay for detection of bovine enteric coronavirus.* J. Clin. Microbiol. 19:388–393.

Danforth, H.D. 1983. *Use of monoclonal antibodies directed against Eimeria tenella sporozoites to determine stage specificity and in vitro effect on parasite penetration and development.* Am. J. Vet. Res. 44:1722–1727.

Davis, W.C., T. Yilma, L.E. Perryman and T.C. McGuire. 1985. *Perspectives on the application of monoclonal antibody and transfection technology in veterinary microbiology.* Prog. Vet. Microbiol. Immunol. 1:1–24.

Davis, W.C., L.E. Perryman and T.C. McGuire. 1984. *The identification and analysis of major functional populations of differentiated cells.* Pp. 121–150 in N.J. Stern and H.R. Gamble, eds., *Hybridoma Technology in Agricultural and Veterinary Research.* Rowman and Allanheld, Totowa, New Jersey.

Dobbelaere, D.A.E., P.R. Spooner, W.C. Barry and A.D. Irvin. 1984. *Monoclonal antibody neutralizes the sporozoite stage of different Theileria parva stocks.* Parasite Immunol. 6:361–370.

Galfre, G. and C. Milstein. 1981. *Preparation of monoclonal antibodies: strategies and procedures.* Meth. Enzymol. 73:3–46.

Gamble, H.R. 1984a. *Screening hybridomas for antibody production.* Pp. 26–32 in N.J. Stern and H.R. Gamble, eds., *Hybridoma Technology in Agricultural and Veterinary Research.* Rowman and Allanheld, Totowa, New Jersey.

Gamble, H.R. 1984b. *Application of hybridoma technology to the diagnosis of parasitic disease.* Vet. Parasitol. 14:251–261.

Gamble, H.R. 1985. *Trichinella spiralis: Immunization of mice using monoclonal antibody affinity-isolated antigens.* Exp. Parasitol. (in press).

Gamble, H.R. and C.E. Graham. 1984. *Monoclonal antibody-purified antigen for the immunodiagnosis of trichinosis.* Am. J. Vet. Res. 45:67–74.

Goldsby, R.A., S. Srikumaran and A.J. Guidry. 1984. *Cell culture and the origins of hybridoma technology.* Pp. 8–14, in N.J. Stern and H.R. Gamble, eds., *Hybridoma Technology in Agricultural and Veterinary Research.* Rowman and Allanheld, Totowa, New Jersey.

Halk, E.L., H.T. Hsu, J. Aebig and J. Franke. 1984. *Production of monoclonal antibodies against three ilarviruses and alfalfa mosaic virus and their use in serotyping.* Phytopathology 74: 367–372.

Haas, J.B. and R.H. Kennett. 1980. *Characterization of hybridoma immunoglobulins by sodium dodecylsulfate-polyacrylamide gel electrophoresis.* Pp. 407–411 in R.H. Kennett, T.J. McKearn, and K.D. Bechtol, eds., *Monoclonal Antibodies: Hybridomas: A New Dimension in Biological Analysis.* Plenum Press, New York.

Hill, E.K., J.H. Hill and D.P. Durand. 1984. *Production of monoclonal antibodies to viruses in the potyvirus group: use in radioimmunoassay.* J. Gen. Virol. 65:525–532.

Hsu, H.T., R.L. Jordan and R.H. Lawson. 1984. *Monoclonal antibodies and plant viruses.* ASM News 50:99–102.

Ikuta, K., S. Veda, S. Kato and K. Hirai. 1984. *Identification with monoclonal antibodies of glycoproteins of Marek's disease virus and herpesvirus of turkeys related to virus neutralization.* J. Virol. 49:1014–1017.

Iorio, R.M. and M.A. Bratt. 1983. *Monoclonal antibodies to Newcastle disease virus: delineation of four epitopes on the HN glycoprotein.* J. Virol. 48:440–450.

Jacobs, M. 1984. *Basal localization of presumptive auxin transport carriers in pea stem cells.* Pp. 239–242, in N.J. Stern and H.R. Gamble, eds., *Hybridoma Technology in Agricultural and Veterinary Research.* Rowman and Allanheld, Totowa, New Jersey.

Jonak, Z.L., V. Braman and R.H. Kennett. 1984. *Transfection of primary mouse lymphocytes with human DNA: production of continuous cell lines producing monoclonal immunoglobulins.* Hybridoma 3:107–114.

Kearney, J.S., A. Radbruch, B. Liesegang and K. Rajewsky. 1979. *A new mouse myeloma cell line which has lost immunoglobulin expression that permits the construction of antibody-secreting hybridomas.* J. Immunol. 123:1548–1550.

Kennet, R.H. 1980. *Freezing of hybridoma cells.* P. 375, in R.H. Kennett, T.J. McKearn, and K.B. Bechtol, eds., *Monoclonal Antibodies; Hybridomas: A New Dimension in Biological Analyses.* Plenum Press, New York.

Kohler, G. and C. Milstein. 1975. *Continuous cultures of fused cells secreting antibody of predefined specificity.* Nature 256:495–498.

Krivi, G.G. and E. Rowold Jr. 1984. *Monoclonal antibodies to bovine somatotropin: immunoadsorbent reagents to mammalian somatotropins.* Hybridoma 3:151–162.

Lampson, L.A. 1980. *Immunoprecipitation with monoclonal antibodies.* Pp. 395–397, in R.H. Kennett, T.J. McKearn, and K.B. Bechtol, eds., *Monoclonal Antibodies: Hybridomas: A New Dimension in Biological Analyses.* Plenum Press, New York.

Lampson, L.A. and C.A. Fisher. 1985. *Immunoblot specificity of monoclonal antibodies assayed against complex extracts.* Anal. Biochem. 144:55–64.

Lankow, R.K., S.H. Woodhead, R.J. Patterson, R. Massey and G. Schochetman. 1984. *Monoclonal antibody diagnostics in plant disease management.* Plant Dis. 68:1100–1101.

Letchworth, G.J. and J.A. Appleton. 1984. *Methods for the production of monoclonal antibodies.* USDA, Agriculture Handbook 630.

Littlefield, J.W. 1964. *Selection of hybrids from mating of fibroblasts in vitro and their presumed recombinants.* Science 145:709–710.

Lunney, J.K. 1984. *Monoclonal antibodies in the analysis of immune responses in swine.* Pp. 298–301 in N.J. Stern and H.R. Gamble, eds., *Hybridoma Technology in Agricultural and Veterinary Research.* Rowman and Allanheld, Totowa, New Jersey.

Magee, W.E., C.F. Beck and S.S. Ristow. 1984. *Monoclonal antibodies specific for Corynebacterium sepedonicum, the causative agent of potato ring rot.* P. 326 in N.J. Stern and H.R. Gamble, eds., *Hybridoma Technology in Agricultural and Veterinary Research.* Rowman and Allanheld, Totowa, New Jersey.

Mammerickx, M., D. Portetelle, C. Bruck and A. Burny. 1984. *Use of an ELISA involving monoclonal antibody for the detection of antibodies against bovine leukemia virus in a herd with a high incidence of enzootic bovine leukosis.* Zbl. Vet. Med. 31:210–218.

Marx, J.L. 1985. *Making antibodies without the antigens.* Science 228:162–165.

Mattingly, J.A. 1984. *An enzyme immunoassay for the detection of all Salmonella using a combination of a myeloma protein and a hybridoma antibody.* J. Immunol. Meth. 73:147–156.

McKearn, T.J. 1980. *Cloning of hybridoma cells by limiting dilution in fluid phase.* P. 374 in R.H. Kennett, T.J. McKearn, and K.B. Bechtol, eds., *Monoclonal Antibodies: Hybridomas: A New Dimension in Biological Analyses.* Plenum Press, New York.

McHugh, Y.E. 1984. *In vitro immunization for hybridoma production.* Pp. 216–221 in N.J. Stern and H.R. Gamble, eds., *Hybridoma Technology in Agricultural and Veterinary Research.* Rowman and Allanheld, Totowa, New Jersey.

Mills, K.W. and K.L. Tietze. 1984. *Monoclonal antibody enzyme-linked immunosorbent assay for the identification of K99-positive Escherichia coli isolates from calves.* J. Clin. Microbiol. 19:498–501.

Morter, R.L. 1984. *Genetically engineered monoclonal antibody for E. coli diarrhea in calves.* Mod. Vet. Prac. June, pp. 427–428.

Morgan, D.O., B.H. Robertson, D.M. Moore, C.A. Timpone and P.D. McKercher. 1984. *Aphthoviruses: control of foot-and-mouth disease with genetically engineered vaccines.* Pp. 135–145 in E. Kurstak, ed., *Control of Virus Diseases,* Marcel Dekker, Inc., New York.

Musoke, A.J., V.M. Nantulya, F.R. Rurangirwa and G. Buscher. 1984. *Evidence for a common antigenic determinant on sporozoites of several Theileria parva strains.* Immunology 52: 231–238.

Newman, M.J., K.H. Beegle and D.F. Antczak. 1984. *Xenogenic monoclonal antibodies to cell surface antigens of equine lymphocytes.* Am. J. Vet. Res. 45:626–632.

Nilsson, K., W. Scheirer, O.W. Merten, L. Ostberg, E. Liehl, H.W.D. Katinger and K. Mosbach. 1983. *Entrapment of animal cells for production of monoclonal antibodies and other biomolecules.* Nature 310:629–630.

Oi, V.T. and L.A. Herzenberg. 1980. *Immunoglobulin-producing cell lines.* Pp. 351–372 in B.B. Mishell and S.M. Shiigi, eds., *Selected Methods in Cellular Immunology.* Freeman, San Francisco.

Paul, P.S., R.A. VanDeusen and W.L. Mengling. 1985. *Monoclonal precipitating antibodies to porcine immunoglobulin M.* Vet. Immunol. Immunopathol. 8:311–328.

Pescovitz, M.D., J.K. Lunney, and D.H. Sachs. 1984. *Preparation and characterization of monoclonal antibodies reactive with porcine PBL.* J. Immunol. 133:368–375.

Pescovitz, M.D., J.K. Lunney and D.H. Sachs. 1985. *Murine anti-swine T4 and T8 monoclonal antibodies: distribution and effects on proliferative and cytotoxic T cells.* J. Immunol. 134: 37–44.

Potocnjak, D., F. Zavala, R. Nussenzweig and V. Nussenzweig. 1982. *Inhibition of idiotype–anti-idiotype interaction for detection of a parasite antigen: a new immunoassay.* Science 215:1637–1639.

Pratt, L.H. 1984. *Application of monoclonal antibodies to the characterization of phytochrome.* Pp. 243–245 in N.J. Stern and H.R. Gamble, eds., *Hybridoma Technology in Agricultural and Veterinary Research.* Rowman and Allanheld, Totowa, New Jersey.

Quinn, R., A.M. Campbell and A.P. Phillips. 1984. *A monoclonal antibody specific for the A antigen of Brucella spp.* J. Gen. Microbiol. 130:2285–2289.

Raybould, T.J.G., C.F. Crouch, L.J. McDougall and T.C. Watts. 1985. *Bovine-murine hybridoma that secretes bovine monoclonal antibody of defined specificity.* Am. J. Vet. Res. 46:426–427.

Robertson, B.H., D.O. Morgan and D.M. Moore. 1984. *Localization of neutralizing epitopes defined by monoclonal antibodies generated against the outer capsid polypeptide, VP1, of foot-and-mouth disease virus A12.* Virus Res. 1:489–500.

Sacks, D.L., K.M. Esser and A. Sher. 1982. *Immunization of mice against African trypanosomiasis using anti-idiotypic antibodies.* J. Expl. Med. 155:1108–1119.

Sadowski, P.L., R.A. Wilson and D.M. Sherman. 1984. *A monoclonal antibody specific for the K99 fimbrial adhesin in Escherichia coli, useful for the reduction of mortality in neonatal pigs.* P. 331 in N.J. Stern and H.R. Gamble, eds., *Hybridoma Technology in Agricultural and Veterinary Research.* Rowman and Allanheld, Totowa, New Jersey.

Schurig, G.C., C. Hammerberg and B.R. Finkler. 1984. *Monoclonal antibodies to Brucella surface antigens associated with the smooth lipopolysaccharide complex.* Am. J. Vet. Res. 45:967–971.

Silberman, L.G., N. Datta, P. Hoops and S.J. Roux. 1985. *Characterization of monoclonal antibodies to oat phytochrome by competitive radioimmunoassays and comparative immunoblots of phytochrome peptides.* Arch. Biochem. Biophys. 236:150–158.

Silberstein, D.S. and D.D. Despommier. 1984. *Antigens from Trichinella spiralis that induce a protective response in the mouse.* J. Immunol. 123:898–904.

Srikumaran, S., A.J. Guidry and R.A. Goldsby. 1982. *Production and characterization of monoclonal antibodies to bovine immunoglobulin G2.* Am. J. Vet. Res. 43:21–25.

Srikumaran, S., A.J. Guidry and R.A. Goldsby. 1983. *Bovine x mouse hybridomas that secrete bovine immunoglobulin G1.* Science 220:522–524.

Thompson, M.R., H. Brandwein, M. Labine-Racke and R.A. Giannella. 1984. *Simple and reliable enzyme-linked immunosorbent assay with monoclonal antibodies for detection of Escherichia coli heat-stable enterotoxins.* J. Clin. Microbiol. 20:59–64.

Uytdehaag, F.G.C.M. and A.D.M.E. Osterhaus. 1985. *Induction of neutralizing antibody in mice against poliovirus type II with monoclonal anti-idiotypic antibody.* J. Immunol. 134:1225–1229.

Van Ness, J., U.K. Laemmli and D.E. Pettijohn. 1984. *Immunization in vitro and production of monoclonal antibodies specific to insoluble and weakly immunogenic proteins.* Proc. Nat. Acad. Sci. 81:7897–7901.

Van Regenmortel, M.H.V. 1984. *Molecular dissection of antigens by monoclonal antibodies.* Pp. 43–82 in N.J. Stern and H.R. Gamble, eds., *Hybridoma Technology in Agricultural and Veterinary Research.* Rowman and Allanheld, Totowa, New Jersey.

Wright, I.G., M. White, P.D. Tracey-Patte, R.A. Donaldson, B.V. Goodger, D.J. Waltisbuhl and D.F. Mahoney. 1983. *Babesia bovis: isolation of a protective antigen by using monoclonal antibodies.* Inf. Immun. 41:244–250.

Zimmerman, V. and J. Vienken. 1984. *Electrofusion of cells.* Pp. 173–199 in N.J. Stern and H.R. Gamble, eds., *Hybridoma Technology in Agricultural and Veterinary Research.* Rowman and Allanheld, Totowa, New Jersey.

Pronuclear transplantation reveals progressive imprinting of the genome during gametogenesis

CLEMENT L. MARKERT
Department of Biology, Yale University, New Haven, CT 06520, U.S.A.

Abstract Attempts to clone mammals by nuclear transplantation or to create homozygotes by diploidizing haploid eggs produced by parthenogenesis or by microsurgical removal of one pronucleus have failed because of prior nuclear differentiation. Moreover, pronuclear transplantation into enucleated eggs has in several laboratories resulted in the construction of eggs equipped with 2 male pronuclei, 2 female pronuclei, or one male and one female pronucleus. Only eggs equipped with both a male and a female pronucleus develop to term. The initial genetic makeup of the gamete nuclei is irrelevant. At least three different microsurgical techniques have been used, and the results all point to the same conclusion: nuclei are "imprinted" during gametogenesis so that male and female pronuclei are functionally different and this difference is required for normal development. This imprinting may be the earliest expression of the nuclear differentiation that occurs during normal cell differentiation and that progressively restricts the developmental capacity of nuclei. The biological behavior of male and female genomes suggests that chromosomal interactions are essential to generating specific patterns of gene function and that members of a homologous pair of chromosomes differ in function in accord with the sex of origin. This difference is replicable, changeable, and erasable, but normally persists until a new cycle of gametogenesis occurs.

Introduction

The removal of pronuclei from fertilized mouse eggs, or the transfer of pronuclei between zygotes offers exciting possibilities for making homozygous mammals, or for manufacturing strains of mammals from a single individual, particularly from males, or perhaps most important, offers the prospect of cloning mammals. Pronuclear transfer is a special case of the more general technology of nuclear transplantation from individual somatic cells of an embryo into enucleated eggs. This procedure equips the egg with the genetic makeup of the cell providing the nucleus and is fundamental to successful cloning of mammals. It must be stated at the outset that the goals of this nuclear transfer technology have not been achieved and at present do not seem achievable without substantial increases in our knowledge of nuclear differentiation during development.

Microsurgical manipulation

First, what has been achieved by microsurgical manipulation of the nuclei of mammalian zygotes or early embryos? Mammalian eggs at first glance seem very fragile, and not likely to survive the kind of treatment experimental embryologists have inflicted on amphibian eggs and nuclei in the classic work on nuclear transplantation. Nuclear transplantation in amphibians began more than 30 years ago and is still a fruitful area of investigation. But even in amphibia successful transplantation of adult nuclei into enucleated eggs does not lead to the development of adult individuals (Briggs, 1977; DiBerardino, 1980; McKinnell, 1978). With mammals the transplantation procedure is far less successful (McGrath and Solter, 1984c) than with amphibians. T.P. Lin (1966) first showed that mouse eggs were tough enough to withstand microsurgical injection of small drops of oil into the egg. Such eggs when fertilized continued to cleave and presumably could develop into fully formed mice. With Lin's achievement as a stimulus, my laboratory began many years ago to develop microsurgical techniques for nuclear transplantation in mouse eggs. Among the first experiments was the removal of one pronucleus from a fertilized mouse egg leaving either the male or female pronucleus behind. The haploid zygote was then treated with cytochalasin B, or more recently with the less toxic cytochalasin D, until the pronucleus replicated equipping the zygote with two identical haploid pronuclei, either gynogenetic or androgenetic. The two pronuclei next came together in the center of the egg, the nuclear membranes broke down, and the two sets of haploid chromosomes assembled on the same spindle to produce a diploid cell which then cleaved to make an early embryo (Markert and Petters, 1977). Such embryos were completely homozygous, and virtually identical to parthenote embryos produced by the LT/Sv strain of mice (Eppig, 1978). These parthenote embryos are also diploid, will proceed through the early developmental stages, and a few will even implant if given the opportunity in a suitably prepared uterus (Kaufman, Barton and Surani, 1977). In our hands none of the hundreds of diploidized homozygous embryos we produced ever succeeded in developing to term after transfer to the uteri of pseudopregnant mice (Markert and Seidel, 1981; Markert, 1982). Such homozygotes did show extensive ability to differentiate many kinds of cells and tissues. Some developed about halfway through gestation (Markert, 1982). These results were disappointing and involved only enucleation and not the transfer of a pronucleus from one zygote to another. Several other laboratories used the same strategy in an effort to produce homozygous mice, but without success (Modlinski, 1980). Since we used inbred strains, lethal genes were scarcely responsible; in fact, our homozygotes were virtually identical to other members of the same strain produced by sexual reproduction through many generations of brother–sister matings. After one published claim (Hoppe and Illmensee, 1977) that these homozygous embryos could in fact develop to term to produce truly homozygous mice that were capable of reproducing, many laboratories attempted to repeat the reported results. All failed, and with much the same results. The early homozygous embryos could develop, they could implant, but none proceeded all the way to term (Surani and Barton, 1983; Surani, Barton and Norris, 1984; Mann and Lovell-Badge, 1984; McGrath and Solter, 1984b).

Pronuclei transfer

Transferring pronuclei from one egg to another is technically difficult. For one thing, the pronuclei are very large, about 25 μm in diameter. It is much easier to transfer nuclei from adult somatic cells into eggs than it is to transfer pronuclei. New, efficient techniques for enucleation and nuclear transfer developed in the laboratory of McGrath and Solter (1983) at the Wistar Institute allowed the transfer of very large nuclei into eggs. Their procedure begins by treating the egg with both cytochalasin and colcemid. Such treatment disaggregates the microfilaments and the microtubules and makes the zygotes very flaccid. It was then possible to suck a pronucleus into a large-bore pipette without rupturing the plasma membrane of the egg, and then to pinch off the pronucleus surrounded by a small amount of cytoplasm and enclosed by plasma membrane. In effect, a very small cell (karyoplast) containing a pronucleus was formed. This karyoplast could then be fused with a zygote, either containing two pronuclei, one pronucleus (one having been removed), or no pronuclei (the zygote having been completely enucleated). Membrane fusion was effected by means of inactivated Sendai virus. We have also produced such fusions with polyethyleneglycol. Microinjection through the plasma membrane is also possible, but the mechanical injection almost always proves lethal in early cleavage stages, if cleavage occurs at all.

The very efficient and simple technique of McGrath and Solter (1983) has made it possible for other laboratories to do many pronuclear transfers and the results are now quite clear. It is possible to equip fertilized mouse eggs with a variety of pronuclear combinations. The two original pronuclei can be removed and replaced with two pronuclei obtained from other fertilized eggs of the same or different genotype. The nuclear manipulations were designed to reveal the capacities for normal development of eggs equipped with either two female pronuclei, two male pronuclei, or with a male and a female pronucleus obtained from different fertilized eggs. In all cases, whenever a fertilized egg is equipped with just female pronuclei or just male pronuclei, development never proceeds to term. Eggs equipped with a male and a female pronucleus, as happens in normal fertilization or by design in microsurgical transplantation, do frequently develop all the way to term (Surani, Barton and Norris, 1984; McGrath and Solter, 1984b). Such development occurs without regard to the genetic makeup of individual pronuclei. Whether the pronuclei are from the same mouse strain or not is irrelevant. What is clearly required is a nucleus from oogenesis and another from spermatogenesis in order for development to proceed to term. The specific nature of the abnormalities evident in the moribund embryos developed from two male or two female pronuclei differ, but lethality is the endpoint in all cases.

Sex specific imprinting of genome

These data compel us to recognize that a sex specific imprinting of the genome must have taken place during gametogenesis. Moreover the consequence of this imprinting to produce functionally different genomes in the male and female germ lines occurs without regard to the specific genetic makeup of the DNA; only the embryonic history of the nucleus is important. It should be emphasized that the homozygotes, embryos equipped with two male or two female pronuclei, do develop readily through early cleavage stages. They do sometimes implant, but they all die by about the middle of gestation. Surani, Barton and Norris (1984), McGrath and Solter (1984b), and Mann and Lovell-Badge (1984) demonstrated decisively that a complementary contribution from the male and female genomes is required for normal development, but not for cell survival. The nature of this male/female complementation in gamete genomes is a profound mystery, but clearly of extraordinary importance to our understanding of the general processes of nuclear differentiation during development. We have not identified the molecular basis for imprinting the genomes of the male and female germ lines in complementary ways. Perhaps sex-imprinted structural arrangements of the chromosomes from the two pronuclei lie at the basis of this complementary requirement (Markert, 1982).

The most dramatic demonstration that the differentiation of identical genomes of eggs and sperm is responsible for the complementary requirements for continued development after fertilization was made by Dr. Wesley Whitten (1975). He worked with a strain of mice that frequently proved to be hermaphroditic. Ovo-testes or sometimes an ovary on one side and a testis on the other were common. He examined many such mice at birth. If both a testis and ovary were present, the ovary was transplanted to a normal female mouse from which the ovaries had been removed. When the female bearing the transplanted ovary reached adulthood, she was mated with the original mouse hermaphrodite, a mouse now equipped only with a testis. Two such mice were mated successfully and produced offspring; thus a single individual, the original hermaphrodite, genetically speaking, reproduced with itself. This experiment demonstrates that identity of genotype is not the root cause of the inability of homozygotes and parthenotes to develop. Rather it is the requirement that the male and female genomes be imprinted differently during gametogenesis.

A final reinforcement for this general conclusion is provided by an abnormal kind of pregnancy in women. Occasionally a pregnant woman produces an hydatidiform mole, a conceptus devoid of any true embryo, but consisting of extra-embryonic tissues that can become malignant. Cytological and chromosomal examination of such moles has proved that the genome is derived entirely from sperm. No chromosomal contribution from the egg is evident. Apparently, since the moles are diploid, the chromosomes of a single sperm were doubled and the female pronucleus disappeared (Check, 1978; Kaji and Ohama, 1977). Such hydatidiform moles are the analogs of diploidized homozygous mice produced by microsurgical removal of the female pronucleus from a fertilized egg followed by doubling of the male genome. Occasionally, hydatidiform moles seem to have been derived from two different sperm. Even so, no female genome is present. The moles derived from two different sperm discourage any belief that injecting two sperm into a mouse egg or into the eggs of any other mammal would lead to normal development (Markert, 1983). If normal development were possible, the use of sperm from a prize bull, for example, could generate a new strain of cattle, derived entirely from the prize bull. Such a strain, of course, would not be a clone, but would represent a subset of the genetic material available in the original bull. Such strains might prove very useful. However, at the present time there is no reason to believe that such a procedure would be successful.

The usefulness of homozygotes in animal breeding and the obvious economic benefit of cloning valuable livestock encourages continued research (Markert, 1984a, 1984b). Although the first efforts with pronuclear transfer have failed to give useful practical results, an attractive possibility beckons us onward. Parthenotes and homozygotes manufactured by diploidizing a single pronucleus (gynogenetic or androgenetic) do not lead directly to cell death, but only to embryonic death. Homozygotes derived entirely from male pronuclei or from female pronuclei cannot develop to term by themselves but with sufficient help from normal diploid cells a substantial number of these homozygous cells should be able to participate in all aspects of development. Whether such cells can produce gametes, however, is still uncertain. Their analogs, parthenote cells, can under similar conditions in chimeras produce gametes, but only rarely. The rescue of parthenotes or diploidized homozygotes in chimeras could prove very important to the manufacture of valuable new strains of mammals provided that gametes can be derived from the parthenote or homozygote cells.

Especially instructive are the results with parthenotes in the LT/Sv strain by mice. In the most dramatic case, parthenote embryos were rescued by combining them with normal diploid embryos to produce chimeras (Stevens, Varnum and Eicher, 1977; Surani, Barton and Kaufman, 1977). At least in one case (Stevens, 1978), such a chimera produced offspring derived from gametes that differentiated from the parthenogenetic component. This particular chimera was reproductively equivalent to a successful parthenote or to a successful diploidized homozygote. Such chimeras could be used for the production of the next generation and would therefore offer the advantages originally anticipated from homozygotes. No one has yet succeeded, however, in rescuing a diploidized gynogenetic or androgenetic microsurgically produced embryo. We are attempting this rescue now, but we have not so far succeeded despite a fairly extensive effort. The rescue of the parthenote was a rare event and rescuing homozygotes may be at least as rare.

Cloning by nuclear transplantation

To simply transfer nuclei from somatic cells of an already proved heterozygous phenotype would generate an identical twin of the individual providing the nucleus, and would be the most welcome achievement of all, cloning by nuclear transplantation. It now seems that the failure to produce normal development with gynogenetic or androgenetic embryos is merely a dramatic illustration of early nuclear differentiation during development. Previously we thought nuclear differentiation began long after the cleavage stages. At least in amphibians nuclei are developmentally totipotent to the gastrula stage of development, when there are more than 10,000 cells in the embryo. But in mammals nuclear differentiation seems to start with the oocyte or spermatocyte. Thus we are forced to re-examine the most fundamental problem of developmental genetics: how is the genome regulated to produce sequential patterns of gene function?

Conclusion

A large amount of additional data can be mobilized from many branches of biology to demonstrate that the male genome is imprinted differently from the female genome, and is recognized to be different by the cytoplasm of the egg (Chandra, 1971). Perhaps the most relevant observation is X-chromosome inactivation. In marsupials it is always the male X-chromosome that is preferentially inactivated throughout the life of the female. X-chromosome inactivation, in other words, does not occur at random as is generally true in eutherian mammals. But even in mice the male and female X chromosome are recognized as different in the extra-embryonic membranes. The male-derived X is preferentially inactivated, and since inbred strains of mice have genetically identical X chromosomes, whether from the male or the female, we must attribute this preferential inactivation to a prior imprinting of the X during spermatogenesis. Yet another chromosome in mice, chromosome 17, functions differently depending upon the sex of origin. A mutant gene on chromosome 17 is lethal in late stages of embryonic development if introduced into the zygote via the maternal genome, but is viable if the same chromosome carrying the same mutant gene is introduced into the genome via the sperm (McGrath and Solter, 1984a). Thus out of the 20 haploid chromosomes of mice at least two already are well known to be imprinted differently by virtue of their sex of origin. One might hazard the guess that all, or nearly all, of the chromosomes are different by virtue of their sex of origin, whether from the sperm or the egg.

Many lines of evidence, such as the data on X-chromosome inactivation and re-activation demonstrate that the genome becomes progessively differentiated not in the nucleotide sequence, but in functional capacity by some as yet unknown mechanism. Whatever the nature of the mechanism, it must imprint genomes so that the imprinting is replicable but changeable under specific circumstances during development and ultimately erasable in the formation of gametes. Methylation of cytosine residues in DNA offers one possibility since we know that these methylation patterns are replicable, changeable, and erasable. When we discover the molecular basis or structural basis for this differentiation of the genome, we will obviously have a better understanding of how we might restore a differentiated genome to the early undifferentiated state, permitting an implanted genome to function in the egg so as to produce a new normal individual. When that day arrives, we will not only clone mammals, but we will know a great deal more about developmental genetics, and should be able to exploit such knowledge and techniques to enhance many aspects of the animal husbandry industry. But most important would be our greatly increased understanding of developmental biology.

Acknowledgments The original research reported here was supported by NSF Grant No. DAR-7910054.

References

Briggs, R. 1977. *Genetics of cell type determination.* Pp. 23–43 in L. Saxen and L. Weiss, eds., *Cell Interactions in Differentiation.* Academic Press, New York.

Chandra, H.S. 1971. *Inactivation of whole chromosomes in mammals and coccids: some comparisons.* Genet. Res. 18:265–276.

Check. W. 1978. *Chromosomes provide new view of hydatidiform moles.* JAMA 239–399.

DiBerardino, M.A. 1980. *Genetic stability and modulation of metazoan nuclei transplanted into eggs and oocytes.* Differentiation 17:17–30.

Eppig, J.J. 1978. *Developmental potential of LT/Sv parthenotes derived from oocytes matured in vivo and in vitro.* Dev. Biol. 65:224–249.

Hoppe, P.C. and K. Illmensee. 1977. *Microsurgically produced homozygous-diploid uniparental mice.* Proc. Nat. Acad. Sci. 74:5656–5661.

Kaji, T. and K. Ohama. 1977. *Androgenetic origin of hydatidiform mole.* Nature 268:633–634.

Kaufman, M.H., S.C. Barton and M.A. Surani. 1977. *Normal postimplantation development of mouse parthenogenetic embryos to the forelimb bud state.* Nature 265:53–55.

Lin, T.P. 1966. *Microinjection of mouse eggs.* Science 151:333–337.

Mann, J.R. and R.H. Lovell-Badge. 1984. *Inviability of parthenogenones is determined by pronuclei, not egg cytoplasm.* Nature 310:66–67.

Markert, C.L. and R.M. Petters. 1977. *Homozygous mouse embryos produced by microsurgery.* J. Expl. Zool. 201:295–302.

Markert, C.L. and G.E. Seidel, Jr. 1981 *Parthenogenesis, identical twins, and cloning in mammals.* Pp. 181–200 in *New Technologies in Animal Breeding*. Academic Press, New York.

Market, C.L. 1982. *Parthenogenesis, homozygosity, and cloning in mammals.* J. Heredity 73: 390–397.

Markert, C.L. 1983. *Fertilization of mammalian eggs by sperm injection.* J. Expl. Zool. 228: 195–201.

Markert, C.L. 1984a. *Cloning mammals: current reality and future prospects.* Theriogenology 21:60–67.

Markert, C.L. 1984b. *Genetic manipulation of mammalian embryos: current techniques and their potential usefulness in livestock improvement.* Pp. 13–19 in *Proceedings of the 10th International Congress on Animal Reproduction and Artificial Insemination. 4:II.*

McGrath, J. and D. Solter. 1983. *Nuclear transplantation in the mouse embryo by microsurgery and cell fusion.* Science 220:1300–1302.

McGrath, J. and D. Solter. 1984a. *Maternal T^{hp} lethality in the mouse is a nuclear, not cytoplasmic, defect.* Nature 308:550–551.

McGrath, J. and D. Solter. 1984b. *Completion of mouse embryogenesis requires both the maternal and paternal genomes.* Cell 37:179–183.

McGrath, J. and D. Solter. 1984c. *Inability of mouse blastomere nuclei transferred to enucleated zygotes to support development in vitro.* Science 226:1317–1319.

McKinnell, R.G. 1978. *Cloning: Nuclear Transplantation in Amphibia.* University of Minnesota Press, Minneapolis.

Modlinski, J.A. 1980. *Preimplantation development of microsurgically obtained haploid and homozygous diploid mouse embryos and effects of pretreatment with cytochalasin B on enucleated eggs.* J. Embryol. Expl. Morphol. 60:153–161.

Stevens, L.C. 1978. *Totipotent cells of parthenogenetic origin in a chimaeric mouse.* Nature 276:266–267.

Stevens, L.D., D.S. Varnum and E.M. Eicher. 1977. *Viable chimaeras produced from normal and parthenogenetic mouse embryos.* Nature 269:515–517.

Surani, M.A.H., S.C. Barton and M.H. Kaufman. 1977. *Development to term of chimaeras between diploid parthenogenetic and fertilized embryos.* Nature 270:601–603.

Surani, M.A.H. and S.C. Barton. 1983. *Development of gynogenetic eggs in the mouse: implications for parthenogenetic embryos.* Science 222:1034–1036.

Surani, M.A.H., S.C. Barton and M.L. Norris. 1984. *Development of reconstituted mouse eggs suggests imprinting of the genome during gametogenesis.* Nature 308:548–550.

Whitten, W.K. 1975. *Chromosomal basis for hermaphrodism in mice.* Pp. 189–205 in C.L. Markert and J. Papaconstantinou, eds., *The Developmental Biology of Reproduction.* Academic Press, New York.

Somaclonal and gametoclonal variation

D.A. EVANS
DNA Plant Technology Corporation, 2611 Branch Pike, Cinnaminson, New Jersey 08077, U.S.A.

Abstract For several years it has been recognized that introduction of plant cells into culture results in genetic changes. These genetic alterations have been recovered in the plants regenerated from cell cultures. More recently it has been recognized that this method of introducing genetic changes into crop plants could be used to develop new breeding lines. The technology of introducing genetic variation by using cell culture has been termed somaclonal and gametoclonal variation. Somaclonal variation has been detected and documented in tomato. In particular, several single gene mutations induced by somaclonal variation have been mapped to specific loci. As this variation represents a new tool for the plant breeder, breeding strategies for the use of this variation are presented and discussed. Somaclonal and gametoclonal variation are new tools for the geneticist and plant breeder that permit reduction in the time period for new variety development and that permit access to new classes of genetic variation.

Introduction

Even from a cursory examination of the tissue culture scientific literature, it is obvious that cell culture can be used to generate clones (genetic carbon copies) or variants. These two types of output represent completely different tools for a breeding program. For cell culture to become an effective breeding tool, it is imperative that, to the extent possible, these phenomena be controlled. Clones have application for asexual propagation of desirable, tested genotype, while variants are primarily used as donor material to develop new varieties or parent lines. Over the past several years, techniques have been developed that can result in either clonal fidelity or high frequency somaclonal variation. These techniques have led to the use of somaclonal and gametoclonal variation in several breeding programs. In this review these new techniques will be outlined with emphasis on the recovery of valuable variants.

Review of literature

Several recent reviews have discussed many of the published reports of somaclonal variation (cf. Evans, Sharp and Medina-Filho, 1984; Scowcroft, Larkin and Brettel, 1983; Orton, 1984). Variation was first detected in cultured cells and regenerated plants several years ago. The first detailed cytological evaluation was completed on regenerated sugarcane plants (Heinz and Mee, 1969). These authors also recognized quite early the value of variant plants regenerated from sugarcane cell

gene changes, and cytoplasmic genetic variants have all been detected among the progeny of somaclones.

Variation in chromosome number, particularly tetraploidy, $2n = 48$, was detected. In addition, among the first 230 regenerated tomato plants, 13 discrete nuclear gene mutations were identified in the R_1 generation. We have since evaluated several thousand regenerated plants from several commercial tomato varieties. A large number of single gene mutations have been identified. These include recessive mutations for male sterility, jointless pedicel, several fruit colors, lethal chlorophyll deficiency, virescence, and mottled leaf appearance, a semi-dominant mutation controlling isozymes of alcohol dehydrogenase, and dominant mutations controlling fruit ripening, growth habit, and disease resistance (Evans et al., 1984). In addition, several regulatory mutants have been detected which seem to control multiple biosynthetic pathways. For example, a virescent mutant defective in chlorophyll biosynthesis was also found to have orange flowers and orange fruit color (Evans and Sharp, 1983). Similarly, a mottled leaf color mutant controlled by a single recessive gene also resulted in mottled fruit with orange spots on red fruit (Evans et al., 1984). Both these nuclear mutations seem to exert control on pigments (chlorophyll and carotenoids) that are accumulated in the plastids, but which are in different biosynthetic pathways.

Genetic analysis was completed by evaluating self-fertilized progeny of selected plants for several generations, and by crossing new somaclones to known mutants when available. The genetic evaluation completed for the tangerine virescent mutant has been published (Evans et al., 1984). Tangerine-virescent plants were first identified among R_1 plants in the field as having orange flowers and fruit. Upon examination it was noted that each R_1 plant with orange fruit was also virescent, i.e. new, young, developing leaves were yellow as opposed to green. Single plant selections were made from these R_1 plants and based on segregation in the R_2 generation the R_1 plants could be classified as homozygous normal, heterozygous, or homozygous mutant (tv–tcl/tv–tcl). The homozygous R_1 plants all bred true in the R_2 and subsequent generations. The heterozygous R_1 plants segregated in a 3:1, normal:mutant, ratio in the R_2 generation. Among the segregating R_2 progeny the flower, fruit and leaf color defects cosegregated suggesting control of the two pigments, chlorophyll and lycopene, by a single, pleiotropic gene. To determine the gene location of tv–tcl, selected mutant R_1 plants were crossed to a known orange flower and fruit mutant (tangerine) in a complementation test. This mutant though, was not peliotropic for the leaf color trait of tv–tcl. All resulting hybrid plants had normal leaves, but contained orange flowers and fruit. Since the two mutants are non-complementing, we concluded that the new somaclone (tv–tc1) is a mutant for a new allele in a previously known gene at position 95 on the long arm of chromosome 10 (Evans and Sharp, 1983). This has been the most precise genetic characterization of a new somaclonal variant to date. Moreover, when the hybrids between tv–tcl/tv–tc1 and t/t were self-fertilized to evaluate the virescence character, a resulting 3:1 ratio for normal:virescence was obtained. Hence, it has been possible to conclude that this locus contains two elements that can mutate independently. The new tv allele is recessive to the t allele for virescence.

Several other genetic characters uncovered by somaclonal variation in tomato have been genetically characterized. These include recessive mutations for jointless pedicel and leaf and fruit pigment, semi-dominant mutation for an alcohol dehydrogenase isozyme, and dominant mutations for fruit color and resistance to *Fusarium oxysporum* race 2. Two of these single gene traits warrant further mention. One variant (j–tc2) was regenerated as a homozygous recessive mutation. The single gene mutation, for jointless pedicel, is economically valuable as it results in plants suitable for mechanical harvesting as the harvested fruit has stem attached. Presumably this somaclone originated by mutation, followed by mitotic recombination and subsequent shoot regeneration. This new mutation complements the known j mutation so that it is not encoded by the same gene and may represent a new mutation in an undescribed gene encoding for jointless pedicel. It is currently being tested for complementation with the other known jointless mutant, j-2. Second, the mottled mutation (m–tc1) appears to be distinct from any of the hundreds of previously reported tomato mutants. the m–tc1 variant is a chlorophyll deficient mutant that was identified in the greenhouse. The variegation is somewhat similar, but distinct, from previously reported mottled nuclear mutants of tomato or from a known plastome mutant of tomato. In addition, it is also possible to discern orange-red mottling on fruits of homozygous (m–tc1/m–tc1) plants. Once again, as with tv–tc1, the m–tc1 mutant has an effect on both chlorophyll and carotenoid pigments.

Based on several years of experimentation the following generalizations can be concluded regarding somaclonal variation in tomato (for detailed elaboration see Evans et al., 1984): (1) Chromosome number variation, particularly polyploidy, can be recovered in regenerated plants; (2) Several single gene mutations have been recovered in several different tomato varieties; (3) Somaclones include dominant, semi-dominant and recessive nuclear mutations; (4) The frequency of single

Table 1. Documented Somaclonal Variation Among Plants Regenerated from Tissue Cultures

Species	Altered Characteristics	Techniques	Inheritance Examined	Reference
Allium sativum	bulb size and shape, clove number, failure to flower, plant height, plant vigor, chromosome number	c	—	Novak, 1980
Ananas cosmosus	spine type, leaf color, leaf waxiness, foliage density, albino stripes	m	—	Wakasa, 1979
Apium graveolens (celery)	isozyme variation; chromosome structure	c	—	Browers and Orton, 1982 Orton 1983
Apium graveolens (celery)	*Fusarium* resistance	c	—	Pullman, et al., 1984
Arachis hypogeae (ground nut)	chromosome number	c	—	Bajaj, 1985
Avena sativa (oat)	chromosome number and structure	c	—	McCoy et al., 1982
Avena sativa	heteromorphic pairs, trisomics, monosomics, interchanges	c	—	McCoy, 1980
Brassica juncea (rapeseed)	seed color, oil content, vigor, fertility	c	+	George and Rao, 1983
Brassica napus	chromosome number	c	—	Sacristan, 1981
Brassica spp.	fasciation, variegation, chromosome number	c	—	Dunwell, 1981
Chrysanthemum × *morifolium*	flower color	c	—	Ben-Jaacov and Langhans, 1972
Chrysanthemum × *morifolium*	plant vigor, leaf morphology and hairness lateral bud growth, phyllotaxy	c	—	Sutter and Langhans, 1981

Daucus carota	erect stems, leaf dissection, leaf thickness, leaf color	c	—	Ibrahim, 1969
Dendrobium (orchid)	flower size and color, petal shape	c	—	Vajrabhaya, 1977
Fragaria × *ananassa* (strawberry)	vigor, leaf area and shape, pubescence density, yield per crown, harvest time, fruit weight	m	—	Swartz et al., 1981
Haworthia setata	chromosome number, vigor, leaf shape, leaf color, esterase, zymogram, chromosome associations, pollen fertility	c	—	Ogihara, 1981
Hordeum spp.	isoenzyme intensity, growth habit, head morphology, auricle size, chromosome	c	—	Orton, 1980
Hordeum vulgare	albinism, leaf shape, fertile tillers	c	+	Deambrogio and Dale, 1980
Lactuca sativa	leaf shape, leaf color, axillary bud, plant vigor	c	+	Sibi, 1976
Lactuca sativa (lettuce)	plant height and vigor, fertility, leaf color, chlorophyll pigment	c	+	Engler and Grogan, 1984
Lilium spp	plant vigor, leaf variegation	c	—	Stimart, Ascher, and Zagorski, 1980
Lolium spp	chromosome number, leaf shape and size, floral development, growth vigor, survival, perreniality	c	—	Alhloowalia, 1978
Lolium spp	leaf shape and size, floral development, plant vigor and survival, chlorophyll pigment	c	—	Ahloowalia, 1983
Lotus corniculatus (bird's foot trefoil)	plant height, leaflet shape, fertility seed yield, stems per plant	c, m	—	Orshinsky and Tomes, 1984

Table 1 (continued)

Species	Altered Characteristics	Techniques	Inheritance Examined	Reference
Lycopersicon esculentum	plant height, vigor, days to flower, leaf shape and size	c	+	Sibi, 1982
L. esculentum (tomato)	fruit color, growth habit, fertility, flower color, chlorophyll content	c	+	Evans and Sharp, 1983
L. esculentum (tomato)	fruit color, disease resistance, alcohol dehydrogenase isozymes	c	+	Evans et al., 1984
Lycopersicon peruvianum	fertility (S-locus)	c	+	Sree Ramulu, 1982
Medicago sativa	chromosome number, leaf shape, petiolule length, herbage yield, plant height, shoot length, rooting response	c	—	Reisch and Bingham, 1981
Medicago sativa	cotyledon number and shape, leaf morphology, plant vigor	c	+	Johnson, Stuteville and Skinner, 1980
Medicago sativa (lucern)	Verticillium wilt resistance, leaf shape and size, leaf thickness, petiole length, stem thickness	c	—	Latunde-dada and Lucas, 1983
Medicago sativa (alfalfa)	yield, fertility	c	+	Pfeiffer and Bingham, 1984
M. sativa (alfalfa)	flower color, leaf morphology, fertility chromosome number	c	+	Groose and Bingham, 1984
M. sativa (alfalfa)	cold susceptibility, yield, crown rot susceptibility, leaflet shape, internode length, fertility	c	—	Johnson et al., 1984

Nicotiana alata	time to flower, flower and leaf morphology, pollen viability, plant height	c	+	Bravo and Evans, 1985
Nicotiana (somatic hybrids)	flower shape and size, leaf shape	c	—	Evans et al., 1982
Nicotiana suaveolens × *N. glutinosa*	CO_2 absorption, chlorophyll content	c	+	Mousseau, 1970
N. sylvestris	leaf shape, photoperiod response, peroxidase isoenzyme	c	—	Maliga et al., 1979
N. sylvestris	plant height, days to flower, leaf shape chlorophyll pigment	c	+	Prat, 1983
N. tabacum	chromosome number, male fertility plant vigor, leaf shape	c	—	Butenko, Shemina and Froloua, 1967
N. tabacum	plant vigor, flower morphology, male and female fertility	c	—	Syono and Furuya, 1972
N. tabacum	chromosome number, leaf shape, leaf color, pollen and seed fertility plant vigor	c	+	Ogura, 1976
N. tabacum	plant vigor, root development, leaf shape, growth habit	c	+	Berlyn, 1980
N. tabacum	leaf color, leaf shape	p	+	Barbier and Dulieu, 1980
N. tabacum	chromosome number, leaf morphology sterility, leaf color	c	—	Sacristan and Melchers, 1969
N. tabacum	chromosome number, plant morphology,	p	—	Takebe, Labib and Melchers, 1971
N. tabacum (tobacco) ($a_1 - a_2$)	leaf color, fertility, leaf morphology	P	+	Dulieu and Barbier, 1982
N. tabacum (hybrid)	leaf color	c	+	Barbier and Dulieu, 1983
N. tabacum	tricotyledons, resistance to hydroxyurea	c	+	Chaleff and Keil, 1982

Table 1 (Continued)

Species	Altered Characteristics	Techniques	Inheritance Examined	Reference
N. tabacum	DNA content, nuclear volume	c	—	Berlyn, 1983
N. tabacum	leaf shape and size; floral development, plant vigor and survival, chlorophyll pigment	c	—	Thanutong, Furusawa and Yamamoto, 1983
N. tabacum	chlorophyll content, leaf shape	p	+	Lorz and Scowcroft, 1983
Oryza sativa (rice)	early heading, chlorophyll content, culm length, fertility	c	+	Fukui, 1983
Oryza sativa	seed fertility, plant height, heading date, morphology, chlorophyll deficiency	c	+	Oono, 1978
Oryza sativa	plant height, chlorophyll pigment, grain weight	c	+	Sun et al., 1983
Panicum spp	plant size, leaf shape, tillering	c	—	Bajaj, Sidhu and Dubey, 1981
Pelargonium spp	plant and organ size, leaf and flower morphology, oil constituents, fasciation pubescence, anthocyanin pigmentation	c	—	Skirvin and Janick, 1976
Pelargonium graveolens	chromosome number, leaf shape and type	c	—	Janick, Skirvin and Janders, 1977
Rubus sp. (blackberry)	leaf size and pattern, fruit size chlorophyll pigment	m	—	Swartz et al., 1983
Saccharum spp. (sugarcane)	Fiji disease and downy mildew resistance, sucrose content	c	—	Krishnamurthi, 1982
Saccharum spp.	eyespot toxin resistance, plant height, chlorophyll content	c	—	Larkin and Scowcroft, 1983
Saccharum spp.	plant morphology, chromosome number, isozyme systems	c	—	Heinz and Mee, 1971

Saccharum spp.	Auricle length, leaf attitude, hairness, sugar content, chromosome number	c	—	Liu and Chen, 1976
Saccharum spp.	leaf length, leaf width, erectness of leaves, stalk diameter, number of stalks/clump, time of flowering, sucrose percent, cane weight, degrees Brix	c	—	Sreenivasan and Jalaja, 1983
Saccharum spp.	leaf and stalk characters	c	—	Irvine, 1984
Saccharum officinarum	resistance to *Drechslera sacchari*	c	—	Heinz et al., 1977
S. officinarum	cane and sugar yield; stalk number, length, diameter, volume and density; fiber percent	c	—	Liu and Chen, 1978
S. officinarum	resistance to *H. sacchari* toxin	c	—	Larkin and Scowcroft, 1981
Solanum tuberosum	resistance to *Alternaria solani*	p	—	Matern, Strobel and Shepard, 1978
S. tuberosum	tuber shape	p	—	Shepard et al., 1980
S. tuberosum	tuber shape, yield, maturity date, photoperiod requirement, plant morphology, resistance to *A. solani* and *Phytophthora infestans*	p	—	Secor and Shepard, 1981
S. tuberosum	leaf morphology, leaf color, glossiness, hairness, plant growth habit	p	—	Thomas, 1981
S. tuberosum	tuber shape, tuber color, leaf shape	c	—	Van Harten et al., 1981
S. tuberosum (potato)	chromosome structure, yield, disease resistance	p	—	Shepard, 1982
S. tuberosum	leaf shape and color, stem pigment, plant height and vigor	p	—	Thomas et al., 1982
S. tuberosum	chromosome number	c	—	Karp et al., 1982

Table 1 (Continued)

Species	Altered Characteristics	Techniques	Inheritance Examined	Reference
S. tuberosum	leaf variegation, leaf shape and color, vigor, anthocyanin content	p	—	Austin and Cassells, 1983
S. tuberosum	vigor, leaf and stem morphology	c	—	Sree Ramulu, Dijkhuis, and Roest, 1983
S. tuberosum	tuber skin color, leaf shape and size	p	—	Bright et al., 1983
S. tuberosum	PVY resistance	p	—	Murakishi, 1984
Sorghum bicolor	seed fertility, leaf morphology plant growth habit	c	—	Gamborg et al., 1977
Trifolium incarnatum	male and female fertility, leaf shape, flower head structure	c	—	Beach and Smith, 1979
Triticale	chromosome number, chlorophyll pigment,	c	—	Nakamura and Keller, 1982
Triticum turgidum	chromosome number	c	—	Bennici and D'Amato, 1978
Triticum aestivum (wheat)	fertility, chlorophyll pigment	c	—	Sears and Deckard, 1982
T. aestivum	plant height, time to flower, morphology	c	—	Maddock et al., 1983
T. aestivum	plant height, awn and tiller number, grain color waxiness, gliadin proteins, amylase regulation, maturity date	c	+	Larkin et al., 1984
T. aestivum	gliadin proteins	c	+	Cooper et al., 1984
Triticum durum (durum wheat)	spike length, chromosome number, fertility	c	—	Lupi et al., 1981
Zea mays (maize)	mitochondrial DNA, toxin response, fertility	c	+	Kemble et al., 1982

Z. mays	leaf arrangement, plant height, node number	c	—	Green, 1977
Z. mays	chromosome number, pollen fertility, endosperm and seedling mutations, e.g., opaque, germless, etched, yellow, yellow green, pale green, viviparous, virescent	c	+	Edallo et al., 1981
Z. mays	plant height, leaf shape, chlorophyll pigment, flowering date, yield	c	+	Beckert, Pollacsek and Caenen, 1983
Z. mays	early tasseling, leaf spotting, curling leaves, clumped tassel, tillering, vigor	c	+	Brookhousen et al., 1984
Z. mays	early flowering, lower grain moisture	c	—	Cummings, 1984

Techniques
m = meristem culture
p = protoplast culture
c = callus or suspension culture or direct regeneration

gene mutation using our procedure is in the neighborhood of 1 mutant in every 20–25 regenerated plants; (5) Some evidence suggests that new single gene mutants not previously reported using conventional mutagenesis have been recovered using somaclonal variation; (6) The occurrence of 3:1 ratios for single gene mutants in R_1 plant suggests that mutants are of clonal origin and that the mutation occurred prior to shoot regeneration (i.e., no mosaics are detected); (7) Evidence suggests that mitotic recombination (reciprocal or non-reciprocal) may also account for some somaclonal variation; (8) Evidence suggests that mutations in chloroplast DNA (detected by both maternal inheritance and restriction enzyme analysis) can also be recovered; and (9) Agriculturally useful variants leading to development of new breeding lines have been recovered via somaclonal variation.

A detailed listing of reports of somaclonal variation is summarized in Table 1. Three general observations are apparent from this table: (1) There have been a large number of reports of somaclonal variation in the past two years, suggesting that activity is increasing rapidly in this area of research; (2) In contrast to early reports, a larger percentage of the more recent reports in this area include some genetic analysis. The importance of genetic analysis for use of somaclones particularly in sexually propagated crops has only recently been realized; (3) Many of the reports on specific crops are being researched in several laboratories. For example, at least five different laboratories are actively studying somaclonal variation in potato. This has resulted in more detailed studies on the phenomenon of somaclonal variation and permitted more comparisons between laboratories. Based on the progress during the past several years, it is clear that somaclonal variation is becoming a more widely explored technique for several crops.

Control of variability

When attempting to produce somaclones for a new crop plant, conditions such as genotype, explant source, duration of culture, and culture conditions all appear to be critical as these factors influence both the ability to regenerate plants and the frequency of variants produced. In some crop plants all of these factors can be varied at will, while in other crops the constraints of regneration may limit the flexibility to manipulate one or more of these critical variables.

The frequency of variants recovered will therefore, no doubt, vary between crop plants based on the constraints of regeneration. For example, in potato several genotypes can be regenerated on several media using several different explants. It is not surprising that in potato some authors have reported somaclonal variants while others have obtained fairly uniform populations of plants.

Before describing each variable that is possible, it is important to recognize that somaclonal variation can result from two sources. Somaclonal variation appears to be the result of both pre-existing genetic variation that is expressed in regenerated plants and cell culture-induced variation that is induced by the process of culturing explants. If a cell culture or genotype related variable influences either of these sources of variation, it can alter the frequency of somaclonal variation. The evidence for two sources of variation is based on observations in several species. In geranium, Skirvin and Janick (1976) compared plants obtained from in vivo stem, root and petiole cuttings. Plants from stem cuttings, a normal method of geranium propagation, were uniform, whereas plants from in vivo root and petiole cuttings were quite variable. Similarly, geranium plants regenerated from callus were also quite variable. Hence, some variability is correlated with donor explant and pre-exists in the tissue used to establish cell cultures. D'Amato (1978) has stated that, based on nuclear genetic differences between differentiated cells and genetic phenomena occurring during the first few mitotic divisions of callus formation, most explants comprise a heterogeneous cell population. Chromosome mosaicism has been frequently reported in many plant species including sugarcane which has been used for many studies on somaclonal variation. The importance of culture condition on the frequency of somaclonal variation has been documented for several crops. Thomas et al. (1982) observed a number of potato plants regenerated from one protoplast and found variation in the plants. Hence, in this system the variation was clearly induced by the process of tissue culturing.

(a) Genotype. The genotype of plants used for somaclonal variation is an important variable. Genotype can influence both frequency of regeneration and the frequency of somaclones. These factors are most evident in potato where a large number of cultivars have been examined in several laboratories. Gunn and Shepard (1981) found differences in the number of regenerated plants in two cultivars of potato using identical conditions. Bright et al. (1983) were able to regenerate plants from protoplasts of only 6 of 10 potato cultivars tested. Evans, Sharp and Flick (1981, Table 16) have summarized the effect of genotype on plant regeneration for several crops. The most dramatic effects were observed between species and varieties of *Triticum* and between genotypes of *Pisum sativum.* The genetic basis for regeneration is best illustrated in alfalfa where two cycles of recurrent selection have been used to increase the frequency of regeneration from 12 to 67% (Bingham et al., 1975). There is also a suggestion from the literature that plants of

different genotypes have different frequencies of somaclonal variation. While Shepard et al. (1980) recorded a high frequency of phenotypic variants for the potato cultivar Russet Burbank, Wenzel et al. (1979) observed very little variation amongst 200 plants regenerated from protoplasts of 5 dihaploid clones of potato. However, it is difficult to compare these two reports as different culture conditions were used in the laboratories. Sun et al. (1983) compared the frequency of polyploid regenerates in 18 varieties of rice. Multiploids were recovered in the indica varieties but not in the japonica varieties of rice. Similarly, the frequency of chlorophyll deficiency mutants varied significantly between the two types of rice varieties.

(b) Explant source. The source of explant has been considered most often as a critical variable for somaclonal variation. As mentioned above, work with geranium has demonstrated that explant source is critical for recovery of somaclonal variants. Variants were recovered from in vivo root petiole cuttings, but not from stem cuttings of geranium. Similarly, it has been well documented that in cultivars of sugarcane which are chromosome mosaics that propagation by stem cuttings can result in plants with variable chromosome number. In addition, since not all explants are equal in terms of regenerability, it is likely that different selective pressures would be exerted against different explants. This could result in different frequencies and spectrums of somaclonal variation among plants from different explants.

The first somaclonal variants of potato were detected among plants regenerated from protoplasts. This resulted in a flurry of efforts in several laboratories to develop procedures for regeneration of plants from potato protoplasts. However, it has been recognized more recently that somaclonal variants of potato could be recovered by direct culture of leaf, rachis, and petiole explants. In a more direct experiment, Barbier and Dulieu (1980) compared the frequency of variation in genetically marked lines of tobacco regenerated from cotyledons and cotyledon-derived protoplasts. Based on comparison of cells that had been in culture for a comparable period of time, there were no differences between protoplast and non-protoplast derived plants in either frequency or type of variants.

Based on several early publications it has generally been concluded that somaclonal variation is restricted to plants derived via organogenesis from long-term cell cultures or to plants derived from protoplasts. In contrast, it has been stated that plants obtained by somatic embryogenesis or by meristem culture would produce clones. This premise should, however, be reexamined. There are several reports of clones recovered via somatic embryogenesis or via meristem culture (cf. Hanna, Lu and Vasil, 1984; Murashige, 1978). However, there are also several reports of somaclonal variants arising from somatic embryos and from cultured meristems. For example, Browers and Orton (1982) reported somaclonal variants in celery derived from somatic embryos. Similarly, *Medicago sativa* (alfalfa) and several of the cereals listed in Table 1 are embryogenic, yet still produce somaclonal variants. The variants recovered from cultured meristems are perhaps of greatest concern. Tissue culture dogma advocates use of mericloning for large scale commercial propagation of desirable genotypes. However, several recent reports suggest that this dogma should be reexamined, particularly for cultivated crops. Orshinsky and Tomes (1984) compared plants of six genotypes of bird's-foot trefoil (*Lotus corniculatus*) derived from node culture with plants propagated by stem cuttings. Plants from node culture were identified that flowered at a lower node, were shorter, had reduced leaflet length:width ratio, lower pollen stainability and/or reduced seed yield. Several of these traits were considered to be epigenetic phenomena due to residual effects of culture conditions. However, some traits such as increased forage yield and stem number persisted into the second year of field testing. Swartz, Galleta and Zimmerman (1981) also found a number of distinct variants among strawberry plants derived from shoot tips. These variants included several general traits that may represent epigenetic changes due to culture conditions as well as several distinct, discrete variants including those with sectorial chlorosis, dwarfs, disease sensitivity, and a female-sterile variant. The epigenetic changes were observed in nearly all strawberry regenerates and included plants with a greater tendency to produce runners, smaller fruit size, and lower yield. The epigenetic changes may be associated with virus elimination during plant regeneration. Swartz, Galletta and Zimmerman (1983) observed similar variation in tissue culture-propagated thornless blackberries. Among tissue culture-derived plants fruit size was smaller but the plants had increased vigor. Some differences were also noted in leaf size and pattern. Perhaps the most dramatic changes in mericlones have been reported in orchids (Vajrabhaya, 1977). In orchids, virus elimination occurs during meristem culture resulting in more vigorous plants. However, in addition to epigenetic changes, Vajrabhaya also reported variation in shape of flowers and in flower color resulting in selection of new novel flower types of *Dendrobium*. Hence, even in tissue culture systems traditionally viewed to be genetically stable, such as somatic embryos and meristems, somaclones have been detected.

(c) Duration of cell culture. Chromosome number variation has been reported for many established plant cell cultures (cf. Bayliss, 1980). This variation in chromosome number is influenced by several factors: (1) Preexisting chromosome mosaicism in plants used for culture initiation; (2) Nuclear fragmentation associated with first cell division of callus initiation (Cionini, Bennici and D'Amato, 1978); (3) Endoreduplication or endomitosis occurring culture initiation (cf. D'Amato, 1978); and (4) Abnormalities of the mitotic process resulting in aneuploid cells. While there is evidence that chromosome stability can be achieved in long-term cell cultures (Evans and Gamborg, 1982), it is widely accepted that most long-term cultures are chromosomally variable. The correlation between duration of culture and accumulation of chromosomal variants was first documented for *Daucus carota* (Smith and Street, 1974). While cell cultures may contain many abnormal cells, regeneration acts as a sieve reducing the frequency of aneuploid plants. In some cases, only or mostly normal diploid plants have been regenerated despite the presence of abnormal cells in the cultures used for regeneration (cf. D'Amato, 1977). On the other hand, polyploid and aneuploid plants have been regenerated in vitro from a large number of plant species. Polyploid plants have been recovered in geranium, ornamental *Nicotiana*, cultivated tobacco, tomato and alfalfa (see Evans et al., 1984). Aneuploid plants have also been recovered following regeneration of plants from cell cultures of several plant species. Aneuploids have been reported more often from plants of polyploid or hybrid origin, e.g., *N. tabacum* (Sacristan and Melchers, 1969) and *Saccharum* hybrids (Heinz and Mee, 1971), where loss or addition of a few chromosomes can be tolerated. Several authors have stated that variation increases with increasing duration of culture. Barbier and Dulieu (1983), using a genetically marked explant source, have shown that while most genetic changes occur in the first few mitoses in culture that some genetic changes increase with the duration of culture. Also using protoplasts of a heterozygous donor, Lorz and Scowcroft (1983) showed that by doubling the duration of culture, the frequency of genetic changes increased from 1.4 to 6%. Fukui (1983) has monitored the occurrence of multiple mutations in plants regenerated from callus cultures of rice. He was able to describe the sequential occurrence of 4 separate mutations during the culture period of a single callus line.

A final indication that extended culture produces an elevated frequency of somaclones is the results obtained with somatic hybrids. In most cases, somatic hybrids require up to 6 months in culture to recover plants. Such protoplast-derived hybrids, even between closely related species result in a very high frequency of somaclonal variation. It has been demonstrated that somatic hybrids are a richer source of variability than comparable sexual hybrids (Evans et al., 1982). In some cases, several unique variants have been detected between clones of somatic hybrids (Evans et al., 1983).

(d) Culture conditions. It has been known for several years that growth regulator composition of the culture medium can influence frequency of karyotypic alterations in cultured cells (Bayliss, 1975). The growth regulator 2,4-D has been most frequently considered to be responsible for chomosome variability (Singh, Kao and Miller, 1975). A high concentration of 2,4-D was associated with increased variability among regenerated *Hordeum* plants (Deambrogio and Dale, 1980). Similarly, use of 2,4-D in place of NAA in culture medium for potato also increased the frequency of abnormal plants (Shepard, 1981). In both these studies the alterations in growth regulators that resulted in greater variability were associated with longer duration of culture. Hence, it is difficult to ascribe the increased variability solely to changes in growth regulator concentrations. On the other hand, Bravo and Evans (unpublished observations) regenerated plants of an ornamental *Nicotiana* directly from leaf explants on a single medium while only varying the concentration of 6BA. Variants were only obtained at the highest two concentrations, 5 and 10 μM 6BA. In addition, the frequency of variants was substantially greater among plants from 10 μM than from 5 μM 6BA.

These four variables are all important in the recovery of somaclones among regenerated plants. In many species, these variables cannot be modified and still permit recovery of plants. However, to the extent possible, these variables should be carefully monitored in both comparing published experiments and in establishing systems for in vitro plant regeneration.

Genetic basis of somaclonal variation

The variation originally detected in regenerated plants of sugarcane and potato was not genetically analyzed. Detailed data on the transmission of variation to sexual progeny is necessary to facilitate the use of somaclonal variation for improvement of sexually propagated crops. Based on experiments with several crop species the following variation has been documented based on evaluation of R_0 or R_1 plants: changes in chromosome number or structure, single gene nuclear mutations, and cytoplasmic DNA changes. In addition, several less-defined changes have been suggested.

Variant plants with altered chromosome number have been reported by several authors (cf. D'Amato, 1977). Polyploidy is the most frequently observed chromosomal abnormality and the mechanisms for its origin have been discussed (D'Amato, 1978). The suggestion has been that in aneuploid species such as sugarcane and potato, when duplicate chromosomes are already present, the addition or loss of one chromosome is usually not lethal. Changes in chromosome number are commonly associated with reduced fertility and with altered genetic ratios in progeny of self-fertilized plants. In addition to changes in chromosome number, chromosome rearrangements have been detected by analyzing meiosis in regenerated plants. Translocations have been reported in potato (Shepard, 1982), ryegrass (Ahloowalia, 1976), and oat (McCoy et al., 1982). The published work also suggests that chromosome deletion, inversions and other minor reciprocal and non-reciprocal rearrangements occur among regenerated plants. Depending on the ploidy of the crop plant used and the extent of chromosome modification, such regenerated plants could have reduced fertility or altered transmission of genetic characters.

In contrast to gross chromosomal changes, more subtle changes resembling single gene mutations have also been detected by several workers. Recessive single gene mutations are suspected if: (1) The variant does not appear in the R_0 plant; and (2) Self-fertilized R_1 progeny segregate in an expected 3:1 Mendelian ratio for the trait of interest. To confirm that the trait of interest is indeed a single gene trait, progeny tests should be completed on selected single R_1 plants to identify segregators and non-segregators and to ensure continued 3:1 transmission in R_2 segregating populations. Moreover, mutant plants should breed true in the R_2 generation. When possible, complementation tests should be completed with known mutants for the crop of interest. This type of detailed genetic analysis to confirm single gene traits has only been completed for tomato (Evans and Sharp, 1983) in the work that is summarized in the Literature Review above. Three to one ratios in R_1 plants, the first step in this type of analysis, have been reported for maize (Edallo et al., 1981), *Nicotiana sylvestris* (Prat, 1983), rice (Fukui, 1983; Sun et al., 1983), and wheat (Larkin et al., 1984).

In some cases, use of specific genetically marked strains has aided in evaluation of plants regenerated from cell culture. Dulieu and Barbier (1982) and Lorz and Scowcroft (1983) have both regenerated plants from *N. tabacum* with specific chlorophyll deficiency markers present in heterozygous tissue, genetic changes could be detected by the appearance of albino or dark-green regenerated plants. Dulieu and Barbier have reported a high frequency (9.6%) of variant regenerates at the a_1 and y loci and have ascribed these genetic changes to the combination of deletion and mitotic recombination. Lorz and Scowcroft (1983) detected genetic changes in 3.7% of morphogenetic cell colonies using the Su locus. When light-green regenerated shoots (Su/su) were self-fertilized, up to 37% of the regenerated plants had segregation ratios that were distorted from the expected 1:2:1 ratio of dark-green: light green:albino. These authors did not speculate on the genetic basis of these altered R_1 segregation ratios. The value of using genetically marked heterozygous donor material was evident in each of these studies as the authors were able to demonstrate that much of the reported variation was proportional to the duration of culture. Hence, a significant proportion of the genetic variability detected in regenerated plants must be induced by the cell culture procedure.

In addition to single nuclear gene changes, cytoplasmic genetic changes have also been detected by somaclonal variation. The most detailed work on the evaluation of cytoplasmic genetic changes via somaclonal variation has been completed by Gengenbach and his colleagues by evaluating plants for two cytoplasmic traits. Sensitivity to host specific toxin of *Drechslera maydis* race T, the causative agent of southern corn leaf blight, is associated with all genotypes containing Texas male sterile (cms-T) cytoplasm. In seed derived plants these two characters are tightly associated and both controlled by mitochondrial DNA (mt DNA). Gengenbach et al. (1977) selected in vitro for resistance to toxin and regenerated resistant plants with the aim of recovering toxin resistant, cms recombinant breeding lines. However, even among tissue culture regenerates resistance was associated with a concomitant reversion to male fertility. When the restriction endonuclease pattern of mt DNA was evaluated (Kemble et al., 1982), it was evident that significant changes had occurred in mt DNA of plants derived from cell culture. Variations in chloroplast DNA have been detected among tomato somaclones (Flick, unpublished observation). It is not surprising that organelle DNA-variants are uncovered as the number of organelles in a developing shoot apex is much smaller than in a mature cell (cf. Bendich and Gauriloff, 1984). Hence, if a mutation occurs in organelle DNA, it is more likely to become the dominant plastid type during sorting out if the mutant occurs in 1 of 10 organelles than in 1 of 100 organelles.

Several authors have suggested that other genetic changes are responsible for the range of variation detected in regenerated plants (cf. Scowcroft and Larkin, 1982; Orton, 1983). The most frequently cited and well documented changes are small chromosome rearrangements. By evaluating meiotic behavior of plants regenerated from cell culture, translocations, deletions and inversions have all been detected (cf.

Ahloowalia, 1978; McCoy et al., 1982). While large changes in chromosome structure have been detected, it is likely that less dramatic structural changes that were not detected, also occur frequently. Small changes in chromosome structure could alter expression and genetic transmission of specific genes. This could include deletion of one copy (copies) of a gene, duplications of one copy (copies) of a gene, or gene conversion during repair processes. In addition, recombination or chromosome breakage may occur in preferential regions or "hot spots" of particular chromosomes, thereby affecting some regions of the genome in a disproportionally higher frequency. With all the phenomena occurring simultaneously, the altered segregation ratios such as those detected by Lorz and Scowcroft (1983) are not at all unexpected.

Mitotic crossing over (MCO) could also account for some of the variation detected in regenerated plants. This could include both symmetric and asymmetric recombination. MCO may account for the recovery of homozygous recessive single gene mutations in some regenerated plants (cf. Evans and Sharp, 1983). As breeders have previously only had access to variation that is normally transmitted through meiosis, the recovery of products of MCO may constitute a unique source of new genetic variation.

Despite the lack of genetic data, several authors have also speculated that transposable elements may also be responsible for somaclonal variation. Variation in the insertion of plasmid-like DNA found in mitrochondria of cms-s corn has been detected in corn cell cultures (Chourey and Kemble, 1982). Heterozygous light-green (Su/su) somaclones with a high frequency of colored spots of the leaf surface have been detected for a clone of both *N. tabacum* (Lorz and Scowcroft, 1983) and a *N. tabacum* + *N. sylvestris* somatic hybrid (Evans et al., 1983). The somatic hybrid has an unstable pattern of inheritance that would be consistent with an unstable gene, although detailed genetic analysis has not yet been completed.

Gametoclonal variation

As described above, the types of genetic changes that are recovered in plants regenerated from cell culture are dependent upon the donor material that is used for plant regeneration. For tomato we used a diploid inbred variety, hence most of the variant regenerated plants were heterozygous resulting in segregation of new mutations in the R_1 generation. Lorz and Scowcroft (1983) used heterozygous material so that mutations could be visually detected in regenerated plants. Among these somaclones several plants had distorted segregation ratios in the R_1 generation. Since the generation of variation will ultimately be used for breeding and crop improvement, it is important to distinguish somatic-derived somaclones and gametic-derived gametoclones. The gametes are products of meiosis, governed by Mendel's laws of segregation and independent assortment. Three genetic differences should be pointed out as evidence that gametoclonal and somaclonal variation are distinct. (1) Both dominant and recessive mutants induced by gametoclonal variation will be expressed directly in haploid regenerated plants from diploid anthers, since only a single copy of each gene is present. Hence, regenerated gametoclones (R) can be analyzed directly to identify new variants. (2) Recombinational events that are recovered in gametoclones would be the result of meiotic crossing over, not mitotic crossing over. While not fully characterized, evidence from *Neurospora* suggests that these two phenomena do not occur at the same frequency along the gene map. For example, MCO may be used to separate genes that are hard to separate by meiotic crossing over. (3) To use gametoclones, chromosome number must be doubled. The most frequently used method to double chromosomes is treatment with colchicine. Colchicine is known to induce mutations (Franzke and Rose, 1952). Hence, by the time gametoclones are genetically analyzed, some of the observed variation may not be due to gametoclonal variation, but may be due to mutagenic effects of colchicine.

The value of gametoclonal variation is evident from the reports of new variety development from Chinese workers (cf. Zeng, 1983). For wheat and rice, anthers of F_1 hybrid plants were introduced into culture and the recombinant microspores produced new doubled haploid lines that contained genetic information from both parents. Anther culture has been used for the recovery of recombinant plants of an F_1 wheat hybrid between Xian nog 5675 (a variety with white glumes, top awn, clavate spike and short stalk) and Jili (a variety with red glumes, awn, fusiform spike and tall stalk). Several doubled haploid plants were recovered that expressed mixed characters of the two parents (e.g., red glume, top awn, clavate spike, and tall stalk). Hence, gametoclonal variation is valuable for hybrid sorting, i.e., development of homozygous diploid recombinant lines from interspecies or intraspecies F_1 hybrids.

Since growth of cells in culture results in single gene mutations (Evans and Sharp, 1983), culture of microspores could result in recessive mutations that are visible in the R_0 generation. Hence, it is likely that gametoclonal variants can be detected in the R_0 generation, while evaluation of R_1 progeny must be completed to detect the full spectrum of somaclonal variants. However, it is likely that mutant cells do not regenerate as well as wild type cells. Hence, the mutation spectrum and frequency obtained from regenerated haploids may be strikingly different from regenerated diploid tissue, in which recessive mutations in heterozygous condition may have no effect on regeneration capacity.

In some cases this gametoclonal variation is no doubt due to uncovering residual heterozygosity. However, in several cases residual heterozygosity cannot be used to explain the variation obtained. Two groups of experiments have been reported that document that the majority of variation was not due to residual heterozygosity. First, several authors have initiated anther cultures using highly inbred donor plants. Arcia, Wernsman and Burk (1978) used tobacco plants that had been self-fertilized for 15 generations prior to anther culture. Among doubled haploid plants they found substantial variation for yield, plant height, number of leaves, and percentage of nornicotine. Since the donor plants were highly inbred, the variation was most likely not due to vesidual heterozygosity. A second approach to preclude residual heterozygosity is to examine the variation that appears in subsequent cycles of androgenesis. That is, to culture anthers from anther-derived doubled haploids. Brown, Wernsman and Schnell (1983) found variation in yield, plant height, days to flower and total alkaloids following a second cycle to anther culture of *N. tabacum*. Similarly, de Paepe, Prat and Knight (1983) found variation for several characters in *N. sylvestris* following several cycles of anther culture. Hence, gametoclonal variation is the result of several factors: meiotic recombination occurring prior to culture, mutations occurring prior to culture, genetic changes associated with use of microspores for regeneration, mutations occurring in culture prior to regeneration and mutations caused by or occurring after the doubling process.

Most of the early reports of gametoclonal variation cited changes in chromosome number. Hu (1983) has described the aneuploids obtained following anther culture of maize, wheat, and a wheat x triticale hybrid. A wide range of aneuploids and mixoploids were recovered as well as numerous genetic changes have been documented that cannot be ascribed to changes in chromosome number. These include changes in both qualitative and quantitative traits.

Examination of the reports of gametoclonal variation suggest that some of the variation is unique to the anther culture procedure. Early reports of anther culture of cereals noted that many of the regenerated shoots were albino. These albinos have proplastids but are defective in thylakoid stacking resulting in non-functional organelles. It has been suggested that these albinos are a result of deletions in chloroplast DNA. These albinos are not normally detected in plants regenerated from somatic tissue. Similarly, using conventional mutagenesis or somaclonal variation, no new variants for the S locus governing self-incompatibility could be detected in *Lycopersicon peruvianum* (Sree Ramulu, 1982). However, one new S allele was detected among 17 or 53 plants regenerated from cultured anthers. These results suggest that the mutation spectrum obtained by gametoclonal variation may differ from that obtained by somaclonal variation. Similarly, de Paepe et al. (1983) noted that variants routinely uncovered following anther culture of *Nicotiana sylvestris* were never recovered among plants regenerated from protoplasts. Several authors have reported yield reduction in doubled haploids of tobacco, particularly the flue cured varieties. Dhillon, Wernsman and Miksche (1983) have suggested that this is the result of changes in heterochromatin among microspore-derived plants that is probably due to gene amplification. The yield reduction is not observed for all tobacco varieties and seems to be a unique consequence of anther culture. Hence, several factors suggest that gametoclonal variation results in a unique array of genetic changes that are distinct from somaclonal variation.

Breeding strategies and future uses

Based on the data summarized above, there appear to be several unique variants produced by somaclonal and gametoclonal variation. These tools, although not fully characterized, appear to represent useful sources for genetic variation that could be of value to plant breeders. Based on the results outlined above, several strategies for the use of somaclonal and gametoclonal variation are evident. These procedures are compared to more conventional approaches such as mutation breeding and backcross breeding in Table 2. In each case attention should be directed to the unique features of this new technology and the interface of this technology with conventional breeding.

As single gene mutations and organelle gene mutations have been produced by somaclonal variation, one obvious strategy is to introduce the best available varieties into cell culture to select for improvement of a specific character. Hence, somaclonal variation could be used to uncover new variants that retain all the favorable qualities of an existing variety while adding one additional trait, such as disease resistance or herbicide resistance. Work with sugarcane and tomato has already suggested that this approach is feasible. Once new R_1 variants are identified, these should be field tested in replicated plots to ascertain genetic stability. Seed should be increased at the same time to permit rapid variety development of promising lines. Reciprocal crosses between desirable R_1's and seed derived controls can be used to gain an understanding of the genetic basis of the somaclonal variants. New promising breeding lines can be reintroduced into cell culture to add an additional

Table 2. Comparison of Procedures for Variety Development

	Somaclonal Variation	Gametoclonal Variation	Mutation Breeding	Backcross Program
Source of variation	Spontaneous and induced	Spontaneous and induced	Induced	Natural populations
Likelihood of success	Undirected variation	Some direction high percentage of success	Undirected variation	Guaranteed except where linkages not broken
Alteration of quantitative traits	Possible	Possible	Possible	Rarely successful
Rate of progress	More than 1 trait per generation	More than 1 trait per generation	More than 1 trait per generation	One trait in 5–7 sexual generations
Chimerism	None or low frequency	None or low frequency	Major problem	None
Species limits	In all species that can be regenerated	In all species that can be regenerated	All species	Only sexually propagated crops
Time for breeding line development	One generation	One generation	Three generations	Up to six generations

character or to improve agronomic performance of a selected somaclonal variant. By using this approach, it is possible to produce new breeding lines with desirable traits in a short period of time.

It is desirable to obtain true breeding homozygous diploid plants at many steps during a conventional breeding program. Hence, the value of anther culture for rapid production of doubled haploid lines has long been recognized (see review Sharp, Reed and Evans, 1984). It is possible to develop strategies for using gametoclonal variation for crop improvement. Two plants each with desirable characteristics can be crossed to produce a hybrid that expresses all the desirable characteristics. Such plants that contain desirable characters of Plant A and Plant B could also include commercially available F_1 hybrids. Anthers from hybrid plants are cultured to produce haploids, then doubled haploids. The plants with expression of the best characteristics of Plant A and Plant B are selected from among the doubled haploids and used to generate new parent lines. The lines can then be self-fertilized to generate new open pollinated breeding lines or can be used in a breeding program to produce new hybrid seed. In most cases, the variation recovered using this approach is the result of meiotic recombination occurring during sexual reproduction of the F_1 hybrid. This could result in transgressive segregation to uncover new unique gene combinations. In addition, it is also likely that variation induced by cell culture could be detected in plants regenerated from cultured anthers.

It is anticipated that in the next 5–10 years somaclonal variation and gametoclonal variation will find widespread application in crop improvement programs. In general, the techniques are relatively simple compared to those of recombinant DNA, and they result in a rich source of genetic variability. In each case, plants can be transferred directly to the field and evaluated as part of an ongoing breeding program. Somaclones have now been produced in a wide variety of crop plants using a wide range of regeneration procedures each resulting in new variants. Several specific developments are anticipated in the future.

(1) To the extent that desirable somaclones can be identified in the test tube rather than in field trials, somaclonal variation can be more efficient and cost effective. This requires a correlation between the cellular and whole plant response to specific chemicals used as selective agents. Promising results have already been obtained by selecting for resistance to host-specific pathotoxins (Gengenbach et al., 1977) and for herbicide resistance (Chaleff and Parsons, 1978). The value of cellular selection of desirable genetic modifications has been elucidated (Chaleff, 1983). It is likely that this approach will become more valuable as a correlation between cell selection and whole plant performance is documented for more traits.

(2) Somaclonal and gametoclonal variation also hold promise as an adjunct to protoplast fusion. The array of plants obtained by protoplast fusion was much more variable than by sexual hybridization for *Nicotiana* interspecific hybrids (Evans et al., 1982). This phenomenon can be used to select somatic hybrid somaclones with the appropriate mixtures of desirable genetic traits of the two parent lines used for protoplast fusion. As protoplast fusion results in a summation chromosome number, anther culture can be used to reduce the chromosome number to that which is consistant with sexual hybridization. This can be accomplished by fusion of haploid protoplasts (Melchers and Labib, 1974) or by taking advantage of gametoclonal variation, it is also possible to culture anthers of the the somatic hybrids. By completing anther culture after protoplast fusion, it is possible to combine protoplast fusion, somaclonal variation from protoplast regeneration and gametoclonal variation to produce an array of plants with maximum variability. It is likely that techniques of gametoclonal and somaclonal variation will be interfaced with protoplast fusion in the next several years.

(3) Finally, it is anticipated that researchers will gain an understanding of gametoclonal variation by characterizing the process in greater detail. An understanding of the production of albinos from anther culture of cereals and the yield reduction associated with anther culture of flue-cured tobacco varieties should be forthcoming. It is possible by culturing immature ovules to obtain maternal (megaspore-derived) haploid plants. A comparison of megaspore and microspore-derived haploids would aid in characterizing much of the variability reported for tobacco dihaploids.

In addition to these specific developments it should be reemphasized that much of the future prospects for developments in somaclonal and gametoclonal variation are dependent upon use of these tools by plant breeders. To this extent, a tight linkage between tissue culturists/geneticists and breeders is imperative to advance this technology.

References

Ahloowalia, B.S. 1976. *Chromosomal changes in para-sexually produced ryegrass.* Pp. 115–122 in K. Jones and P.E. Brandham, eds., *Current Chromosome Research.* Elsevier Press, Amsterdam.

Ahloowalia, B.S. 1978. *Novel ryegrass genotypes regenerated from embryo-callus culture.* P. 62 (abst.). Fourth International Association Plant Tissue Culture Meeting, Calgary, Canada.

Ahloowalia, B.S. 1983. *Spectrum of variation in somaclones of triploid ryegrass.* Crop Sci. 23: 1141–1147.

Arcia, M.A., E.A. Wernsman and L.G. Burk. 1978. *Performance of anther-derived dihaploids and their conventionally inbred parents as lines in F_1 hybrids and F_2 generations.* Crop Sci. 18:413–418.

Austin, S. and A.C. Cassells. 1983. *Variation between plants regenerated from individual calli produced from separated potato stem callus cells.* Plant Sci. Lett. 31:107–114.

Bajaj, Y.P.S. 1985. *Peanut.* in D.A. Evans, W.R. Sharp and P.V. Ammirato, eds., *Handbook of Plant Cell Culture*, Vol. 4. MacMillan Press, New York (in press).

Bajaj, Y.P.S., B.S. Sidhu and V.R. Dubey. 1981. *Regeneration of genetically diverse plants from tissue cultures of forage grass-Panicum spp.* Euphytica 30:135–140.

Barbier, M. and H.L. Dulieu. 1980. *Genetic changes observed on tobacco (Nicotiana tabacum) plants regenerated from cotyledons by in vitro culture.* Ann. Amelio. Plant. 30:321–344.

Barbier, M. and H. Dulieu. 1983. *Early occurrence of genetic variants in protoplast cultures.* Plant Sci. Lett. 29:201–206.

Bayliss, M.W. 1975. *The effects of growth in vitro on the chromosome complement of Daucus carota (L.) suspension cultures.* Chromosoma 51:401–411.

Bayliss, M.W. 1980. *Chromosomal variation in plant tissues in culture.* Pp. 113–144 in I.K. Vasil, ed., *Perspectives in Plant Cell and Tissue Culture.* Int. Rev. Cytol. Suppl. 11A. Academic Press, New York.

Beach, K.H. and R.R. Smith. 1979. *Plant regeneration from callus of red clover and crimson clover.* Plant Sci. Lett. 16:231–238.

Beckert, M., M. Pollacsek and M. Caenen. 1983. *Study of genetic variability in maize (Zea mays L) after in vitro regeneration of plants from callus culture.* Agronomis 3:9–18.

Bendich, A.J. and L.P. Gauriloff. 1984. *Morphometric analysis of cucurbit mitochondria: the relationship between chondriome volume and DNA content.* Protoplasma 119:1–7.

Ben-Jaacov, J. and R.W. Langhans. 1972. *Rapid multiplication of Chrysanthemum plants by stem tip proliferation.* Hort. Sci. 7:289–290.

Bennici, A. and F. D'Amato. 1978. *In vitro regeneration of durum wheat plants. I. Chromosome numbers of regenerated plants.* A. Pflanzenzuecht. 81:305–311.

Berlyn, M.B. 1980. *Isolation and characterization of isonicotinic acid hydrazide resistant mutants of Nicotiana tabacum.* Theor. Appl. Genet. 58:19–26.

Berlyn, M.B. 1983. *Patterns of variability in DNA content and nuclear volume in regenerating cultures of Nicotiana tabacum.* Can. Jour. Genet. Cytol. 25:354–360.

Bingham, E.T., L.V. Hurley, D.M. Kaatz and J.W. Saunders. 1975. *Breeding alfalfa which regenerates from callus in culture.* Crop Sci. 15:719–721.

Bright, S., V. Jarrett, R. Nelson, G. Creissen, A. Karp., J. Franklin, P. Norbury, J. Kueh, S. Rognes and B. Miflin. 1983. *Modification of agronomic traits using in vitro technology.* Pp. 251–265 in S.H. Mantell and H. Smith, eds., *Plant Biotechnology.* Cambridge University Press, Cambridge.

Brook Houzen, P.D., J.P. Cook, C.R. Cowley and M.D. Hollingsworth. 1984. *Analysis of corn mutants derived from in vitro culture.* P. 60 (abst.). Amer. Soc. Agron., Las Vegas, Nevada.

Browers, M.A. and T.J. Orton. 1982. *Transmission of gross chromosomal variability from suspension cultures into regenerated celery plants.* Jour. Hered. 73:159–162.

Brown, J.S., E.A. Wernsman and R.J. Schnell. 1983. *Effect of a second cycle of anther culture on flue-cured lines of tobacco.* Crop Sci. 23:729–733.

Butenko, R.G., Z.B. Shemina and L.V. Frolova. 1967. *Induced organogenesis and characteristics of plants produced in tobacco tissue culture.* Genetika 3:29–39.

Chaleff, R.G. 1981. *Genetics of Higher Plants.* Cambridge University Press, Cambridge.

Chaleff, R.S. 1983. *Isolation of agronomically useful mutants from plant cell cultures.* Science 219:676–682.

Chaleff, R.S. and M.F. Parsons. 1978. *Direct selection in vitro for herbicide-resistant mutants of Nicotiana tabacum.* Proc. Nat. Acad. Sci. USA 75:5104–5107.

Chaleff, R.S. and R.L. Keil. 1982. *Origins of variability among cultured cells and regenerated plants of Nicotiana tabacum.* Pp. 175–187 in E.D. Earle and Y. Demarly, eds., *Variability in Plants Regenerated from Tissue Culture.* Praeger Press, New York.

Chourey, P.S. and R.J. Kemble. 1982. Transposition event in tissue cultured cells of s-cms genotype of maize. Maize Genet. Coop. Newsl. 56:70.

Cionini, P.G., A. Bennici and F. D'Amato. 1978. *Nuclear cytology of callus induction and development in vitro. I. Callus from Vicia faba cotyledons.* Protoplasma 96:101–112.

Cooper, D.B., R.G. Sears, A.C. Guenzi, N.L. Lapitan, B.L. Jones and G.L. Lookhart. 1984. *Heritable somaclonal variation in wheat gliadin proteins.* P. 62 (abst.). Amer. Soc. Agron., Las Vegas, Nevada.

Cummings, D.P. 1984. *Somaclonal variation in corn: a field selection experiment.* P. 63 (abst.). Amer. Soc. Agron., Las Vegas, Nevada.

D'Amato, F. 1977. *Cytogenetics of differentiation in tissue and cell culture.* Pp. 343–357 in J. Reinert and Y.P.S. Bajaj, eds., *Applied and Fundamental Aspects of Plants Cell, Tissue, and Organ Culture.* Springer-Verlag, Berlin.

D'Amato, F. 1978. *Chromosome number variation in cultured cells and regenerated plants.* Pp. 287–295 in T.A. Thorpe, ed., *Frontiers of Plant Tissue Culture.* IAPTC/University of Calgary, Calgary.

Deambrogio, E. and P.I. Dale. 1980. *Effect of 2,4-D on the frequency of regenerated plants in barley (Hordeum vulgare) cultivar 'Akka' and on genetic variability between them.* Cereal Res. Commun. 8:417–424.

De Paepe, R., D. Prat and J. Knight. 1983. *Effects of consecutive androgeneses on morphology and fertility in Nicotiana sylvestris.* Can. J. Bot. 61:2038–2046.

Dhillon, S.S., E.A. Wernsman and J.P. Miksche. 1983. *Evaluation of nuclear DNA content and heterochromatin changes in anther-derived dihaploids of tobacco (Nicotiana tabacum) cv. Coker 193.* Can. J. Genet. Cytol. 25:169–173.

Dulieu, H. and M. Barbier. 1982. *High frequencies of genetic variant plants regenerated from cotyledons of tobacco.* Pp. 211–229 in L. Earle and Y. Demarly, eds., *Variability in Plants Regenerated from Tissue Culture.* Praeger Press, New York.

Dunwell, J.M. 1981. *In vitro regeneration from excised leaf-disks of 3 Brassica species.* J. Expl. Bot. 32:789–799.

Edallo, S., C. Zuccinalli, M. Perenzin and F. Salamini. 1981. *Chromosomal variation and frequency of spontaneous mutation associated with in vitro culture and plant regeneration in maize.* Maydica 26:39–56.

Engler, D.E. and R.G. Grogan. 1984. *Variation in lettuce plants regenerated from protoplasts.* J. Hered. 75:426–430.

Evans, D.A. and O.L. Gamborg. 1982. *Chromosome stability of cell suspension cultures of Nicotiana supp.* Plant Cell Rep. 1:104–107.

Evans, D.A. and W.R. Sharp. 1983. *Single gene mutations in tomato plants regenerated from tissue culture.* Science 221:949–951.

Evans, D.A., W.R. Sharp, and C.E. Flick. 1981. *Growth and behavior of cell cultures: embryogenesis and organogenesis.* Pp. 45–113 in T.A. Thorpe, ed., *Plant Tissue Culture: Methods and Applications in Agriculture.* Academic Press, New York.

Evans, D.A., C.E. Flick, S.A. Kut and S.M. Reed. 1982. *Comparison of Nicotiana tabacum and Nicotiana nesophila hybrids produced by ovule culture and protoplast fusion.* Theoret. Appl. Genet. 62:193–198.

Evans, D.A., J.E. Bravo, S.A. Kut and C.E. Flick. 1983. *Genetic behavior of somatic hybrids in the genus Nicotiana: N. otophora + N. tabacum and N. sylvestris + N. tabacum.* Theoret. Appl. Genet. 65:93–101.

Evans, D.A., W.R. Sharp and H.P. Medina-Filho. 1984. *Somaclonal and gametoclonal variation.* Amer. J. Bot. 71:759–774.

Everett, N.P. and S.M. Flashman. 1984. *Plants from culture.* Nature 307:115.

Franzke, C.J. and J.G. Rose. 1952. *Colchicine induced variants in sorghum.* J. Hered. 43:107–115.

Fukui, K. 1983. *Sequential occurrence of mutations in a growing rice callus.* Theoret. Appl. Genet. 65:225–230.

Gamborg, O.L., J.P. Shyluk, D.S. Brar and F. Constabel. 1977. *Morphogenesis and plant regeneration from callus of immature embryos of sorghum.* Plant Sci. Lett. 10:67–74.

Gengenbach, B.G., C.E. Green and C.M. Donovan. 1977. *Inheritance of selected pathotoxin resistance in maize plants regenerated from cell cultures.* Proc. Nat. Acad. Sci. 74:5113–5117.

George, L. and P.S. Rao. 1983. *Yellow seeded variants in in vitro regenerants of mustard (Brassica juncea coss var. RAI-5).* Plant Sci. Lett. 30:327–330.

Green, C.E. 1977. *Prospects for crop improvement in the field of cell culture.* Hort. Sci. 12: 131–134.

Groose, R.W. and E.T. Bingham. 1984. *Variation in plants regenerated from tissue culture of tetraploid alfalfa heterozygous for several traits.* Crop Sci. 24:655–658.

Gunn, R.E. and J.F. Shepard. 1981. *Regeneration of plants from mesophyll-derived protoplasts of British potato (Solanium tuberosum L.) cultivars.* Plant Sci. Lett. 22:97–101.

Hanna, W.M., C. Lu and I.K. Vasil. 1984. *Uniformity of plants regenerated from somatic embryos of Panicum maximum Jacq (Guinea grass).* Theor. Appl. Genet. 67:155–159.

Heinz, D.J. and G.W.P. Mee. 1969. *Plant differentiation from callus tissue of Saccharum species.* Crops Sci. 9:346–348.

Heinz, D.J. and G.W.P. Mee. 1971. *Morphologic, cytogenetic, and enzymatic variation in Saccharum species hybrid clones derived from callus tissue.* Amer. J. Bot. 58:257–262.

Heinz, D.J., M. Krishnamurthi, L.G. Nickell and A. Maretzki. 1977. *Cell, tissue, and organ culture in sugarcane improvement.* Pp. 3–17 in J. Reinert and Y.P.S., eds., *Applied and Fundamental Aspects of Plant Cell, Tissue, and Organ Culture.* Springer-Verlag, Berlin.

Hu, H. 1983. *Genetic stability and variability of pollen-derived plants.* Pp. 145–157 in S.K. Sen and K.L. Giles, eds., *Plant Cell Culture in Crop Improvement.* Plenum Press, New York.

Ibrahim, R.K. 1969. *Normal and abnormal plants from carrot root tissue cultures.* Can. J. Bot. 47:825–826.

Irvine, J.E. 1984. *The frequency of marker changes in sugarcane plants regenerated from callus culture.* Plant Cell Tiss. Org. Cult. 3:201–209.

Janick. J., R.M. Skirvin and R.B. Janders. 1977. *Comparison of in vitro and in vivo tissue culture systems in scented geranium.* J. Hered. 68:62–64.

Johnson, L.B., D.L. Stuteville and D.Z. Skinner. 1980. *Phenotypic variants in progeny of selfed alfalfa regenerated from calli.* Plant Physiol. (suppl.) 65:36.

Karp, A., R.S. Nelson, E. Thomas and S.W.J. Bright. 1982. *Chromosome variation in protoplast-derived potato plants.* Theoret. Appl. Genet. 63:265–272.

Kemble, R.J., R.I.S. Brettell and R.B. Flavell. 1982. *Mitochondrial DNA analyses of fertile and sterile maize plants from tissue culture with the Texas male sterile cytoplasm.* Theoret. Appl. Genet. 62:213–217.

Krishnamurthi, M. 1981. *Sugarcane improvement through tissue culture and review of progress.* Pp. 70–77 in A.N. Rao, ed., *Tissue Culture of Economically Important Plants.* Asian Network for Biological Sciences, Singapore.

Krishnamurthi, M. 1982. *Disease resistance in sugarcane developed through tissue culture.* Pp. 769–770 in A. Fujiwara, ed., *Plant Tissue Culture.* Jpn. Assoc., Plant Tissue Culture, Tokyo.

Larkin, P.J. and W.R. Scowcroft. 1983. *Somaclonal variation – a novel source of variability from cell culture for plant improvement.* Theoret. Appl. Genet. 60:197–214.

Larkin, P.J. and W.R. Scowcroft. 1981. *Somaclonal variation and eyespot toxin tolerance in sugarcane.* Plant Cell Tiss. Org. Cult. 2:111–121.

Larkin, P.J., S.A. Ryan, R.I.S. Brettell and W.R. Scowcroft. 1984. *Heritable somaclonal variation in wheat.* Theoret. Appl. Genet. 67:443–455.

Liu, M.C. and W.H. Chen. 1978. *Tissue and cell culture as aids to sugarcane breeding.* II. *Performance and yield potential of callus-derived lines.* Euphytica 27:273–282.

Lorz, H. and W.R. Scowcroft. 1983. *Variability among plants and their progeny regenerated from protoplasts of Su/su heterozygotes on Nicotiana tabacum.* Theoret. Appl. Genet. 66:67–75.

Lupi, M.C., A. Bennici, S. Baroncelli, D. Gennai and F. D'Amato. 1981. *In vitro regeneration of Durum wheat plants.* II. *Diplomatic selection in aneusomatic plants.* Z. Pflanzenzuchtg. 87: 167–171.

Maddock, S.E., V.A. Lancaster, R. Risiott and J. Franklin. 1983. *Plant regeneration from cultured immature embryos and inflorescences of 25 cultivars of wheat (Triticum aestivum).* J. Expl. Bot. 34:915–926.

Maliga, P., Z.R. Kiss, P.J. Dix and G. Lazar. 1979. *A streptomycin resistant line of Nicotiana sylvestria unable to flower.* Mol. Gen. Genet. 172:13–15.

McCoy, T.J. 1980. *Cytogenetic ability of tissue culture regenerated plants of Avena sativa: high frequency of chromosome loss.* Can. J. Genet. Cytol. 24:37–50.

McCoy, T.J., R.L. Phillips and H.W. Rines. 1982. *Cytogenetic analysis of plants regenerated from oat (Avena sativa) tissue culture: high frequency of partial chromosome loss.* Can. J. Genet. Cytol. 24:37–50.

Melchers, G. and G. Labib. 1974. *Somatic hybridization of plants by fusion protoplasts. I. Selection of light resistant hybrids of "haploid" light sensitive varieties of tobacco.* Mol. Gen. Genet. 135:277–294.

Mousseau, J. 1970. *Fluctuation induite par la neoformation de bourgeons in vitro.* Colloq. Int. CNRS 193:234–239.

Murakishi, H.H. 1984. *Reaction of protoclones of potoato to potato virus Y.* Phytopath. 74:801.

Murashige, T. 1978. *The impact of tissue culture on agriculture.* Pp. 15–26 in T.A. Thorpe, ed., *Frontiers of Plant Tissue Culture.* University Calgary Press, Calgary.

Nakamura, C. and W.A. Keller. 1982. *Callus proliferation and plant regeneration from immature embryos of hexaploid Triticale.* Z. Pflanzenzuchtg. 88:137–160.

Novak. F.J. 1980. *Phenotype and cytological status of plants regenerated from callus cultures of Allium sativum L.* Z. Pflanzenzuecht. 84:250–260.

Ogihara, Y. 1981. *Tissue culture in Haworthia. IV. Genetic characterization of plants regenerated from callus.* Theor. Appl. Genet. 60:353–363.

Ogura, H. 1976. *The cytological chimeras in original regenerates from tobacco tissue cultures and in their offspring.* Jpn. J. Genet. 51:161–174.

Oono, K. 1978. *Test tube breeding of rice by tissue culture.* Proceedings Symposium Methods In Crop Breeding. Tropical Agriculture Research Series No. 11. Ministry of Agriculture and Forestry, Ibaraki, Japan.

Orshinsky, B.R. and D.T. Tomes. 1984. *Comparison of plants derived from cuttings, node cultures, and ethyl methanesulfonate treated node cultures of bird's foot trefoil (Lotus corniculatus).* Can. J. Bot. 62:1501–1504.

Orton, T.J. 1980. *Chromosomal variability in tissue cultures and regenerated plants of Hordeum.* Theor. Appl. Genet. 56:101–112.

Orton, T.J. 1983. *Spontaneous electrophoretic and chromosomal variability in callus cultures and regenerated plants of celery.* Theoret. Appl. Genet. 67:17–24.

Orton, T.J. 1984. *Genertic variation in somatic tissue: mthod or madness.* Pp. 153–189 in D.S. Ingram and P.H. Williams, eds., *Advances in Plant Pathology.* Academis Press, New York.

Pfeiffer, T.W. and E.T. Bingham. 1984. *Comparisons of alfalfa somaclonal and sexual derivatives from the same genetic source.* Theoret. Appl. Genet. 67:263–266.

Prat, D. 1983. *Genetic variability induced in Nicotiana sylvestris by protoplast culture.* Theoret. Appl. Genet. 64:223–230.

Pullman, G., L. Rappaport and S. Heath-Pagliuso. 1984. *Somaclonal variation in celery: towards selection for resistance to Fusarium wilt.* Hort. Sci. 19:589.

Reisch, B. and E.T. Bingham. 1981. *Plants from ethonine-resistant alfalfa tissue cultures: variations in growth and morphological characteristics.* Crop Sci. 21:783–788.

Sacristan, M.D. 1981. *Regeneration of plants from long-term callus cultures of haploid Brassica napus.* Z. Pflanzenzuecht. 86:248–253.

Sacristan, M.D. and G. Melchers. 1969. *The caryological analysis of plants regenerated from tumorous and other callus cultures of tobacco.* Mol. Gen. Genet. 105:317–333.

Scowcroft, W.R., and P.J. Larkin. 1982. *Somaclonal variation: a new option for plant improvement.* Pp. 159–178 in I.K. Vasil, W.R. Scowcroft, and K.J. Frey, eds., *Plant Improvement and Somatic Cell Genetics.* Academic Press, New York.

Scowcroft, W.R., P.J. Larkin and R.I.S. Brettell. 1983. *Genetic variation from tissue culture.* Pp. 139–162 in J.P. Helgeson and B.J. Deverall, eds., *Use of Tissue Culture and Protoplasts in Plant Pathology.* Academic Press, New York.

Sears, R.G. and E.L. Deekard. 1982. *Tissue culture variability in wheat: callus induction and plant regeneration.* Crop. Sci. 22:546–550.

Secor, G.A. and J.F. Shepard. 1981. *Variability of protoplast-derived potato clones.* Crop Sci. 21:102–105.

Shahin, E. 1984. *Isolation, culture, and regeneration of potato leaf protoplasts from plants preconditioned in vitro.* Pp. 381–390 in I.K. Vasil, ed., *Cell Culture and Somatic Cell Genetics of Plants.* Academic Press, New York.

Sharp, W.R., S.M. Reed and D.A. Evans. 1984. *Production and application of haploid plants.* Pp. 347–381 in P.B. Vose and S.G. Blixt, eds., *Crop Breeding: A Contemporary Basis.* Pergamon Press, Oxford.

Shepard, J.F. 1981. *Protoplasts as sources of disease resistance in plants.* Ann. Rev. Phytopath. 19:145–155.

Shepard, J.F. 1982. *The regeneration of potato plants from leaf-cell protoplasts.* Sci. Amer. 246:154–166.

Shepard, J.F. and R.E. Totten. 1977. *Mesophyll cell protoplasts of potato: isolation, proliferation, and plant regeneration.* Plant Physiol. 60:313–316.

Shepard, J.F., D. Bidney and E. Shahin. 1980. *Potato protoplasts in crop improvement.* Science 208:17–24.

Sibi, M. 1976. *Genetic program in higher plants.* II. *Experimental aspects, production of variants by in vitro tissue culture of Lactuca sativa. Increase in vigor in outcrosses.* Ann. Amelior. Plant 26:523–547.

Sibi, M. 1982. *Heritable epigenetic variations from in vitro tissue culture of Lycopersicon esculentum (var. Monal bo).* Pp. 228–244 in E.D. Earle and Y. Demarly, eds., *Plants Regenerated from Tissue Culture.* Praeger Press, New York.

Singh, B.D., K.N. Kao and R.A. Miller. 1975. *Karyotypic changes and selection pressure in Haplopappus gracilis suspension cultures.* Can. J. Genet. Cytol. 17:109–116.

Skirvin, R.M. and J. Janick. 1976. *Tissue culture-induced variation in scented Pelargonium spp.* J. Amer. Soc. Hort. Sci. 101:281–290.

Smith, S.M. and H.E. Street. 1974. *The decline of embryogenic potential as callus and suspension cultures of carrot (Daucus carota L.) are serially subcultured.* Ann. Bot. 38:233–241.

Sreenivasan, J.V. and N.C. Jalaja. 1983. *Sugarcane varietal improvement through tissue culture.* Pp. 371–376 in S.K. Sen and K.L. Giles, eds., *Plant Cell Culture in Crop Improvement.* Plenum Press, New York.

Sree Ramulu, K. 1982. *Genetic instability at the S-locus of lycopersicon peruvianum plants regenerated from in vitro culture of anthers: generation of new S-specificities and S-allele reversions.* Heredity 49:319–330.

Sree Ramulu, K.P. Dijkhuis and S. Roest. 1983. *Phenotypic variation and ploidy level of plants regenerated from protoplasts of tetrapolid potato (Solanum tuberosum L. cv. Bintje).* Theoret. Appl. Genet. 65:329–338.

Stimart, D.P., P.D. Ascher and J.S. Zagorski. 1980. *Plants from callus of the interspecific hybrid Lilium 'Black Beauty'.* Hort. Sci. 15:313–315.

Sun, Z.X., C.Z. Zhao, K.L. Zheng, H.F. Qi and Y.P. Fu. 1983. *Somaclonal genetics of rice, Oryza sativa L.* Theoret. Appl. Genet. 67:67–73.

Sutter, E. and R.W. Langhans. 1981. *Abnormalities in Chrysanthemum regenerated from long-term cultures.* Ann. Bot. 48:559–568.

Swartz, H.J., G.J. Galletta and R.H. Zimmerman. 1981. *Field performance and phenotypic stability of tissue-culture-propagated strawberries.* J. Amer. Soc. Hort. Sci. 106:667–673.

Swartz, H.J., G.J. Galletta and R.H. Zimmerman. 1983. *Field performance and phenotypic stability of tissue-culture-propagated thornless blackberries.* J. Amer. Soc. Hort. Sci. 108: 285–290.

Syono, L. and T. Furuya. 1972. *Abnormal flower formation of tobacco plants regenerated from callus cultures.* Bot. Mag. Tokyo 85:273–284.

Takebe, I., G. Labib and G. Melchers. 1971. *Regeneration of whole plants from isolated mesophyll protoplasts of tobacco.* Naturwissenschaften 58:318–320.

Thanutong, P., I. Furusawa and M. Yamamoto. 1983. *Resistant tobacco plants from protoplast-derived calluses selected for their resistance to Pseudomonas and Alternaria toxins.* Theor. Appl. Genet. 66:209–215.

Thomas, E. 1981. *Plant regeneration from shoot culture-derived protoplasts of tetraploid potato (Solanum tuberosum cv. 'Maris Bard').* Plant Sci. Lett. 23:81–88.

Thomas, E., S.W.J. Bright, J. Franklin, V.A. Lancaster, B.J. Miflin and R. Gibson. 1982. *Variation amongst protoplast-derived potato plants (Solanum tuberosum cv. 'Maris Bard').* Theoret. Appl. Genet. 62:65–68.

Vajrabhaya, T. 1977. *Variations in clonal propagation.* Pp. 179–201 in J. Arditti, ed., *Orchid Biology: Reviews and Perspectives.* Columbia University Press, New York.

Van Harten, A.M. and C. Broertjes. 1981. *In vitro adventitious bud techniques for vegetative propagation and mutation breeding of potato (Solanum tuberosum L.).* II. *Significanca for mutation breeding.* Euphytica 30:1–8.

Wakasa, K. 1979. *Variation in the plants differentiated from tissue culture of pineapple.* Jpn. J. Breed. 29:13–22.

Wenzel, G., O. Schieder, T. Przewozny, S.K. Sopory and G. Melchers. 1979. *A comparison of single cell-derived Solanum tuberosum plants and a model for their application in breeding programs.* Theoret. Appl. Genet. 55:49–55.

Zeng, J.Z. 1983. *Application of anther culture technique to crop improvement in China.* Pp. 351–363 in S.K. Sen and K.L. Giles, eds., *Plant Cell Culture in Crop Improvement.* Plenum Press, New York.

Part 2

Genetic modifications

Lectin genes and a transposable element in soybean

LILA O. VODKIN*, PATSY R. RHODES*, JOEL M. CHANDLEE* and ROY W. HARDING**
**Plant Molecular Genetics Laboratory, Agricultural Research Service, U.S. Department of Agriculture, Beltsville, Maryland 20705, U.S.A.*
***On sabbatical leave from Smithsonian Environmental Research Center, 12441 Parklawn Drive, Rockville, Maryland 20852, U.S.A.*

Abstract Biotechnology can contribute to solving agricultural problems by providing the means to probe the organization and function of plant genomes. One major question is why certain genes are expressed only in particular plant tissues and at specific times during development. This knowledge is essential for creating improved varieties by genetic engineering. Some current research aimed at using gene transfer to study the developmental regulation of plant genes is discussed. Our specific approach to this goal is to examine the expression of a developmentally regulated soybean gene which produces seed lectin only during embryo formation. Before genes of agronomic significance can be manipulated by genetic engineering, they must be identified and isolated. Knowledge of the genes and biochemical pathways that control most plant traits of agronomic importance is very limited. In this regard, the use of transposons to isolate or tag genes whose protein products are unknown holds promise. While a large amount of work exists on transposable elements in *Zea mays*, little is known in soybean. We describe an element, Tgm1, that interrupts the soybean lectin gene and blocks its expression. Tgm1 has a familial relationship to the En/Spm transposable element of maize. Attempts to correlate the soybean element to other mutable traits or mutations in soybean are described.

Introduction

The widespread use of recombinant DNA and molecular biology techniques in plant biology and agriculture generates excitement for several reasons. One is the potential of these techniques for commercial application, and another is the powerful opportunity they provide for probing the organization and function of plant genomes. These aims are not mutually exclusive but are likely to be closely interdependent. Advances in one area are likely to aid and stimulate progress in the other.

The present aims of our laboratory focus on two broad, long term objectives: (a) developmental regulation of gene activity and (b) the nature and action of mobile elements in soybean. While these aims are scientific goals in basic research, there are potential benefits to applied agricultural biotechnology from knowledge obtained in these areas. Routine use of genetic engineering techniques for crop improvement will require a means to control developmental and tissue specific gene action. For example, a gene that controls a seed product should be expressed in the seed and not in the leaves or other organs of the recipient plant. Defined regulatory sequences which can be coupled to

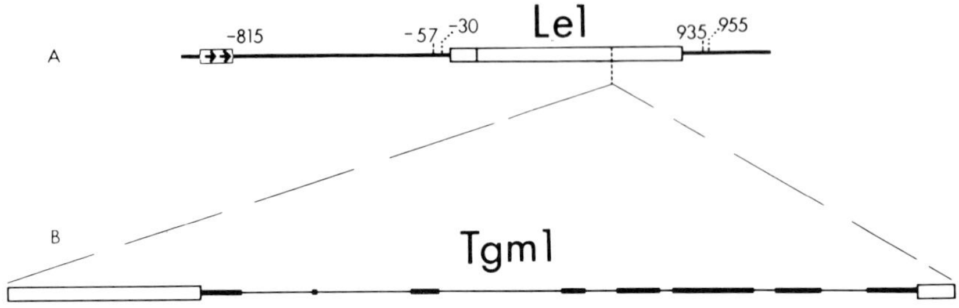

Figure 1. Schematic drawing of the soybean seed lectin gene (A) and the lectin insertion element (B). The coding region of *Le1* is denoted by a short open box representing the signal peptide sequence and a longer open box denoting the coding frame for the mature lectin protein. The promoter (TATAAATA) at −57 from the initiation codon is indicated as are the 5′ cap site at −30 and the polyadenylation sites at +935 and +955. *Le1* contains no introns. Two tandem repeats of 58 bp each in the 5′ flanking region are designated by arrows enclosed by the box at −815. The position at which the 3.5 kb Tgm1 insertion splits the lectin coding region in the lectin negative cultivar is indicated by the dashed line at +600. The open boxes of Tgm1 delineate 726 bp of the left arm and 144 bp of the right arm which consist of tandem repeating units. The relative position of 80–90% A + T rich areas (bold lines) and 60–70% A + T areas (thin lines) are also presented. (Adapted from Rhodes and Vodkin, 1985.)

the foreign gene of interest and thereby confer proper developmental expression to the transferred gene must be available.

Transposable elements have proven extremely useful in some organisms to identify and isolate genes whose protein products are unknown. Many genes important to the plant breeder fall into this category. These include genes that confer disease resistances, mineral utilization, and morphological characteristics such as organ shapes, branching, and plant height. Development of approaches in soybean to identify such genes using transposable elements would enhance our ability to isolate, understand, and utilize agronomically or developmentally important genes.

Developmental regulation of gene activity

For the purpose of understanding the regulation of plant genes, we have examined the molecular basis for a naturally occurring mutation that affects expression of a specific seed protein, soybean lectin. Lectins are carbohydrate binding proteins commonly found in legume seed. Purified plant lectins are able to bind specific carboyhydrates on membrane surfaces and agglutinate cells; however, the functions they perform in plants are unclear. Whereas lectin normally accounts for up to 5% of the total seed protein, a survey of *Glycine max* lines revealed that several lacked lectin protein (Pull et al., 1978). Seed lectin protein is encoded by a single gene, *Le1.* We determined that a lectin negative (Le^-) soybean variety contains this lectin gene but its expression is blocked by the presence of a large insertion which has the structural features of a transposable element (Vodkin, Rhodes and Goldberg, 1983; Goldberg, Hoschek and Vodkin, 1983).

Figure 1 shows a schematic diagram of the DNA sequence of *Le1*, its flanking regions, and the insertion which interrupts this gene in the Le$^-$ variety. *Le1* lacks intervening sequences and produces a 1.0 kilobase (kb) mRNA. The protein encoded by the gene contains a 32 amino acid signal sequence which directs packaging of lectin into protein bodies of the cotyledon cells. Each subunit polypeptide of 253 amino acids (26,700 daltons) contains about 2000 daltons of carbohydrate, and four identical subunits associate to form a mature tetrameric protein.

Expression of lectin mRNA and protein is developmentally regulated. Cotyledon cells of developing soybean seeds normally synthesize about 4000 lectin mRNA molecules per cell during a period known as mid-maturation, a time of maximum accumulation of seed storage proteins in soybean (Goldberg et al., 1983). The 32,000 dalton seed lectin polypeptide is not detected in other plant parts such as leaves or roots although there are proteins with immunological cross reactivity to the seed lectin found in roots and other soybean tissues (Vodkin, unpublished observations).

Presumably, regulatory sequences which reside 5′ (upstream) of the lectin protein coding region are involved in specifying that this particular gene is expressed only in the developing seed. Methods to test the structure–function relationships of gene sequences in living cells must involve the introduction of genes into cells and whole plants. Extensive research on *Agrobacterium tumefaciens*, a soil bacterium, has led recently to its successful use as a vector for gene transfer in plants (Fraley et al., 1983). *A. grobacterium tumefaciens* can infect a broad range of dicotyledonous plants. *Nicotiana* is a commonly used host species since tobacco plants are easy to manipulate in culture and regenerate whole plants from single cells. When the bacteria enter a wounded area of the plant, they produce a neoplastic growth known as a crown gall tumor. The tumor mass can be excised from the plant and grown in vitro without the normal hormone supplements needed to maintain plant cell cultures and without the infecting bacterium. A large, resident plasmid (Ti-plasmid) within *A. tumefaciens* carries the necessary functions for infectivity and a 25 kb region of this plasmid (known as T-DNA) is transferred stably into DNA of the infected plant cells (Chilton et al., 1977). The T-DNA region contains genes which code for products that have hormone-like effects on plant tissue. Thus, the transformed plant tissue is propagated as a tumorous growth which is normally unable to regenerate shoots and roots.

Ti-plasmid vectors are currently being used to examine developmental regulation of gene activity. A gene coding for the small subunit of RuBP carboxylase from pea was introduced into petunia tumors and found to be expressed under its own promoter in this new plant host (Broglie et al., 1984). The small subunit promoter when fused to a bacterial enzyme will direct the synthesis of the enzyme within tobacco tumors (Herrera-Estrella et al., 1984). More interestingly, the activity of the foreign gene is light regulated in the new host tumor cells in a manner similar to the effect of light on the activity of this gene in pea. Thus, one might expect to obtain expression of a number of plant genes in tumor cells of unrelated plant species provided that these genes are normally expressed in callus tissue. This opens the way to designing deletions or other mutations in the regulatory regions and then assaying for the effect of the mutation on the expression of the gene after it is transferred back into host plant tissues. Such an analysis has recently been undertaken for the small subunit RuBP carboxylase and the results demonstrate that a light responsive sequence of 33 bp resides very close to the TATA box promoter region (Morelli et al., 1985).

Unfortunately, one disadvantage of the Ti-plasmid system is that properties which confer hormone independent tumor growth, also prevent the normal regeneration of shoots from transformed cell lines. This barrier is being overcome, however, by construction and use of several types of "disarmed" Ti-plasmids. In disarmed vectors the genes which cause hormone production and thus oncogenic growth of transformed plant cells have been eliminated or inactivated (Caplan et al., 1983; de Frammond, Barton and Chilton, 1983). This permits regeneration of whole plants from infected tissue and subsequent examination of how the transferred gene is expressed during the normal life cycle of the modified plant.

Using Ti plasmid vectors, we have transferred the lectin gene into tobacco tumor cells. A 5.7 kb Bgl II fragment that contains the seed lectin gene and approximately 2.9 kb of flanking 5′ sequence and 1.8 kb of flanking 3′ sequence has been cloned into the Bam HI site of *Escherichia coli* shuttle vector, pDA17 (Riedel and Austen, 1984). The resulting plasmid, pLe1-B8 was then moved into the *A. tumefaciens* Ti-plasmid by a triparental mating procedure (Ditta et al., 1980). pDA17 carries regions of homology to the T-DNA which allow homologous recombination with kanamycin resistance as a selectable marker to detect the recombination events. Exconjugates from the mating procedure which are resistant to kanamycin and which were confirmed to contain the lectin gene by Southern hybridization were used to infect wounded *Nicotiana tabacum* or *N. plumbaginifolia* whole

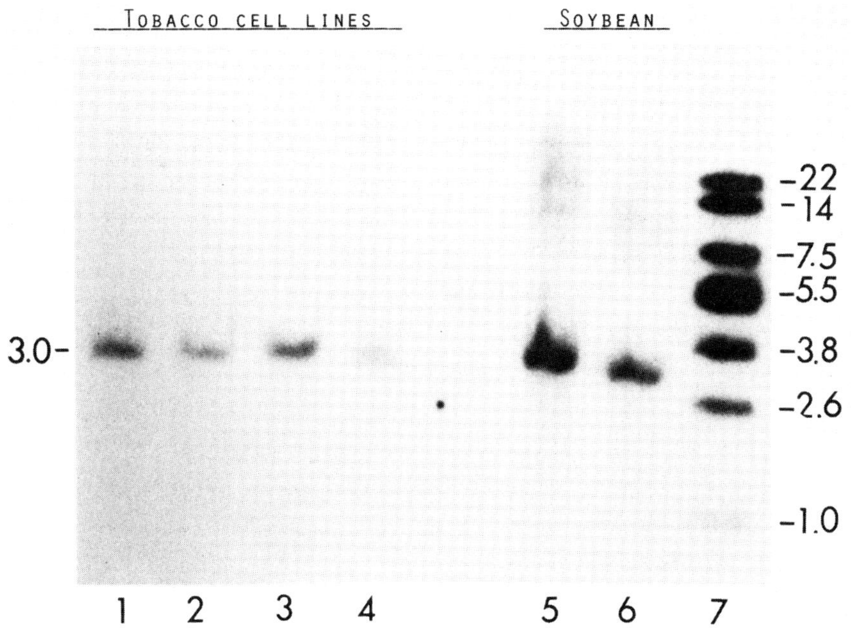

Figure 2. DNA blot indicating the presence of the soybean lectin gene within tobacco tumor cell lines. DNA was extracted from hormone-independent and octopine-producing tobacco cell lines that had been established from tumors induced by *Agrobacterium* strains carrying the lectin gene. The DNA was digested with XbaI, and the Southern blot was probed with an M13 subclone containing the lectin sequence. Lanes 1–4 are independently transformed cell lines of tobacco; lanes 5 and 6 are soybean DNA samples as controls; and lane 7 is a marker standard of the indicated molecular weights. A single 3.0 kb XbaI fragment is expected and was found in all of the tobacco tumor preparations. Control tobacco DNAs from non-transformed tissues do not show any hybridization with the lectin sequence (not shown).

plants or inverted stems grown axenically. After several weeks, the small tumors which formed were excised and transferred to plates containing standard Murashige-Skoog media plus carbenicillin to kill *Agrobacterium* cells and aminoethylcysteine to select against non-transformed plant cells (see Riedel, this symposium volume).

Southern blots of the callus tumors are shown in Fig. 2. When Xba I digests of DNA extracted from the tobacco cell lines are probed with an M13 subclone containing a 300 bp region of the lectin coding sequence, a 3 kb fragment is detected which corresponds to that expected for the soybean lectin gene sequence. Although we have not yet examined border fragments, all of the axenic tumor cultures from several independent infection experiments show the presence of the soybean lectin sequence which is not found in control tobacco DNA. The tobacco tumor lines were also examined for lectin mRNA by S1 analysis of poly A + mRNA using M13 subclones of the lectin gene. Assays for lectin polypeptides were made using a monospecific antibody to probe immunoblots of tumor proteins separated by sodium dodecyl sulfate polyacrylamide electrophoresis. No lectin mRNA or protein was detected in the transformed tobacco tumors, but this was not unexpected. It is unlikely that the *Le1* gene would be expressed in a tumor culture of nondifferentiated tobacco cells since it is not normally expressed in soybean callus culture but only in developing seeds. Presently, we are using non-oncogenic strains of Ti-plasmid to transfer the lectin gene and to obtain regenerated tobacco plants. Unlike soybean, tobacco does not produce a high protein seed, but it is possible that the soybean lectin gene will be developmentally regulated in this unrelated species. A recent report has demonstrated that a seed protein gene, phaseolin, from *Phaseolus vulgaris* is expressed and developmentally regulated in tobacco (Sengupta-Gopalan et al., 1985). A single copy of the phaseolin gene containing 863 bases of flanking 5′ sequence appeared to be expressed in tobacco seeds at nearly equivalent levels per genome copy as in bean seeds.

Obviously, a full understanding of the mechanisms by which a DNA sequence controls how a gene is expressed in certain tissues and times of development will require many other innovative approaches in addition to gene transfer techniques. Specific protein interactions with DNA sequences are undoubtedly involved. Changes in DNA methylation or the organization of genes within chromatin could also influence developmental regulation. At this time, the inability of soybeans to be routinely regenerated from callus or suspension cultures imposes limitations on using gene transfer in soybean for crop improvement and for probing gene regulation.

Uses of transposable elements

Mobile elements are pieces of DNA capable of changing their locations within the genome. During this process they insert into or near other genes and often lead to a mutant phenotype (trait). Although the genetic behavior of transposable elements was first recognized in corn in the 1950's, molecular descriptions of plant elements has emerged only within the past two years. In the intervening time, basic studies on the nature of transposable elements in bacteria and *Drosophila* have shown that they are more than just interesting pieces of DNA. As discussed below, they have exceptional utility as a means to isolate genes whose protein products are unknown.

Several reviews of plant transposable elements have appeared in recent years (Fedoroff, 1983; Freeling, 1984; Doring and Starlinger,

1984). Maize controlling elements often occur in pairs, as the Ac (Activator) and Ds (Dissociation) elements which are members of the same family. Generally, one can think of these elements as being autonomous or non-autonomous with respect to whether or not the element is capable of independent transposition. Both presumably affect gene activity by inserting into or existing from gene loci; however, the non-autonomous elements can transpose only if an active element is also present elsewhere in the genome. Presumably, the non-autonomous element is defective in some function which is supplied by a trans-acting protein from the autonomous element. Isolation and sequencing of the autonomous Ac element shows that there exist several open reading frames which have the potential to code for a transposase that would be necessary to mediate movement of the element (Pohlman, Fedoroff and Messing, 1984).

Plant elements are recognized by their effects on gene expression and can be physically isolated if they have inserted into a gene which has been cloned. In addition to members of the Ac/Ds family, several other elements have been isolated and characterized from corn. These include Robertson's mutator (Barker et al., 1984) and the En/Spm element (Pereira et al., 1985). The genetic effects of Suppressor-mutator (Spm) were described extensively by McClintock (as reviewed by Fedoroff, 1983). It is the same element that was found originally by Peterson and referred to as the Enhancer (En) element (Peterson, 1953). This element shows a range of effects on gene expression which are not yet explained at the molecular level.

In the absence of some knowledge of the protein encoded by a DNA segment, as is the case for genes affecting many complex plant traits, isolation of these genes from a genomic library is impossible without a way to equate a phenotypic change to a cloned DNA sequence. In this regard, plant transposable elements should be very useful. The basic process of "transposon tagging" begins with a genetic cross in which one of the parent lines contains an active element. Subsequently, the progeny are screened for recessive mutations occurring at a high frequency in the trait of interest. These are likely to be due to an insertion of the element into or near a gene controlling that trait. A genomic library of random sequence is then constructed from the mutant plant DNA. Assuming that the transposable element has previously been isolated and characterized, a clone of the element is available. This plasmid containing the transposable element can be radioactively labeled and used as a specific DNA hybridization probe to identify the genomic clone containing the element from the library of mutant plant DNA. This genomic clone will also contain some of the gene into which the element has inserted. In this manner, an unknown gene controlling a plant trait can be identified and isolated without prior knowledge of the information it potentially encodes.

The isolation of genes in bacteria has been greatly facilitated by transposon tagging. This approach has also been used to isolate *Drosophila* genes (Searles et al., 1982). The first two examples of transposon tagging in higher plants are the isolation of the bronze locus in corn (Fedoroff, Furtek and Nelson, 1984) using an Ac element probe and of the *a1* locus using an En probe (O'Reilly et al., 1985). Both of these genes are involved in synthesis of anthocyanin pigments. Many more examples of genes isolated in this manner in corn are likely to follow in the next few years, including some genes for disease resistance and other potentially valuable agronomic traits.

Unusual structure of the soybean element, TGM1

In a soybean variety which does not produce any lectin protein, the seed lectin gene is present but is interrupted by an insertion element at a position about two-thirds into its coding region. DNA sequence analysis has shown that the insertion termini are 30 base pair (bp) inverted repeats and that a 3 bp sequence of the lectin gene has been duplicated so that it flanks both ends of the insertion termini (Vodkin et al., 1983). Inverted repeat termini and small duplications of target site material are structural evidence for transposable element action. Thus, these features of the lectin insertion element demonstrate that it arose by a transposition mechanism. For simplicity, we designate the soybean lectin gene insertion as Tgm1 (transposable element, *Glycine max*).

A remarkable sequence similarity exists between the ends of elements from three unrelated plant species (Fig. 3). Tam1, a 17 kb element which interrupts a chalcone synthase gene in snapdragon (Bonas, Sommer, and Saedler, 1984), Tgm1 from soybean, and the En/Spm element isolated from the *waxy* locus of corn (Schwarz-Sommer et al., 1984) are similar at 11 of the 13 terminal inverted-repeat nucleotides. All three of the elements duplicate a three base segment of the target gene also indicating a similarity in the transposition mechanism.

The complete sequence of Tgm1 is complex and highly structured (Rhodes and Vodkin, 1985). As diagrammed in Fig. 1, extended borders exist beyond the termini of the element. The borders are comprised of a tandem series of sequences which appear to be variations of a prototype 54 bp repeating unit found at nt 673–726. These tandem repeating units occupy a total of 726 bases at the 5′ end of the element and 300

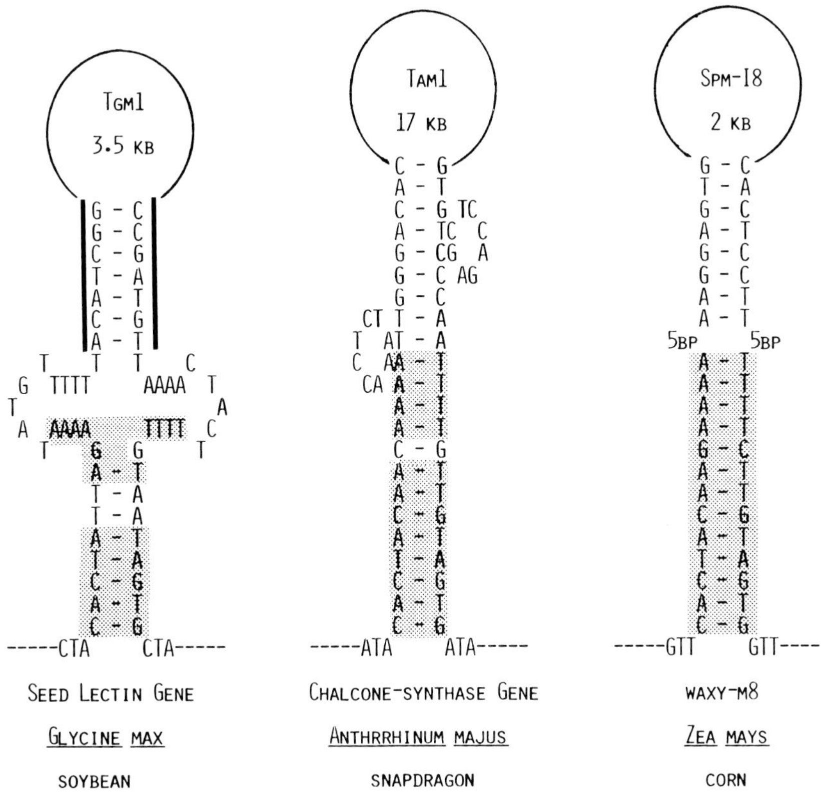

Figure 3. Transposable elements from three unrelated plant species have sequence similarity at their termini. Terminal inverted repeats for the soybean element, Tgm1, which interrupts the seed lectin gene are compared to those of Tam1 a 17 kb element found in the chalcone synthase gene in snapdragon (Bonas et al., 1984) and to the En/Spm element of corn isolated from the *waxy* locus which encodes a UDP-glucose starch transferase (Schwarz-Sommer et al., 1984; Pereira et al., 1985). The shaded bases are identical as compared to the Spm element. All three elements duplicate three bases of the target gene as shown. The seven base sequence of Tgm1 termini which has homology to inverted repeats within the lectin gene is marked by the bold lines.

bases at the 3′ end of the element. Inverted repeats are found within the tandem repeating unit and can be drawn as hairpin structures which have a 16 bp conserved stem. While degenerate forms of the basic structure appear progressively towards the 5′ and 3′ ends of the element, the stem area remains conserved.

Features of the repeating unit and its similarities to the termini of Tgm1 are shown in Fig. 4. The 16 bp stem consists of a Hpa I restriction site at its base and a prominent feature is the presence of a seven nucleotide sequence, ACATCGG, and its inverted complement, CCGATGT. This 7 bp sequence occurs in the 30 bp inverted termini of

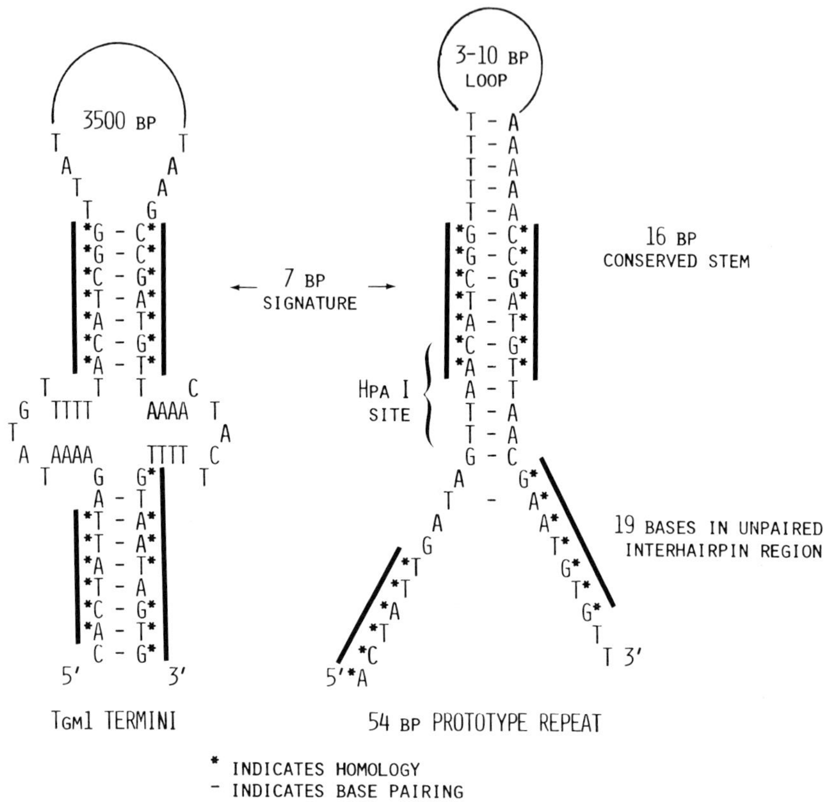

Figure 4. Diagram illustrating similarity between Tgm1 termini and the 54 bp prototype repeat unit of the Tgm1 extended border region. The drawings are intended to point out sequence similarities and do not imply that the structures exist in vivo. The prototype repeat unit consists of a 3–10 base loop and a 16 base pair conserved stem (hairpin) having a Hpa I restriction site at the base. The seven base signature which has homology to inverted repeats within the lectin gene is found in the conserved stem and in the upper region of the Tgm1 termini as indicated by the arrows and the asterisks denoting sequence homology. The sequence similarity between the first 9 bases of the Tgm1 termini and the prototype repeat unit is also denoted by bold lines and asterisks. Since this sequence homology is not exact and the bases are "slipped" with respect to one another, precise pairing in the prototype repeat unit cannot occur. This area thus forms the interhairpin region that separates the tandem hairpin units.

the Tgm1 element and is also identical to seven nucleotides which occur within 17 bp inverted repeats present in the lectin coding region at approximately 80 bp to either side of the target site (Vodkin et al., 1983). Such a symmetrical arrangement may indicate a site specificity for the insertion element. It was surprising to find these seven bp conserved in the internal repeating unit of Tgm1. Including all derivative forms, this sequence appears a total of 33 times in the element. Because of the prevalence of these seven bp throughout the borders, we consider it to be a molecular signature for Tgm1 in the sense that it is an identifying characteristic of the element. There is also similarity between the 13 bp termini of the element and the interhairpin sequence separating the hairpin repeat units of the border region. The homology is striking though not perfect as the bases are "slipped" with respect to one another so that base pairing would not occur in this region of the repeat unit.

We do not know the significance of this unusual border structure, however, we do suspect that the numerous occurrences of the 7 bp signature sequence arranged in an inverted repeat format may play a role in selection of the target gene. The En/Spm and Tam elements also have an extended border region that consists of inverted repeat sequences. While the termini of these three elements have sequence similarity, the conserved stems within the inverted repeat hairpin structures of the borders are not homologous (not shown).

We are testing the possibility that the inverted repeats of the Tgm1 border may play a role in selection of the target gene by isolating sequences related to Tgm1 from the soybean genome. Southern blots of soybean DNA probed with a border region of Tgm1 show a large number of dispersed repeat sequences. When an internal region of the Tgm1 is used to probe genomic DNA, a smaller number of about 10–20 related sequences is found. A number of clones having homology to Tgm1 have been isolated and are being compared by DNA sequencing. These analyses will determine how conserved or prevalent the 13 bp termini and the repeating border sequence arrangement is in the Tgm1 related soybean fragments. We especially want to answer whether the other gene sequences which flank Tgm1 relatives will have at some distance from the target site an inverted repeat similar in sequence to the Tgm1 border. Alternatively, these related sequences may be similar to Tgm1 in having the same basic pattern, i.e., a border consisting of inverted repeats, but the stem portion of the repeat may be different in sequence from the 7 bp signature found in Tgm1 and in the target lectin gene.

Can Tgm1 or related sequences be correlated with other mutant traits in soybean?

Although Tgm1 has the structural features of a transposable element and undoubtedly arose at some point by a transposition mechanism, there is presently no evidence for revertants in which Tgm1 has excised from the lectin gene or for movement of the element to other locations in the soybean genome. Whether it is now a defective element no longer capable of transposition or whether it is activated at low frequencies only in certain genetic backgrounds or under certain physiological conditions is unknown.

Although we are not able to genetically manipulate the lectin insertion sequence as an element for transposon tagging, we are using another approach to attempt to correlate the insertion element with other soybean mutant phenotypes. A number of isolines carrying single gene mutations are available from the USDA soybean germplasm collection in Urbana, Illinois. These include genes which affect chlorophyll, morphological features, pigmentation, disease reactions, and others. They are all in the genetic background of the cultivars Clark or Harosoy and represent about 38 nuclear genes. Generally, these isolines result from backcrossing the recurrent Clark or Harosoy parent for about five generations. Although this leads to a 98% recovery of the genome from the recurrent parent, these lines may still differ at loci other than the one in question.

As discussed in the previous section, sequences related to Tgm1 are present elsewhere in the soybean genome. When genomic DNA is probed with a clone containing the border regions of Tgm1, a large number of fragments are obtained on Southern blots. However, when a middle region is used, a smaller number of bands (approximately 10–15) are apparent. We are comparing genomic DNAs from the recurrent parent and from the mutant isolines using a number of restriction enzyme digests. One interpretation for any observed differences in Southern patterns between the two would be that a sequence related to Tgm1 is present in a restriction fragment that contains the mutant gene in question.

While there is no a priori reason to suspect the involvement of a Tgm1-related element in any of these mutant isolines, there are several loci in soybean whose behavior is likely to be due to mobile elements because they result in variegation or pattern formation. One of these is ringed mutable (r^{m}) which is an allele of the *R* locus, one the genes for seed coat color in soybean (Bernard and Weiss, 1973). In seeds containing r^{m} the black pigment is distributed in a variegated fashion over the seed coat surface often in concentric rings. Another locus of interest is the *I* gene (inhibitor of color) which controls distribution of pigment over the seed coat surface. This gene appears to mutate frequently, and one allele results in a distinct formation of color on the seed coat known as saddle pattern. A third locus of interest is $Y18^{m}$, a yellow mutable allele described by Peterson and Weber (1969). It is a nuclear gene which affects chloroplast function and results in variegated yellow and green sectors in the leaves. We have detected differences between the Southern profiles of DNAs isolated from yellow versus green tissue of the variegated $Y18^{m}$ plants (Chandlee and Vodkin, unpublished observations) indicating a possible association between this mutant gene and an element having homology to Tgm1.

These loci are of particular interest because of their mutable nature. In addition to analysis of DNA blots, crosses are being made between these mutable alleles and some of the Le^- lines to determine whether any interactions might occur. For example, we have observed some unexpected segregation patterns in outcrosses of an Le^- cultivar to other cultivars but have not yet analysed these at the molecular level. Experiments to determine whether active elements can be recovered from soybean in response to mutagens are in progress.

While there are some reports on the spontaneous origin of controlling elements in maize stocks, the numerous mutable alleles studied by McClintock were derived from plants that had undergone a breakage-fusion-bridge cycle on chromosome 9 during meiosis. The drastic structural modifications imposed on chromosome 9 by these events resulted in a "burst of newly arising mutable loci" in the progeny of these plants (McClintock, 1950). It is often assumed that movement of cryptic or endogenous mobile elements is normally suppressed or highly controlled and that stresses which disrupt normal function of the genome may release mobile elements from these control factors (Freeling, 1984). Such stresses may be mutagenic agents such as chemicals or radiation, infection by viruses or other pathogenic agents, and perhaps the process of plant cell tissue culture.

Summary and conclusions

The action of transposable elements constitutes a source of genetic diversity because of the effects that they have on the function and expression of the genes they invade. The study of transposable elements may provide an entry into aspects of differentiation. Rearrangements promoted by "naturally functioning" mobile elements may be involved in the processes of differentiation during clonal development. On the other hand, these elements may be molecular symbionts, or parasites, which survive in the genome as long as they do not place too great a demand on the host organism. In either case, they can be exploited for our own purposes. In corn, it is now feasible to use transposable elements which have been cloned as an approach to characterize genes controlling developmental events, complex traits, disease resistances, or other genes which we might not be able to isolate and identify by any other means. In addition, knowledge of the structure and function of transposable elements in higher plants may lead to the construction of improved vector systems for transformation and modification of plant genomes.

Before the directed modification of plant genomes can become routine, there are fundamental areas which need considerable research in the upcoming years. Assuming that we have decided upon a useful gene to transfer to a plant, we must have efficient vector and delivery systems and the ability to regenerate the plant from cell culture. A critical factor is to know how to control the activity of a transferred gene so that it will be expressed as desired in the plant's developmental program. This requires knowledge of the fundamental processes of gene regulation during differentiation and development. Vectors composed of modules which include the structural gene of interest as well as sequences required to regulate the quantity and time of expression of the protein product will be needed. Ideally, integration should be a non-random process, directing the foreign gene to a certain chromosomal or extrachromosomal site with some degree of reliability. Knowledge of plant replication origins, enhancer elements, tissue specificity control sequences, genetic rearrangements and transposition mechanisms should aid in the design of such vector systems. Advances in our basic knowledge of these processes will enhance the development of vectors for applied purposes. Such vector systems, in turn, will increase our capacity to probe further into the function and regulation of nuclear DNA in higher plants.

Acknowledgements We gratefully acknowledge the work of Robert Goeken and Krystyna Dobrowalska. We thank Dr. Gerard Riedel, Deborah Marin, and Douglas Austen for the pDA17 shuttle vector and for assistance with *Agrobacterium tumefaciens* infections.

References

Barker, R.F., D.V. Thompson, D.R. Talbot, J. Swanson and J.L. Bennetzen, 1984. *Nucleotide sequence of the maize transposable element Mu1.* Nucl. Acids Res. 12:5955–5967.

Bernard, R.L. and M.G. Weiss. 1973. *Qualitative genetics.* Pp. 117–154 in B.E. Caldwell, ed., *Soybeans: Improvement, Production, and Uses.* American Society of Agronomy, Madison, Wisconsin.

Bonas, U., H. Sommer and H. Saedler. 1984. *The 17-kb Tam1 element of Antirrhinum majus induces a 3-bp duplication upon integration into the chalcone synthase gene.* EMBO J. 3: 1015–1019.

Broglie, R., G. Coruzzi, R.T. Fraley, S.G. Rogers, R.B. Horsch, J.G. Niedermeyer, C.L. Fink, J.S. Flick and N-H. Chua. 1984. *Light-regulated expression of a pea ribulose-1, 5-biphosphate carboxylase small subunit gene in transformed plant cells.* Science 224:838–843.

Caplan, A., L. Herrera-Estrella, D. Inze, E. Van Haute, M. Van Montagu, J. Schell and P. Zambryski. 1983. *Introduction of genetic material into plant cells.* Science 222:815–821.

Chilton, M-D., M.H. Drummond, D.J. Merlo, D. Sciaky, A.L. Montoya, M.P. Gordon and E.W. Nester. 1977. *Stable incorporation of plasmid DNA into higher plant cells: the molecular basis of crown gall tumorigenesis.* Cell 11:263–271.

Ditta, G., S. Stanfield, D. Corbin and D.R. Helinski. 1980. *Broad host range DNA cloning system for Gram-negative bacteria: Construction of a gene bank of Rhizobium meliloti.* Proc. Nat. Acad. Sci. USA 77:7347–7351.

Doring, H-P. and P. Starlinger. 1984. *Barbara McClintock's controlling elements: now at the DNA level.* Cell 39:253–259.

Fedoroff, N.V. 1983. *Controlling elements in maize.* Pp. 1–63 in J. Shapiro, ed., *Mobile Genetic Elements.* Academic Press, New York.

Fedoroff, N.V., D.B. Furtek and O.E. Nelson, Jr. 1984. *Cloning of the bronze locus in maize by a simple and generalizable procedure using the transposable controlling element Activator (Ac).* Proc. Nat. Acad. Sci. USA 81:3825–3829.

Fraley, R.T., S.G. Rogers, R.B. Horsch, P.R. Sanders, J.S. Flick, S.P. Adams, M.L. Bittner, L.A. Brand, C.L. Fink, J.S. Fry, G.R. Galluppi, S.B. Goldberg, N.L. Hoffmann and S.C. Woo. 1983. *Expression of bacterial genes in plant cells.* Proc. Nat. Acad. Sci. USA 80:4803–4807.

de Framond, A.J., K.A. Barton and M-D. Chilton. 1983. *Mini-Ti: a new vector strategy for plant genetic engineering.* Bio/Technology 1:262–269.

Freeling, M. 1984. *Plant transposable elements and insertion sequences.* Ann. Rev. Plant Physiol. 35:277–298.

Goldberg, R.B., G. Hoschek and L.O. Vodkin. 1983. *An insertion sequence blocks the expression of a soybean lectin gene.* Cell 33:465–475.

Herrera-Estrella, L., G. Van den Broeck, R. Maenhaut, M. Van Montagu and J. Schell. 1984. *Light-inducible and chloroplast-associated expression of a shimaeric gene introduced into Nicotina tabacum using a Ti-plasmid vector.* Nature 310:115–120.

McClintock, B. 1950. *The origin and behavior of mutable loci in maize.* Proc. Nat. Acad. Sci. USA 36:344–355.

Morelli, G., F. Nagy, R.T. Fraley, S.G. Rogers and N-H. Chua. 1985. *A short conserved sequence is involved in the light-inducibility of a gene encoding ribulose 1, 5-biphosphate carboxylase small subunit pea.* Nature 315:200–204.

O'Reilly, C., N.S. Shepherd, A. Pereira, Z. Schwarz-Sommer, I. Bertram, D.S. Robertson, P.A. Peterson and H. Saedler. 1985. *Molecular cloning of the al locus of Zea mays using the transposable elements En and Mul.* EMBO J. 4:877–882.

Pereira, A. Z. Schwarz-Sommer, A. Gierl, I. Bertram, P.A. Peterson and H. Saedler. 1985. *Genetic and molecular analysis of the Enhancer (En) transposable element system of Zea mays.* EMBO J. 4:17–23.

Peterson, P.A. 1953. *A mutable pale green locus in maize.* Genetics 38:682–683.

Peterson, P.A. and C.R. Weber. 1969. *An unstable locus in soybeans.* Theor. Appl. Genet. 39:156–162.

Pohlman, R.F., N.V. Fedoroff and J. Messing. 1984. *The nucleotide sequence of the maize controlling element Activator.* Cell 37:635–643.

Pull, S.P., S.G. Pueppke, T. Hymowitz and J.H. Orf. 1978. *Soybean lines lacking the 120,000 dalton seed lectin.* Science 200:1277–1279.

Rhodes, P.R. and L.O. Vodkin. 1985. *Highly structured sequence homology between an insertion element and the gene in which it resides.* Proc. Nat. Acad. Sci. USA 82:493–497.

Riedel, G.E. and D.A. Austen, 1984. *E. coli – A. tumefaciens shuttle vectors.* Pp. 74–86 in W. Arber et al., eds., *Impact on Man and Society.* ICSU Press, Cambridge.

Schwarz-Sommer, Z., A. Gierl, R.B. Klosgen, U. Wienand, P.A. Peterson and H. Saedler. 1984. *The spm (En) transposable element controls the excision of a 2-kb DNA insert at the wx $^{m-8}$ allele of Zea mays.* EMBO J. 3:1021–1028.

Searles, L.L., R.S. Jokerst, P.M. Bingham, R.A. Voelker and A.L. Greenleaf. 1982. *Molecular cloning of sequences from a Drosophila RNA polymerase II locus by P element transposon tagging.* Cell 31:585–592.

Sengupta-Gopalan, C., N.A. Reichert, R.F. Barker, T.C. Hall and J.D. Kemp. 1985. *Developmentally regulated expression of the bean b-phaseolin gene in tobacco seed.* Proc. Nat. Acad. Sci. USA 82:3320–3324.

Vodkin, L.O., P.R. Rhodes and R.B. Goldberg. 1983. *A lectin gene insertion has the structural features of a transposable element.* Cell 34:1023–1031.

Microinjection for polygenic traits

R.J. GRIESBACH
Florist and Nursery Crops Laboratory, BARC-West, Beltsville, MD 20705, U.S.A.

Abstract Chromosome-mediated transformation offers the geneticist a unique means for introducing small amounts of foreign DNA into the genome of host cells. Several polygenic traits are found as a block of genes on a chromosome arm. Therefore, one should be able to transfer polygenic traits via chromosome-mediated transformation.

One serious problem with chromosome-mediated transformation is the lack of suitable methods for the introduction of the foreign chromosomes into host cells. We have developed a system for first removing the vacuoles of host cells and then microinjecting those cells with foreign chromosomes. Approximately 25% of the injected protoplasts are capable of further development, while 40% of the control non-injected protoplasts are capable of further growth and development.

Introduction

Many of our agricultural and horticultural important characteristics are encoded by multiple genes or polygenes. Each polygene contributes a small additive effect towards the total characteristic. Because the effect of each polygene is so small, it is impossible to describe polygenic characteristics in terms of individual genes or genotype. In some cases, the individual polygenes are completely independent and spacially separated; while in other instances the single polygenes can occur clustered within a polygenic block. Several polygenic characteristics are salt and drought tolerance, cold resistance, fruit weight, height or size (Briggs and Knowles, 1967).

Most of the current gene transfer biotechnology is very successful for transferring single genes or very small groups of genes. For example, the Ti plasmid of *Agrobacterium tumefaciens* has been successfully used to stability transfer up to 14 kilobases of foreign DNA (Caplan et al., 1983). It is clear that this plasmid as well any other plasmid, could not transfer all of the individual genes encoding a single polygenic trait. In addition, no one has been able to identify, much less isolate individual polygenes.

One technique might be useful for the transfer of some polygenic characteristics. If individual chromosome arms can be stably moved between species, then those polygenic characteristics which occur as polygenic blocks on those chromosomes might be transferred.

There are several methods for isolating plant chromosomes in a physiologically active state (Griesbach, Maimberg and Carlson, 1982a;

Microneedles were pulled from capillary tubes on a pipet puller. The tubes had an outer diameter of 0.85 mm and an inner diameter of 0.6 mm. The diameter of the microneedle was 2.5 μm. The evacuolated protoplasts were 20 μm in diameter. The needles were loaded by placing the tip in the chromosome suspension and applying a vacuum. The chromosomes were injected via pressurized nitrogen. Two nitrogen tanks were connected to a gas burst regulator. One of tank was maintained at 25 lb/in^2 and the other at 40 lbs/in^2. The low pressure tank was used to balance capillary action or to prevent liquid or air from entering the needle during the manipulations. The high pressure tank was used for forcing the chromosome out of the needle during injection. A 0.1 μl suspension of chromosomes was injected into each protoplast.

Results and discussion

Approximately 40% of the evacuolated protoplasts were viable and underwent continued development. A new vacuole reformed within hours and after 16 hours, the evacuolated protoplasts were completely vacuolated. At this time, starch was also seen accumulating in the chloroplasts. After 4–5 days, the first cell division occurred and within 2–3 weeks calli large enough to transfer to differentiation medium were obtained.

Evacuolation leads to a 30% reduction in viability. Microinjection reduces the viability of evacuolated protoplasts another 40%. Thus, only 25% of the injected evacuolated protoplasts are capable of continued development. This frequency is still higher than that obtained when using normal vacuolated protoplasts because microinjection reduces the viability of these protoplasts by 60%.

In addition, there are other problems in that the vacuole can be a physiological barrier to successful microinjection. It is very difficult to inject a protoplast such that the needle does not damage the vacuole. Vacuoles are a rich reservoir of secondary metabolites, hydrolytic enzymes, inorganic molecules and many toxic waste products (Matile, 1978). Any damage to the vacuolar membrane could release these chemicals into the cytoplasm and cause cell death. In addition, the autophagic activity of the vacuole could make it difficult to introduce foreign material into cells. The cell usually recognizes this material as foreign and sequesters it within the vacuole. Another bothersome phenomenon is senescence. The onset of senescence can be triggered by any stress (e.g. microinjection, osmotic shock, or any protoplast manipulation). One of the first steps in senescence is a change in the permeability of the vacuolar membrane which leades to autolysis (Thomas and Stoddart, 1980).

Small, 5 to 10 cell samples from growing microcalli have been analyzed using polyacylamide gel electrophoresis. Several qualitative and quantitative protein differences have been detected between injected and non-injected protoplasts. These differences, however, did not remain constant, but changed with time. Further data needs to be collected as the calli get older. Once the calli are large enough, they will be transferred to differentiation medium. The developing shoots could then be tested for drought tolerance.

It appears that chromosome-mediated transformation via microinjection does occur. Whether or not polygenic characteristics can be transferred is still a question. Chromosome-mediated transformation via microinjection should alleviate many of the problems plant breeders will face in the future, for it would allow the breeder to overcome many of the difficulties of developmental incompatibility and extensive introgression associated with either somatic or sexual wide hybridization.

References

Briggs, F. and P. Knowles. 1967. *Introduction to Plant Breeding.* Reinhold Publishing Co., New York.

Caplan, A., L. Estrella, D. Inze, E. Haute, M. Montagu, J. Schell and P. Zambryski. 1983. *Introduction of genetic material into plant cells.* Science 222:814–820.

Griesbach, R. 1983. *Protoplast microinjection.* Plant Mol. Biol. Rep. 1:32–37.

Griesbach, R. and K. Sink. 1983. *Evacuolation of mesophyll protoplasts.* Plant Sci. Lett. 30: 297–301.

Griesbach, R., R. Malmberg and P. Carlson. 1982a. *Uptake of isolated lily chromosomes by tobacco protoplasts.* J. Hered. 73:151–152.

Griesbach, R., R. Mamberg and P. Carlson. 1982b. *An improved technique for the isolation of higher plant chromosomes.* Plant Sci. Lett. 24:55–60.

Hadlaczky, G., G. Bisztray, T. Praxnoszky and P. Dudits. 1983. *Mass isolation of plant chromosomes and nuclei.* Planta 157:278–285.

Matile, P. 1978. *Biochemistry and function of vacuoles.* Ann. Rev. Plant Physiol. 29:193–318.

Steinbiss, H. and P. Stabel. 1983. *Protoplast derived tobacco cells can survive capillary microinjection of lucifer yellow.* Protoplasma 116:223–227.

Szabados, L., G. Hadlaczky and D. Dudits. 1982. *Uptake of isolated plant chromosomes by plant protoplasts.* Planta 151:141–145.

Thomas, H. and J. Stoddard. 1980. *Leaf senescence.* Ann. Rev. Plant Physiol. 31:83–111.

Flow cytometry: analysis and sorting of X and Y chromosome-bearing spermatozoa

LAWRENCE A. JOHNSON
U.S, Department of Agriculture, Reproduction Laboratory, Animal Science Institute, Beltsville Agricultural Research Center, Agricultural Research Service, Beltsville, Maryland 20705, U.S.A.

Abstract Preselection of sex requires manipulation of spermatozoa because of the sex-determining role of the X or Y chromosome within the cell. Using flow cytometry (FCM), a relatively new technology, one is able to measure small differences in DNA content among individual cells. Utilization of FCM for the analysis and separation of X and Y chromosome-bearing spermatozoa is based upon labeling of DNA with a DNA-specific fluorescent stain and subsequent excitation of the labeled cell with a laser beam at the appropriate wavelength to give maximum fluorescence. Since the X chromosome is larger and contains at least 3.5% (in most species) more DNA than the Y chromosome, the FCM can differentiate the two sperm populations. Controlling the orientation of spermatozoa passing in front of the laser beam on the FCM is essential for high resolution DNA analysis because the fluorescence emitted is affected by the high refractive index and flat shape of the sperm head. A flow cytometer was modified to overcome optical artifacts caused by changes in orientation as spermatozoa pass the laser beam. This modification has allowed us to determine the ratio of X to Y spermatozoa in ejaculated, epididymal and in cryopreserved spermatozoa from several different species with high precision (CV = < 1.5%). Semen from bulls, ram, boars, chinchillas, turkeys, hamsters, rabbits, and voles has been analyzed. The Chinchilla is an excellent model for sorting X and Y spermatozoa because of the wide difference in DNA content (7.5%). We have developed procedures to separate X and Y Chinchilla spermatozoa populations with a purity approaching 95%.

Introduction

Flow cytometry (FCM) is useful in separating closely related yet functionally distinct types of cells. An excellent review of FCM and its application to cell biology is available (Melamed, Mullaney and Mendelsohn, 1979).

Quantitative measurement of the properties of individual cells can be used to investigate many cell types. Spermatozoa can be prepared for flow analysis with relative ease because spermatozoa are free from adhering tissue. Chromosomal constitution is the only established difference between X and Y chromosome-bearing spermatozoa. The amount of deoxyribonucleic (DNA) per spermatozoan is fixed (including the X and Y difference) and is nearly always the same from spermatozoan within a species. Flow cytometric analysis of the DNA content of

Mention of a trade name, proprietary product, or specific equipment does not constitute a guarantee or warranty by the USDA and does not imply its approval to the exclusion of other products that may be suitable.

individual spermatozoa can distinguish X from Y chromosome-bearing spermatozoa.

Insemination of females with semen that has been enriched for one sex or the other (X or Y) has been a subject of keen interest to scientists and non-scientists for generations. The lack of credible data on X–Y sperm separation is probably due to the difficulty in detecting minute differences between X and Y spermatozoa. The cattle industry has been plagued by many get-rich-quick schemes that involve attempts to market purportedly enriched semen. This probably due to the high value of cattle and their fairly long generation interval.

The potential usefulness of sexed semen to the livestock industry has been documented on several occasions (Kiddy and Hafs, 1971; Amann and Seidel, 1982; Kiddy, unpublished observations). These reports have shown that if sexed semen were available in the livestock industry, it would be used extensively, especially by cattle producers. For example, the dairy producer would select semen to provide a preponderance of heifers for milk production, beef and sheep producers would select semen for males because males grow faster than females, and swine producers would likely select semen for females, also for their higher growth rate. Genetic progress would be faster in all species because of the increased selection pressure resulting from preselection for sex.

Controlling the prenatal sex ratio of livestock is the ultimate objective of these studies. This paper will deal with the modification and utilization of an orthogonal flow cytometer/cell sorter (FCM) in the analysis and sorting of spermatozoa.

Instrumentation

Flow cytometric analysis of individual cells offers precision, high sensitivity and rapid analysis or relatively large numbers of individual cells. Two basic flow cytometric systems have been used for analysis of spermatozoa: an epi-illumination flow system using a mercury lamp, and an FCM of orthogonal optical geometry, employing a laser as the light energy source. In the former, the spermatozoa are labeled with a DNA-specific fluorescent dye and passed along a single optical axis toward a microscope objective of high numerical aperture. Hydrodynamic forces cause each spermatozoa to orient its longitudinal axis parallel to the flow. Random rotational orientation does not affect the emitted light reaching the photomultiplier because the optics are radially symmetric. The fluorescent signal is the basis for the photometric measurement. The usefulness of this system for resolving the X and Y spermatozoa population of several species based on DNA content has been reported (Garner et al., 1983). This system, however, is not designed to sort cells of any type.

An FCM using an orthogonal optical system contains no intrinsic protection against optical orientation artifacts. Therefore, commercial instruments that use orthogonal systems must be modified to compensate for orientation artifacts (Pinkel et al., 1979; Johnson and Pinkel, 1985; Fig. 1). The conventional commercial orthogonal FCM contains one optical detector situated at a 90° angle to the laser beam and one light scatter photodiode-type detector in the path of the laser beam (0° angle). Orientation of cells is not controlled in the conventional FCM and the cells pass at random through the laser beam.

To standardize spermatozoal orientation, two modifications were made to an EPICS V FCM (Coulter Corporation, Hialeah, FL). The first modification involved re-tooling the sample injection tube. The tip of a standard cylindrical injection tube was beveled to produce a rectangular ribbon stream rather than a cylindrical stream at the exit (Fig. 1). The hydrodynamic forces of the ribbon type stream cause the flat ovoid sperm heads to line up individually in the plane of the stream (Pinkel et al., 1979; Johnson and Pinkel, 1985) and exit the flow chamber through a 76 μm orifice with uniform orientation.

The second modification to the EPICS V was the addition of a second optical fluorescence detector to replace the light scatter detector that is normally situated in the path of the laser beam. The sample injection tube was then positioned to allow the laser beam to hit the flat side of the ribbon stream (Fig. 1) and thus the flat surface of the sperm. The fluorescence was collected from the opposite flat surface of the spermatozoa; at that angle (0°) the emitted light is less sensitive to a slight angular change than is the case for the detector at the 90° position (Pinkel et al., 1982a).

In the modified flow cytometer/cell sorter, sperm nuclei which had been labeled with a fluorescent DNA-specific dye pass into the flow chamber, are oriented, and exit from the sample injection tube in single file. The sperm nuclei then pass in front of, and are excited by, the laser beam, and emit light which is collected by the fluorescence detectors at a right angle (90°) and in the path of (0° angle) the laser beam. The excitation wavelength is removed with spectral filters and the fluorescent light is directed to the photomultiplier tubes, where the light is changed to an electronic signal, amplified, and sent to the data acquisition system. The fluorescence from each nucleus appears as a pulse on an oscilloscope, where pulse height is proportional to emitted fluorescence and, therefore, DNA content. Digital representations of the pulses are fed to a computer which generates a histogram of pulse heights.

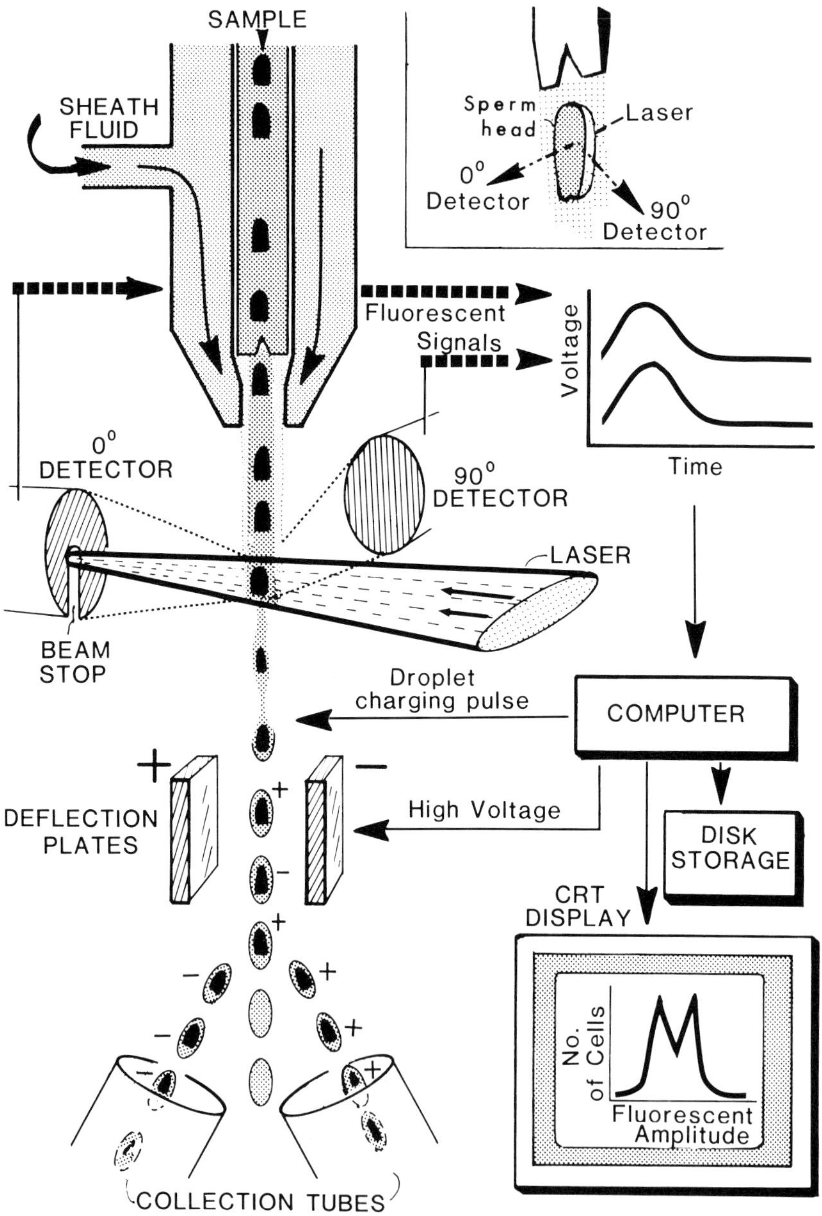
SAMPLE
SHEATH FLUID
Sperm head
Laser
0° Detector
90° Detector
Fluorescent Signals
Voltage
Time
0° DETECTOR
90° DETECTOR
LASER
BEAM STOP
Droplet charging pulse
COMPUTER
DEFLECTION PLATES
High Voltage
DISK STORAGE
CRT DISPLAY
No. of Cells
Fluorescent Amplitude
COLLECTION TUBES

Sorting sperm nuclei

To sort the spermatozoa, the flow chamber is vibrated at high frequency (32 KHz), causing the descending stream carrying the sperm nuclei away from the laser to break into small, uniform droplets (Fig. 1). At sample flow rates used for sorting sperm nuclei, approximately 0.8% of the droplets contain a nucleus. Nuclei can be sorted on the basis of pulse height and thus DNA content. When a nucleus to be sorted is identified by its respective pulse height, timing and charging circuits are activated. As the stream breaks into droplets (Fig. 1), the droplets containing the desired nucleus are given a positive or negative electrical charge based on pulse height. The charged droplet falls through an electrostatic field (generated between two charged plates) and is deflected into one of two collection tubes. The empty, uncharged droplets are not deflected and fall into the discard tube (Fig. 1).

Sample preparation and staining

Epididymal or ejaculated semen for FCM analysis or sorting is washed in a buffer solution followed by fixation in 80% ethanol (Johnson et al., 1984). Cryopreserved semen should be washed in increasing concentration of DMSO (5 to 50%) which is helpful to remove the egg yolk or milk particles that are common to semen extenders (Garner et al., 1983). Samples that have been prepared by either of these techniques can be stored after fixation at 5°C for more than two years without change (L.A. Johnson, unpublished observations).

Several staining regimes can be used. Originally, fixed spermatozoa were stained for DNA with a combination of ethidium bromide plus mithramycin (EBMI) or with 4′-6 diamidino-2-phenylindole (DAPI). Both staining systems were effective (Pinkel et al., 1982a). EBMI is advantageous because its optimum excitation wavelength is in the visible range thus avoiding the use of ultraviolet light (UV). However, further studies showed that DAPI, which is excited in the UV range, stained the

---

Figure 1. Flow chamber, optics and signal processing of modified EPICS V flow cytometer/cell sorter. The hydrodynamic forces exerted on the flat ovoid mammalian sperm nuclei as they exit the beveled sample injection tip cause spermatozoa to be oriented in the plane of the flattened sample stream. Fluorescence emitted from the stained sperm nuclei as it passes through the laser beam is simultaneously measured by optical detectors positioned 0° and 90° to the laser beam. The fluorescent signals are processed, amplified, converted from analog to digital and stored by the computer as histograms. If sperm nuclei are to be sorted, the sample stream is broken into droplets by an ultrasonic transducer. Droplets containing single nuclei of desired fluorescence intensity are electrostatically deflected into collection tubes. Inset: Shows orthogonal relationship of laser, oriented sperm, and the two fluorescence detectors. The flat shape and condensed chromatin of mammalian sperm nuclei cause higher intensity fluorescence to be emitted from the edge of the sperm nucleus, than is emitted from the flat side of the nucleus.

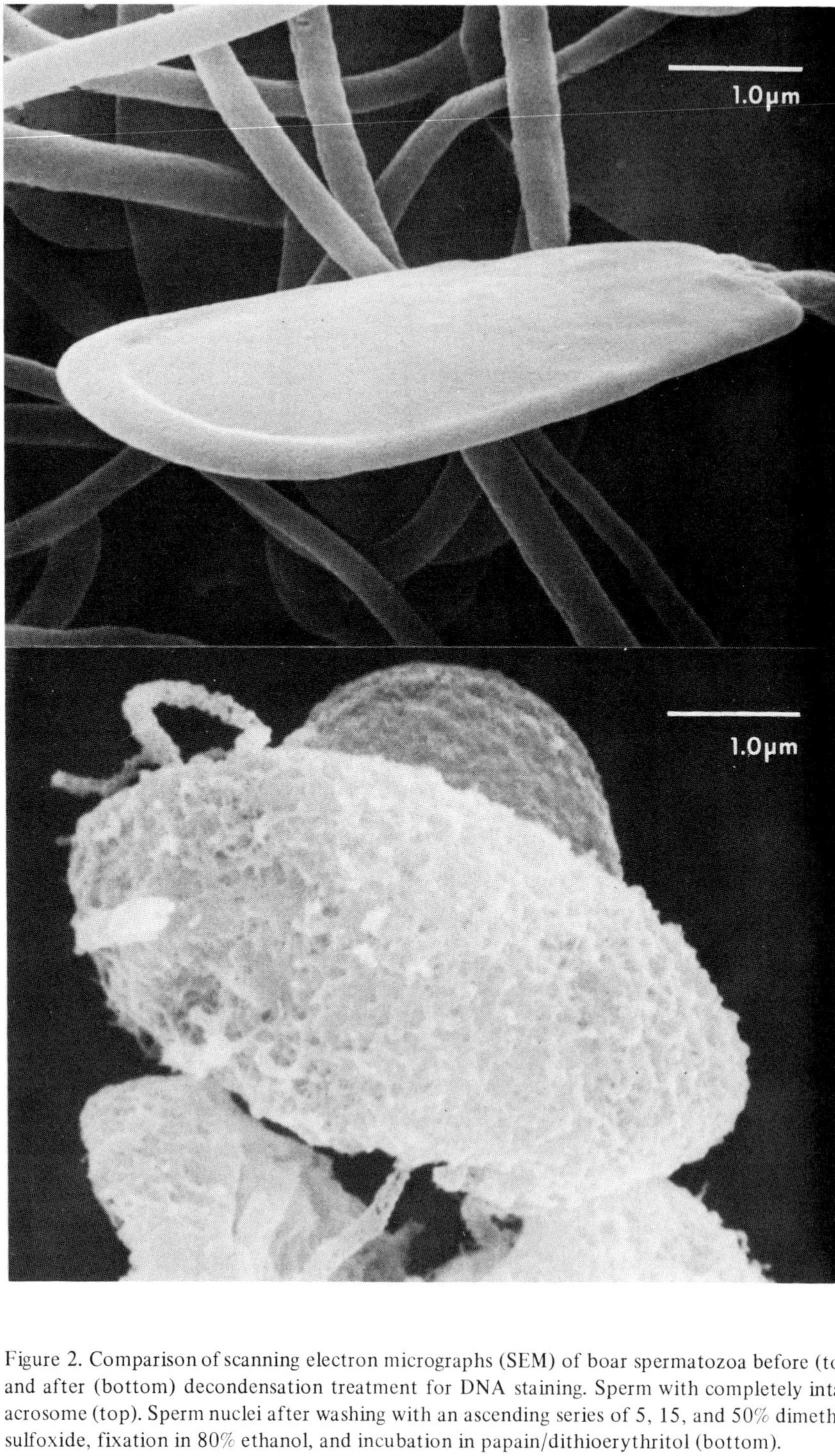

Figure 2. Comparison of scanning electron micrographs (SEM) of boar spermatozoa before (top) and after (bottom) decondensation treatment for DNA staining. Sperm with completely intact acrosome (top). Sperm nuclei after washing with an ascending series of 5, 15, and 50% dimethylsulfoxide, fixation in 80% ethanol, and incubation in papain/dithioerythritol (bottom).

DNA of spermatozoa more effectively (Garner et al., 1983). Recently, we have found Hoechst 33342 to be superior to DAPI (Johnson, 1984). Usually, about 2 million sperm are incubated in a solution of papain and dithioerythritol (Garner et al., 1983) for 15 minutes to loosen sperm membranes. Transmission and scanning electron microscopy studies in our laboratory indicate some DNA decondensation (Fig. 2) takes place during this incubation. This enhances stain penetration. The spermatozoa are then pelleted and resuspended in a solution of 9 mM Hoechst 33342. Staining equilibrium is reached in about 30 minutes. The stained spermatozoa are filtered through 37 μm nylon mesh and analyzed in the modified flow cytometer/cell sorter.

Orientation of sperm in flow systems

Measurement of fluorescence in sperm nuclei is very sensitive to the orientation of the nuclei relative to the laser beam. The modifications discussed in the instrumentation section are designed to minimize the number of misoriented nuclei passing through the system. However, even a system with a modified orthogonal geometry of measurement modified EPICS V, such as described above) cannot completely eliminate the presence of some misoriented sperm nuclei.

The capability of the instrument to select data in a precise fluorescence range (gating) is a key factor in eliminating misoriented sperm nuclei from analysis. Pulses are collected from two fluorescent detectors oriented at right angles to each other. Sperm nuclei passing out of the flow tip are oriented with their flat surface parallel to the ribbon stream and their thin edge perpendicular to the detector positioned at a 90° angle. This detector collects fluorescence emitted by the edge of the sperm nuclei. The flat side of the nuclei receives a direct hit (Fig. 1, inset) from the laser beam and emits fluorescence which is collected by the detector positioned at 0°. Since the fluorescence collected by the 90° detector is the brightest when the edge of the nucleus is exactly perpendicular to the detector, oriented nuclei will show a peak of high fluorescence on the histogram. If the edges of the sperm nuclei are slightly angled, less fluorescence is emitted and the nuclei show up as a broad peak on the leading edge of the histogram. These sperm nuclei are classed as misoriented. Because inclusion of their pulses in a histogram of the 0° detector would reduce the sensitivity of the system in determining differences in DNA content, an electronic gate is placed on signals from the 90° detector to exclude all but the most brightly fluorescing nuclei. Only correlated signals from within this gate are included in the 0° detector histogram. The 0° detector is used to make the final and exact measurement of fluorescence because it is less sensitive than the 90° detector to slight variation in angle of the sperm nucleus (Pinkel et al., 1982a). The effects of cell orientation have been described in detail (Fulwyler, 1977; Dean, Pinkel and Mendelsohn, 1978; Stovel et al., 1978; Pinkel et al., 1982a).

Data acquisition and analysis

The sperm nuclei in suspension are measured at a slow flow rate in comparison to cells that are not subject to optical orientation artifacts, such as most blood cells. For analysis of DNA content, 50 to 300 sperm nuclei per second pass the detector, depending on the magnitude of the X–Y DNA differential. The more similar the DNA complement of X and Y sperm populations, the lower the flow rate needed for precise measurement of DNA. However, sample quality and uniform specimen staining are critical also. A multichannel analyzer (Multiparameter Data Acquisition and Display System, MDADS, Coulter Corporation, Hialeah, FL) records the number of nuclei with a specific fluorescent intensity. The fluorescent intensity scale is divided into 256 discrete channels. Pulse information is accumulated during each analysis until a minimum of 5,000 nuclei are measured. The pulse information is used to construct a fluorescence frequency distribution (histogram). The data from the region of the histogram containing the two peaks were fitted by least squares techniques to a pair of Gaussian distributions, whose means, coefficients of variation (CV), and relative areas are estimated. The relative areas of the two peaks were assumed to reflect the relative proportion of X and Y sperm nuclei in the sample and the difference between the means was assumed to give the difference in DNA content between X and Y chromosome-bearing sperm nuclei. The relative difference in DNA content (percentage) of the two peaks is calculated by difference (Δ):

$$\Delta = 100 \frac{(\bar{x} - \bar{y})}{0.5\,(\bar{x} + \bar{y})}$$

where $\bar{x}$ and $\bar{y}$ represent means of the X and Y peaks.

DNA measurement of X and Y chromosome-bearing spermatozoa

Homogametic males. Turkeys are homogametic and thus their sperm should show a single sharply defined peak of DNA fluorescence (Fig. 3). The unimodal distributions consistently had a CV of between 1 to 2%.

Heterogametic males. Sperm nuclei from more than a dozen species have been analyzed. Analyses of sperm nuclei from several of the species are illustrated in Figs. 4–7. Of the farm animals, boar spermatozoa showed the least difference in DNA content between X and Y nuclei

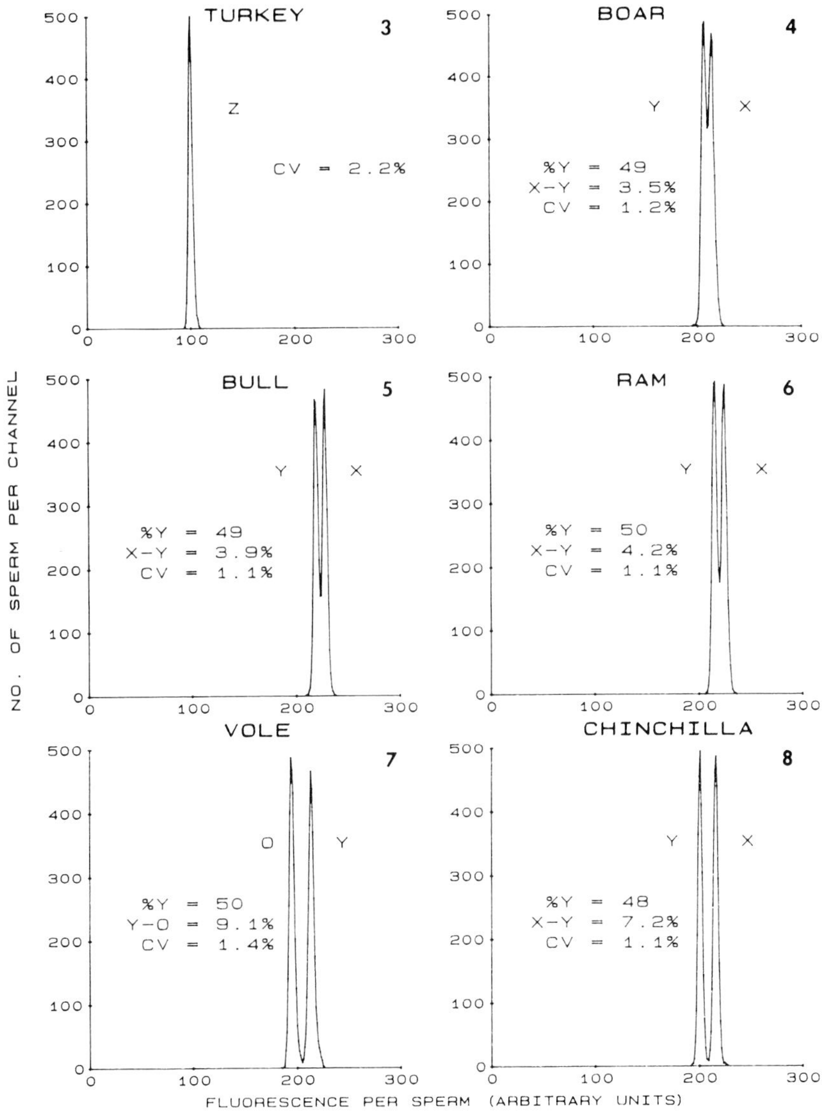

Figures 3–8. Fluorescence histograms of DNA of flow cytometric analyses stained sperm nuclei from various species. All histograms represent pulse signals from the 0° fluorescence detector of a modified EPICS V FCM and do not include signals from misoriented nuclei. All but the homogametic turkey spermatozoa show a bimodal distribution representing the X and Y spermatozoan populations. The vole is unusual in that it produces sperm with either no sex chromosome (0) or the Y chromosome (Y). Data given include the proportion of Y-bearing spermatozoa in the sample analyzed, the percent difference in DNA content between the two sperm types (measured by Hoechst 33342 fluorescence), and the average coefficient of variation (CV) of the two peaks. The data were taken from a computer fit of the histogram to a pair of Gaussian distributions, whose means, CV's and relative areas were adjusted to give the best least squares fit to the data.

(Fig. 4). Measurements of ejaculated, epididymal and cryopreserved spermatozoa give similar bimodal distributions showing a mean X–Y difference of 3.4%. The usual bimodal distribution seen with boar sperm illustrates the limitation of the technique. That is, the less difference in DNA between X and Y spermatozoa, the less precise the analysis. For bull spermatozoa (Fig. 5), the X–Y DNA difference is somewhat larger (3.9%) than for the boar. Again, the X–Y difference is similar for epididymal, ejaculated and cryopreserved spermatozoa. The ram (Fig. 6) has the largest difference in DNA content between X and Y sperm of the farm animals (4.2%). The CV on the histograms for each species is $<1.5\%$.

Moruzzi (1979) measured chromosomal lengths in karyotypes from many species and calculated the percentage DNA difference between X and Y chromosome-bearing sperm. He listed 25 species with a calculated difference in DNA content of more than 6% between X and Y chromosomes. Near the top of the list is the microtine *Microtus oregoni*, a creeping vole that inhabits the coastal range of Oregon, with a calculated difference of 8.8%. The large difference results from this vole carrying no X chromosome so that spermatozoa are classified as "0" and Y rather than X and Y. The histogram in Fig. 7 illustrates the difference between spermatozoa which carry no germ cell chromosome (least fluorescence) and spermatozoa that carry the Y chromosome (greater fluorescence).

Another animal listed by Moruzzi (1979) was the *Chinchilla laniger*, a rodent known well for its soft fur. Calculation of DNA content of spermatozoa from chromosome lengths indicated an X–Y difference of 6.9%. Flow cytometric analysis of the sperm nuclei from the chinchilla (Fig. 8) showed a 7.2% difference in DNA content between X and Y spermatozoa. Because of the distinctive bimodal distribution of DNA between X and Y spermatozoa of the vole and the chinchilla, these species make superb models for the separation of sperm into X and Y populations. The CV for the chinchilla and vole spermatozoa is consistently $<1.5\%$.

Flow sorting of X and Y chromosome-bearing spermatozoa

Nuclei of chinchilla spermatoza have been sorted into X and Y chromosome-bearing populations (Figs. 9, 10 and 11). Epididymal spermatozoa, prepared as described earlier, were stained with Hoechst 33342 and sorted into separate X and Y populations based on DNA content (Fig. 9). The resultant sorted populations of about 200,000 sperm nuclei were then reanalyzed (Figs. 10 and 11). The same conditions were used for the initial analysis, the sort and the reanalysis, so

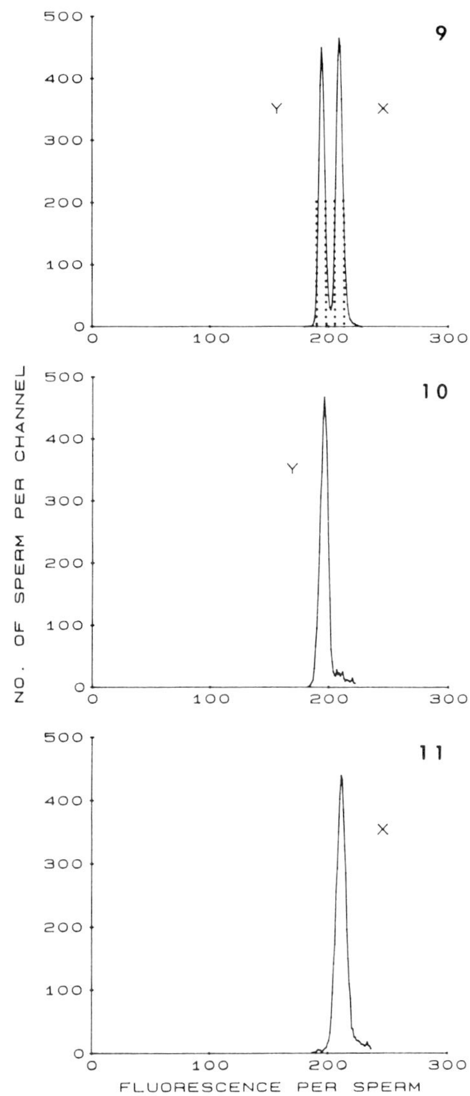

Figures 9–11. Fluorescence histograms of flow cytometric analysis of unsorted (Fig. 9) and sorted, reanalyzed Y (Fig. 10) and X (Fig. 11) chromosome-bearing chinchilla spermatozoa that had been strained for DNA with Hoechst 33342. All data are from the 0° detector. Figure 9 does not include misoriented cells, whereas Figs. 10 and 11 include both oriented and misoriented cells. The two pairs of vertical lines in Fig. 9 represent the areas sorted for the Y (left) and X (right) chromosome-bearing spermatozoa. Flow cytometric analysis of the Y peak sort (Fig. 10) and X peak sort (Fig. 11) show high degree of purity achieved by sorting of the sperm nuclei.

that the histograms shown (Figs. 10 and 11) are directly superimposable on one another.

Actual purities of the individual X and Y populations were estimated by locating the area of the histogram where the opposite peak would have fallen, integrating the counts in that area, and calculating percent of histogram. No grating was used for the reanalysis, and the histograms show a typical skew, or tail, representing misoriented cells measured by the 0° detector. Note that although only oriented sperm nuclei were sorted, upon reanalysis (Figs. 10 and 11), some sperm were misoriented (trailing edge, Fig. 10). The effect of the trailing portion of the histogram on the calculated purity of the Y peak sort was estimated and subtracted from the integrated counts to produce a final value. Based on this method of estimation, the reanalysis of the sorted peaks indicated a purity of about 95% (n = 12).

Although we currently lack fertility data to determine the sex ratio of offspring, the evidence, based on two peaks indicating different DNA content of two spermatozoal populations, is very strong that X and Y spermatozoa were actually separated. We have measured spermatozoa with expected DNA differences between X and Y spermatozoa of 0 to over 9% (turkey, 0%; vole 9.6%). Others have reported data for the mouse (Pinkel et al., 1982a) and for cattle, sheep and swine (Garner et al., 1983). The differences between X and Y spermatozoa in DNA content calculated by Moruzzi (1979) agree with the flow cytometric measurements in the present study. Sperm have been sorted in one other instance. Pinkel et al. (1982b) sorted sperm from the vole (*Microtus oregoni*) by using a specially constructed orienting flow cytometer and reanalyzing the sorted nuclei with an epi-illumination flow system. Our data with the chinchilla corroborates that original data.

Conclusions

Evidence has been presented that a commercially built orthogonal FCM can be suitably modified to overcome optical orientation artifacts peculiar to spermatozoa, that the modified system is capable of analyzing DNA in mammalian spermatozoa with high precision, and that X and Y chromosome-bearing spermatozoa can be differentiated with a coefficient of variation of less than 1.5%. In addition, X and Y chromosome-bearing spermatozoa from the *Chinchilla laniger* have been sorted into two separate populations.

We conclude that the modified system is an effective tool to determine the DNA content of spermatozoa in order to test the efficacy of procedures to separate X from Y chromosome-bearing sperm. Testing "separated" spermatozoa by DNA content avoids the necessity of obtaining large numbers of offspring to determine if the expected sex ratio has been altered. Thus, costs can be reduced, and research designed to produce a practical means of sexing semen can be accelerated. Pinkel et al. (1985) used an epi-illumination FCM to test the efficacy of various enrichment procedures. No changes in sperm sex ratios were seen. We used the modified EPICS V FCM to test the efficacy of four different spermatozoal enrichment procedures. In no case has the expected sperm ratio deviated significantly from the expected 50:50.

The X and Y chromosome bearing spermatozoa from the Chinchilla can be isolated into separate populations at purities of about 95%. Definitive proof that we are indeed separating X and Y sperm must await fertilization data to verify the sex chromosome composition of the isolated spermatozoal populations. Such verification data is currently being acquired.

Isolation of individual X and Y sperm populations is essential for the ultimate development of a sperm marker on which a practical procedure for sexing spermatozoa can be based. It is unlikely that FCM/cell sorting instrumentation will be useful as a practical means of separating spermatozoa into separate populations for subsequent insemination of females because of the millions of spermatozoa required per insemination in most species. However, it is possible that FCM-sorted spermatozoa could be used to fertilize ova in vitro by microinjecting one sperm nucleus into each ovum.

Acknowledgments The author acknowledges with gratitude the contribution of colleagues and associates to the research reported here. In particular, I am grateful to James Flook, Mary Look, and Dr. Daniel Pinkel for their significant contributions to the success of this work, to Dr. Duane Garner for helpful suggestions, and to Dr. Caird Rexroad, Jr. for his assistance in the development of the computer fitting program.

References

Amann, R.P. and G.E. Siedel. 1982. *Prospects for Sexing Mammalian Sperm.* Colorado Associated University Press, Boulder, Colorado.

Dean, P.N., D. Pinkel and M.L. Mendelsohn. 1978. *Hydrodynamic orientation of sperm heads for flow cytometry.* Biophys. J. 23:7–13.

Fulwyler, M.J. 1977. *Hydrodynamic orientation of cells.* J. Histochem. Cytochem. 25:781–783.

Garner, D.L., B.L. Gledhill, D. Pinkel, D. Stephenson, M.A. VanDilla and L.A. Johnson. 1983. *Quantification of the X and Y chromosome-bearing spermatozoa of domestic animals by flow cytometry.* Biol. Reprod. 28:312–321.

Johnson, L.A. 1984. *Relative DNA content of X and Y chromosome-bearing chinchilla and porcine spermatozoa stained with Hoechst 3342,* Page A-11 (abst.). Proc. 10th Int. Congr. Anal. Cytol.

Johnson, L.A., J. Flook, M. Look and D. Pinkel. 1985. *Flow sorting of X and Y chromosome-bearing Chinchilla spermatozoa into two populations.* J. Androl. (Suppl.) 6:128P.

Johnson, L.A. and D. Pinkel. 1986. *Modification of a laser-based flow cytometer for high resolution DNA analysis of mammalian spermatoza.* Cytometry 7: (in press).

Kiddy, C.A. and H.D. Hafs. 1971. *Sex Ratio at Birth – Prospects for Control.* American Society of Animal Science, Urbana, Illinois.

Melamed, M.R., P.F. Mullaney and M.L. Mendelsohn. 1979. *Flow Cytometry and Sorting.* John Wiley and Sons, New York.

Moruzzi, J.F. 1979. *Selecting a mammalian species for the separation of X and Y chromosome-bearing spermatozoa.* J. Reprod. Fertil. 57:319–323.

Pinkel, D., P. Dean, S. Lake, D. Peters, M. Mendelsohn, J. Gray, M.A. VanDilla and B. Gledhill. 1979. *Flow cytometry of mammalian sperm progress in DNA and morphology measurement.* J. Histochem. Cytochem. 27:353–358.

Pinkel, D., D.L. Garner, B.L. Gledhill, S. Lake, D. Stephenson and L.A. Johnson. 1985. *Flow cytometric determination of the proportions of X and Y chromosome-bearing sperm in samples of purportedly separated bull sperm.* J. Anim. Sci. 60:1303–1307.

Pinkel, D., S. Lake, B.L. Gledhill, M.A. VanDilla, D. Stephenson and G. Watchmaker. 1982. *High resolution DNA measurements of mammalian sperm.* Cytometry 3:1–9.

Pinkel, D., B.L. Gledhill, S. Lake, D. Stephenson and M.A. VanDilla. 1982b. *Sex Preselection in mammals: separation of sperm bearing Y and 'O' chromosomes in the vole Microtus oregoni.* Science 218:904–906.

Stovel, R.T., R.G. Sweet and L.A. Herzenberg. 1978. *A means for orienting flat cells in flow systems.* Biophys. J. 23:1–5.

Modification of the germ line in animals

JON GORDON, M.D., Ph.D.
The Department of Obstetrics, Gynecology and Reproductive Science, The Mount Sinai School of Medicine, 1 Gustave L. Levy Place, New York, NY 10029, U.S.A.

Abstract In recent years, techniques have been developed for introducing new genetic material into intact organisms. These procedures, which allow the transfer of genes at the earliest stages of development, make possible the insertion of new genetic material into the germline cells of the developing embryo. This capability thus provides the potential to permanently alter the genetic characteristics of a strain of animals. Whatever new characteristics are supplied by foreign genes, they can be maintained indefinitely in subsequent generations through breeding. Advances of this kind have profound implications for agriculture. Reproducible expression of foreign genes in recipient animals can result in the appearance of new and highly advantageous phenotypes. The most obvious applications of gene transfer technology to farm animals involve manipulation of the immune system or the endocrine system. However, as greater knowledge of the genetic regulation of mammalian development is achieved, and as more genes are cloned, isolated, and studied, the range of applicability of this technology will broaden considerably. Therefore, it must be anticipated that gene transfer will profoundly change the theory and practice of animal breeding.

Introduction

The controlled manipulation, or "genetic engineering" of animals and plants is not a new science. Systems of selective breeding, in use for thousands of years, have provided a diverse variety of highly specialized organisms whose agricultural productivity far exceeds their wild ancestors. This powerful technique is based on the simple realization that many phenotypic traits are based fundamentally on the combination of genes resident within the fertilized zygote. Selective breeding draws its power from the fact that whatever complex of genes is required to generate a specific phenotype, that complex can be selected for by simply enhancing the reproductive capacity of those individuals within a population whose phenotypic characteristics are most desirable and by limiting the reproduction of all others. Thus, though a desired phenotype may result from the coordinated expression of hundreds of genes, all of these genes can be selectively maintained within animals by simple phenotypic selection. Testimony to the power of this approach lies in the observation of the thousands of animal and plant strains which have been selected over the centuries.

Despite the power of selective breeding as a genetic engineering technique, the approach has its limitations. The genetecist employing this strategy must be limited to the genes resident in the species with which he is working. Thus, it is not possible to select for genes in one fish species whose expression in another would provide a more desirable phenotype. Introduction of genes even between closely related mammalian species is also complicated by the fact that hybrids for crosses between species are frequently sterile. Therefore, it is not possible to maintain the hybrid phenotype of choice through subsequent generations.

Until a few years ago, the limitations on genetic selection were accepted as intrinsic to the laws of evolution and speciation. However, with the advent of the recombinant DNA technology, it became possible to purify single genes in large quantities. The advent of gene cloning led to experiments in gene transfer. These studies were first carried out in tissue culture. Cells selected for various genetic deficiencies were exposed to foreign purified DNA (Graham and Van der Eb, 1973), after which they were placed in selective media requiring expression of the deleted gene. Then, by virtue of uptake and expression of the transferred gene, cells survived selection. Thus, cells lacking a gene for thymidine kinase (TK) could be induced to survive selective media requiring the expression of this gene after exposure to DNA from Herpes virus which coded for this enzyme (Bachetti and Graham, 1977; Maitland and MacDougall, 1977; Wigler et al., 1977). These experiments established a powerful new technique of gene transfer. Gene transfer into tissue culture cells has been applied to the cloning of genes and to a wide variety of studies of mammalian gene regulation. Although the uptake and expression of foreign DNA was found to be a relatively rare event, various selection protocols could be utilized to unequivocally identify cells which expressed donor genes.

The development of micromanipulation equipment (Diacumakos, 1973) subsequently led to more varied gene transfer strategies. With these techniques, individual cells could be microinjected with DNA and subsequently grown in clonal populations for further examination. These experiments revealed that a high rate of gene transfer was possible (Graessmann et al., 1979; Capecchi, 1980; Anderson et al., 1980). In all such experiments, however, selective procedures were still utilized to identify expressing populations. However, the high transformation rates when compared with previous approaches suggested that even in the absence of selection, a significant number of cells could be identified which had at least retained transferred DNA. Of course, one would anticipate that in the absence of selection, expression of donor genes was not obligatory.

The observation that microinjected cells were transformed with high efficiency led investigators to consider using cells of the cleaving embryo as recipients for gene transfer. Investigators had previously shown that exposure of cleaving embryos to RNA tumor viruses led to integration of proviral DNA at an early stage of development. Moreover, these viruses expressed their genes later in development as evidenced by the development of leukemia in the animals (Jaenisch, 1976, 1979). Early integration of proviral DNA was demonstrated by the Mendelian transmission of that DNA to progeny (Jaenisch, 1976). Thus, it was clear that foreign DNA could be inserted into early embryos and persist throughout development without significant impairment of the developmental process.

With these observations in mind, efforts were initiated to insert individual cloned genes into early embryos. In 1980, a successful protocol was devised (Gordon et al., 1980). This methodology involved the microinjection of cloned DNA into one pronucleus of the fertilized zygote. Reimplantation of the embryos into pseudopregnant females led to the birth of offspring which carried the foreign genes. The presence of foreign DNA was demonstrated by Southern blot hybridization techniques (Gordon et al., 1980). Persistence of the donor DNA into adult life (Brinster et al., 1981) and to subsequent generations of progeny (Costantini and Lacy, 1981; Gordon and Ruddle, 1981) thus established that foreign DNA cloned from any source could be transferred into the genomes of mammals. These foreign genes could then persist through all subceeding generations. Subsequent studies, to be described below, have established that foreign genes can also be efficiently expressed in adult animals. Data regarding the tissue specificity and efficiency expression will be described, after which the application of gene transfer to practical problems of agriculture will be outlined. The problems with this genetic engineering approach will then be discussed.

Foreign gene expression in transgenic animals

Expression is often tissue-specific. Early experiments with transgenic mice immediately established that despite their integration into apparently random sites from the genome, tissue-specific expression was frequently observed. The mouse metallothionein-1 (Mt-1) promotor, when linked to the Herpes thymidine kinase gene, elicited expression of the Herpes gene in liver and kidney of transgenic animals, while very little expression was detected in other tissues (Brinster et al., 1981). This pattern of expression is characteristic of the endogenous Mt-1 gene which, though ubiquitiously expressed, is preferentially transcribed in liver and kidney. Moreover, the Mt-1 promotor, normally inducible by cadmium or zinc, was also able to further stimulate Herpes thymidine kinase expression in transgenic mice, even though the site of integration within the host genome was not that of the endogenous Mt-1 gene (Palmiter et al., 1982a).

The tissue specificity supplied by gene promoters was further demonstrated by experiments in which the Mt-1 promoter was linked to other coding sequences. In particular, linkage of the Mt-1 promoter to the rat (Palmiter et al., 1982b) or the human (Palmiter et al., 1983) growth hormone gene resulted in rat and human growth hormone production from the livers of transgenic mice. Moreover, expression was quite efficient, often resulting in growth hormone levels in transgenic animals which far exceeded those seen in normal mice (Palmiter et al., 1982b, 1983). These high levels of growth hormone so suppressed production of the endogenous product that atrophy of the pituitary cells normally responsible for growth hormone synthesis was marked (Palmiter et al., 1983).

A series of further experiments reinforced the tissue-specific nature of gene promoters. The chicken transferrin gene, normally synthesized in oviduct and liver of chickens, was found to be preferentially expressed in the livers of transgenic mice (McKnight et al., 1983). Introduction of immunoglobulin genes similarly resulted in expression of those genes primarily in mature B-cells (Brinster et al., 1983; Grosschedl et al., 1984). The pancreatic elastase-1 gene was also found to be localized to pancreatic cells (Swift et al., 1984). These experiments thus established that gene promoters provide tissue specificity and that expression in transgenic mice is frequently comparable to that expected from an endogenous gene in its native chromosomal position. Finally, it has been observed that genes with enhancer sequences are able to further increase tissue-specific gene expression (Brinster et al., 1983). In the case of the immunoglobulin gene this expression can result in the production of functional immunoglobulin molecules (Grosschedl et al., 1984) as well as the ellicic exclusion of expression from counterpart genes within the host genome (Ritchie, Brinster and Strob, 1984).

Expression is stable over generations. Another important characteristic of foreign gene expression in transgenic mice is that it is very stable. With the exception of the Mt-1-Herpes TK fusion gene, foreign genes in transgenic mice appear to be stably expressed even when their tissue distribution is abberrant (Lacy et al., 1983). This finding thus indicates that reproducible phenotypic alteration of animals by gene transfer is feasible. At the present time, it is not understood why the mammalian-viral fusion gene was unpredictably expressed while mammalian genes are stably expressed. However, it has been suggested that "organizing sequences" exist within mammalian genes. These sequences purportedly stabilize gene expression by dictating a specific level of such expression later in development (Palmiter et al., 1984). This intriguing hypothesis remains to be tested, but indicates that introduction of mammalian genes into embryos is likely to result in stable, reproducible, and predictable expression over generations. The likelihood that foreign gene expression will remain stable from the time of its insertion suggests many practical applications for gene transfer, particularly for the agricultural industry. In the ensuing discussion, several examples of practical applications of gene transfer will be described.

Transfer of genes into farm animals

Introduction of genes coding for hormones. As noted previously, the rat and human growth hormone genes have been introduced to mouse embryos. When linked to the Mt-1 promoter, these genes were expressed from the livers of transgenic mice. Expression was in many cases inducible by heavy metal exposure, though in some cases the gene appeared to be expressed constitutively (Palmitter et al., 1982b). The result of growth hormone gene expression in the mice was a dramatic accelerated growth of young pups (Palmiter et al., 1982b, 1983). These animals attained adult size in approximately half the time as that of their control littermates. This observation indicates that the transfer of the growth hormone gene might result in accelerated growth of farm animals. If this proves to be the case, animals could be reared to a size suitable for slaughter in a much shorter time than previous to the advent of this technology. This happenstance would have profound economic ramifications for the agricultural industry.

Transfer of antibiotic-resistance genes. The bacterial gene coding for neomycin resistance has been cloned and well characterized. This gene codes for resistance to neomycin, a highly toxic but potent antibiotic. Because of its toxicity, neomycin is not administered systematically to animals or human beings (it is poorly absorbed, and can therefore be administered orally).

The transfer of the neomycin resistance gene into animals could possibly make them systemically resistant to the antibiotic. Thus, the toxicity of the antibiotic would be negated. Should expression be efficient, enormous doses of neomycin might be administered to animals without significant side effects. Thus, the insertion of such a gene into animals would broaden the spectrum of useful antibiotics available for treatment of various infections and diseases. Although the neomycin gene is the best studied and serves as a paradigm for such efforts, it is entirely conceivable that other genes coding for resistance for various toxic agents could be introduced into animal embryos. These experiments would similarly allow the treatment of adults with high doses of agents which would be very toxic under ordinary circumstances. It is presently unclear whether genes normally resident on bacterial plasmids and which encode resistance to a variety of other antibiotics will function in transgenic animals. However, it must be presumed that success could be achieved with such experiments. Therefore, it is not inconceivable that genetically engineered animals might be highly resistant to a very broad spectrum of antibiotics. This achievement would clearly improve the survivability of agricultural livestock.

Insertion of immunoglobulin genes into animals. Experiments with transgenic mice have already demonstrated that the introduction of immunoglobulin genes into embryos results in effect tissue-specific expression in adults. Moreover, functional antibody has been detected in such animals. Thus, animals can be born with circulating antibodies against antigens to which they had never been exposed. This accomplishment raises the prospect of introducing genes coding for antibody molecules to antigenic determinants present on pathogenic organisms. If such experiments succeeded, an animal could be born already immune to diseases to which it might otherwise have succumbed had vaccination not been available. Thus, the introduction of immunoglobulin genes might serve as a mode of "genetic vaccination". The advantage of this approach would be that repeated vaccination of newborn animals would be unnecessary. Instead, when the animals were born, the foreign gene would simply begin to function and produce circulating antibodies. This approach has advantages other than simple convenience. For example, some pathogenic organisms do not elicit vigorous antibody responses in vivo. However, under appropriate conditions, experiments can be conducted in vitro which will lead to the production of monoclonal antibodies with high specificity and binding affinity to the antigenic determinants resident on these pathogens. Production of monoclonal antibodies can shortly lead to molecular cloning of the genes encoding them. Then, these genes might be introduced into animals to produce circulating monoclonal antibodies against important proteins on infectious foreign organisms. This kind of genetic engineering might actually prove more effective than standard vaccination procedures. It is important to recognize that there is no limit to the number of times foreign genes may be inserted into animals.

Over generations, animals might be engineered which would be born resistant to the most prevalent pathogens in their environment. In fact, continued insertion of foreign genes could produce animals resistant even to rare though dangerous infections. Thus, genetic vaccination is a potentially significant application of embryo gene transfer.

The use of gene transfer for gamete selection. A number of genes have been identified which can serve as dominant selectable markers in gene transfer systems. Dominant selectable genes may be defined as those which can be transferred into genetically normal cells but whose expression within recipient cells can provide a selective advantage under certain conditions. An example of such a gene is Xanthine phosphoribosyl transferase (XPRT). This gene, when introduced into mammalian cells, will rescue those cells from selection in mycophenolic acid and xanthine (Mulligan and Berg, 1980). Another important dominant selectable gene is dihydrofolate reductase (DHFR). Exposure of cultured cells to high levels of the dihydrofolate reductase-inhibiting drug methotrexate (MTX), has led to identification of mutant forms of the enzyme which have high resistance to MTX (Flintoff, Davidson and Siminovitch, 1976). These genes can be transferred into genetically normal tissue culture cells, after which their expression can lead to preferential survival of cells when high doses of MTX are added to the medium. Another advantage of the DHFR gene as the dominant selectible marker is that it is amplifiable (Alt et al., 1978; Haber et al., 1981). That is, as doses of MTX are raised, this gene will undergo amplification with consequent production of increased amounts of enzyme.

Our laboratory has been introducing such a mutant gene into mouse embryos to produce transgenic mice. We have now generated 16 lines of animals which retain the gene, and several of these lines have been shown to express high quantities of the mutant gene product. Because of its dominant selectable characteristics, this gene has the potential for gamete selection. When these transgenic mice are bred, they will transmit the new gene to half of their progeny. The cleaving embryos which inherit the gene can be grown in culture for several days, after which they may be reimplanted to complete development. We are currently engaged in experiments wherein these cultured embryos are grown in medium containing MTX. We are investigating whether cells which inherit the mutant, dominant selectable DHFR gene can preferentially survive in such culture conditions. If this is the case, then preferential survival of embryos inheriting the chromosome on which the new gene has been inserted will be achieved. Success in this effort can lead to protocols for selecting gametes which retain chromosomes on which dominant selectable markers have been inserted. Although the integration of genes into the host genome is an apparently random event, integration into large chromosomes is undoubtedly statistically favored. Because the X-chromosome is particularly large, it is likely to receive foreign genetic material relatively frequently. If a dominant selectable marker is introduced onto the X-chromosome, and if such a marker's expression can be selected for during culture of cleaving embryos, then a system can be devised wherein the X-chromosome is always present in the progeny of individual crosses. Where such an X-chromosome is transmitted through a male parent, this protocol would result in the birth of all female offspring. This kind of gamete selection would be of great value in animal husbandry, where an excess of female progeny over male is frequently advantageous. Thus, gene transfer may be applied to the selection of specific linkage groups within animal populations.

Problems with gene transfer into the germ line

Insertional mutagenesis. One difficulty with gene transfer into the embryo is that insertion of foreign genetic material into the host genome must inevitably disrupt host sequences. Insertion of foreign DNA into a host gene with consequent interruption of the recipient's coding information can have adverse consequences. Introduction of retroviral proviral DNA into early mouse embryos has led to identification of a recessive insertional lethal mutation (Jaenisch et al., 1983). In this case, the α_1 (1) collagen gene was interrupted by the viral DNA with consequent loss of function. Animals homozygous for the retroviral insertion died in mid-gestation (Jaenisch et al., 1983). In a series of experiments involving gene microinjection, two of six transgenic mouse lines studied also were found to have recessive lethal mutations resulting from gene insertion (Wagner et al., 1983). The nature of these mutations has not been characterized. The possibility that gene transfer can interfere with host coding functions can cause problems when applying genetic engineering technology to farm animals. When a new gene is inserted into the germ line of an animal, the most efficient way of maintaining it is to produce homozygotes. Then, all offspring in subsequent generations will carry the gene. However, the aforementioned examples illustrate that the generation of homozygotes is not always possible. The problem of insertional mutagenesis can be partially overcome by crossing animals derived from separate gene transfer experiments. If a single gene is inserted onto a number of different chromosomes, an animal can be multiply heterozygous for the gene at separate insertion sites. Then, the frequency with which the gene is transmitted will be very high, without encountering the problem of insertional mutagenesis. Clearly, however, this approach is less efficient than the production of true homozygotes.

Another form of insertional mutagenesis can cause problems even in heterozygous animals. One line of transgenic mice has been described in which males were unable to transmit the foreign gene (Palmiter et al., 1984). Apparently, the donor sequence disrupted a gene which functions in haploid spermatids and thereby rendered the sperm which carried it nonfunctional (Palmiter et al., 1984).

Translocations. Our laboratories have identified a line of transgenic mice which have sustained a translocation as a result of gene insertion. The initial transgenic mouse produced in these experiments was a female and was fertile. However, all male offspring which inherited the gene were sterile. Subsequent analyses demonstrated a reciprocal translocation between chromosomes 2 and 12. Gene mapping study subsequently localized the foreign gene sequence to one of the translocation chromosomes. The mechanism by which insertion of this material led to a translocation is presently unknown. It is possible that the microinjection procedure mechanically disrupted the chromosomes. We consider it more likely that gene insertion destabilized the chromatin structure and led to spontaneous breakage. Regardless of the mechanism, however, the gene transfer event was clearly associated with chromosome breakage. It must be anticipated, therefore, that transfer of genes into any mammalian species may occasionally be associated with chromosome breakage and loss of fertility. Fortunately, this event appears rare, as this transgenic mouse line is the only one yet identified with a problem of this nature. However, when introducing genes into embryos by microinjection, this hazard must be taken under consideration.

Endocrinopathies. The insertion of genes coding for growth hormone appears at first to be an attractive strategy for inducing rapid growth of livestock. This approach has an advantage over growth hormone injection, however, only if the rapid growth characteristics transmitted to subsequent progeny. In mice, insertion and genetic transmission of growth hormone has been demonstrated (Palmiter et al., 1983). However, succeeding experiments showed that the gene may impair fertility. Injection of growth hormone genes into mice which lack an endogenous functional growth hormone gene corrected the small phenotype, and improved the fertility of males (males of this strain are normally subfertile). However, fertility of females was markedly impaired (Hammer, Palmiter and Brinster, 1984). This negative side effect is probably the consequence of the continuous production of growth hormone from the new gene. The endogenous hormone is normally secreted in a pulsatile manner and mediates sex-specific differentiative processes in several organ systems (Norstedt and Palmiter, 1984).

Foreign genes may act as dominant mutations. As noted above, our laboratory has been transferring a dominant selectable DHFR gene into mouse embryos. We have observed that a number of these transgenic mice, although heterozygous for the foreign gene, manifest a variety of developmental abnormalities. These include reduced fertility, abnormal skeletal development, cranial malformations, anemias, pigment cell defects, and severely stunted growth. Because independently generated lines of mice which carry this gene at a single chromosomal site manifest a similar array of abnormalities, we must conclude that gene expression underlies the appearance of these defects. Thus, in this case, the foreign gene is acting as a dominant mutation. We are not as yet certain why insertion of this gene has had such an effect. However, we presume that insertion and expression of other genes may eventually lead to similar findings. Some of the problems evident in these animals would clearly be highly undesirable in farm animals. We currently hypothesize that it is embryonic expression of the mutant DHFR gene which causes the developmental anomalies seen. Therefore, it may be advisable to link promoters whose function is specific to adult tissues to genes which are to be transferred. Then, deleterious effects upon the developing embryo will not be encountered.

Conclusions

Gene transfer into the germ line of animals clearly has great potential for use in agriculture. Animals may be induced to grow more rapidly, to synthesize antibodies to environmental pathogens and to be highly resistant to various drugs. A greater knowledge of the regulation of intermediary metabolism and how it is affected by abnormal gene products may also lead to more sophisticated gene transfer experiments than those thus far attempted. For example, animals might be engineered to store more calories as protein rather than fat, or to preferentially synthesize unsaturated fatty acids. These changes could yield agricultural products which were both safer and more nutritious.

However, the deleterious effects of foreign gene insertion seen in the few mice made to date, while of interest to the developmental biologist, are certainly undesirable to the agricultural genetecist. Further study of this gene transfer system is needed not only to broaden the variety of experiments with practical value, but also to minimize the untoward effects of gene insertion. To the latter end, exploration of the gene transfer system so as to develop approaches for directing foreign genes to specific sites in the host genome should be encouraged. If this goal were achieved, gene transfer could be safer, more reliable, and more applicable to practical problems in agriculture.

Acknowledgements This work was supported in part by private contributions to the Mount Sinai School of Medicine.

References

Alt, F. W., R. E. Kellems, J. R. Bertino, and R. T. Schimke. 1978. *Selective multiplication of dihydrofolate reductase genes in methotrexate-resistant variants of cultured murine cells*. J. Biol Chem. 253:1357–1370.

Anderson, W. F., L. Killos, L. Sanders-Haigh, P. J. Kretchmer and E. Diacumakos. 1980. *Replication and expression of thymidine kinase and human globin genes microinjected into mouse fibroblasts*. Proc. Nat. Acad. Sci. USA 77:5399–5403.

Bachetti, S. and F. L. Graham. 1977. *Transfer of the gene for thymidine kinase to thymidine kinase deficient human cells by means of purified Herpes simplex viral DNA*. Proc. Nat. Acad. Sci. USA 74:1590–1594.

Brinster, R. L., H.-Y. Chen, M. Trumbauer, A. W. Senear, R. Warren and R. D. Palmiter. 1981. *Somatic expression of Herpes thymidine kinase in mice following injection of a fusion gene into eggs*. Cell 27:223–231.

Brinster, R. L., K. A. Ritchie, R. E. Hammer, R. L. O'Brien, B. Arp and U. Storb. 1983. *Expression of microinjected immunoglobulin gene in the spleen of transgenic mice*. Nature 306:332–336.

Capecchi, M. R. 1980. *High efficiency transformation by direct microinjection of DNA into cultured mammalian cells*. Cell 22:479–488.

Costantini, F. and E. Lacy. 1981. *Introduction of a rabbit B-globin gene into the mouse germ line*. Nature 294:92–94.

Diacumakos, E. 1973. *Methods for micromanipulation of human somatic cells in culture*. Meth. Cell Biol. 7:287–311.

Flintoff, W. F., S. V. Davidson and L. Siminovitch. 1976. *Isolation and partial characterization of three methotrexate-persistent phenotypes from Chinese hamster ovary cells*. Somat. Cell Genet. 2:245–261.

Gordon, J. W. and F. H. Ruddle. 1981. *Integration and stable germ line transmission of genes injected into mouse pronuclei*. Science 214:1244–1246.

Gordon, J. W., G. A. Scangos, D. J. Plotkin, J. A. Barbosa and F. H. Ruddle. 1980. *Genetic transformation of mouse embryos by microinjection of purified DNA*. Proc. Nat. Acad. Sci. USA 77:7380–7384.

Graessmann, A., M. Graessmann, W. C. Topp and M. Botchan. 1979. *Retransformation of a simian virus 40 revertant cell line, which is resistant to viral and DNA infections, by microinjection of viral DNA*. J. Virol. 32:989–994.

Graham, F. L. and A. J. Vander Eb. 1973. *A new technique for the assay of infectivity of human adenovirus V DNA*. Virology 52:456–467.

Grosschedl, R., D. Weaver, D. Baltimore and R. Costantini. 1984. *Introduction of a u immunoglobulin gene into the mouse germ line: specific expression in lymphoid cells and synthesis of a functional antibody*. Cell 38:647–658.

Haber, D. A., S. M. Beverley, M. L. Kiely and R. T. Schimke. 1981. *Properties of an altered dihydrofolate reductase encoded by amplified gene in cultured mouse fibroblasts*. J. Biol. Chem. 256:9501–9510.

Hammer, R. E., R. D. Palmiter and N. L. Brinster. 1984. *Partial correction of murine hereditary growth disorder by germ-line incorporation of a new gene*. Nature 311:65–67.

Jaenisch, R. 1976. *Germline integration and Mendelian transmission of the exogenous Moloney leukemia virus*. Proc. Nat. Acad. Sci. USA 73:1260–1264.

Jaenisch, R. 1979. *Moloney leukemia virus expression and gene amplification in preleukemic and leukemic Balb/Mo mice*. Virology 93:80–90.

Jaenisch, R., K. Harberg, A. Schnieke, J. Lohler, I. Chumakov, D. Jahner, D. Grotkopp and E. Hoffman. 1983. *Germline integration of Moloney murine leukemia virus at the Mov13 locus leads to recessive lethal mutation and early embryonic death*. Cell 32:209–216.

Lacy, E., S. Roberts, E. P. Evans, M. D. Burtenshaw and F. E. Costantini. 1983. *A foreign B-globin gene in transgenic mice: integration at abnormal chromosomal positions and expression in inappropriate tissues.* Cell 34:343–358.

Maitland, H. and J. McDougall. 1977. *Biochemical transformation of mouse cells by fragments of Herpes simplex virus DNA*. Cell 11:233–241.

McKnight, G. S., R. E. Hammer, E. A. Kuenzel and R. L. Brinster. 1983. *Expression of the chicken transferrin gene in transgenic mice*. Cell 34:335–341.

Mulligan, R. C. and P. Berg. 1980. *Expression of a bacterial gene in mammalian cells*. Science 209:1422–1427.

Norstedt, G. and R. Palmiter. 1984. *Secretory rhythm of growth hormone regulates sexual differentiation of mouse liver*. Cell 36:805–812.

Palmiter, R. D., R. L. Brinster, R. E. Hammer, M. E. Trumbauer, M. G. Rosenfeld, N. C. Birnberg and R. M. Evans. 1982b. *Dramatic growth of mice that develop from eggs microinjected with a metallothionein-growth hormone fusion gene*. Nature 300:611–615.

Palmiter, R. D., H.-Y. Chen and R. L. Brinster. 1982a. *Differential regulation of metallothionein-thymidine kinase fusion genes in transgenic mice and their offspring.* Cell 29:701–710.

Palmiter, R. D., G. Norstedt, R. E. Gelinas, R. E. Hammer and R. L. Brinster. 1983. *Metallothionein-human GH fusion genes stimulate growth of mice*. Science 222:809–814.

Palmiter, R. D., T. M. Wilkie, H.-Y. Chen and R. L. Brinster. 1984. *Transmission distortion mosaicism in an unusual transgenic mouse pedigree*. Cell 36:869–877.

Ritchie, K. A., R. L. Brinster and U. Storb. 1984. *Allelic exclusion and control of endogenous immunoglobulin gene rearrangement in k transgenic mice*. Nature 312:517–520.

Swift, G. H., E. R. Hammer, R. J. MacDonald and R. L. Brinster. 1984. *Tissue-specific expression of the rat pancreatic elastase I gene in transgenic mice*. Cell 38:639–646.

Wagner, E. F., L. Covarrubias, T. A. Stewart and B. Mintz. 1983. *Prenatal lethalities in mice homozygous for human growth hormone gene sequences integrated in the germline*. Cell 35:647–655.

Wigler, M., S. Silverstein, L. S. Lee, A. Pellicer, T. Cheng and R. Axel. 1977. *Transfer of purified Herpes virus thymidine kinase gene to cultured mouse cells.* Cell 11:223–232.

Root culture system useful in the study of biotrophic root pathogens in vitro

JACQUES MUGNIER, PAUL W. READY and GERARD E. RIEDEL*
Genetics Institute, 87 Cambridge Park Drive, Cambridge Massachusetts 02140, U.S.A.

Abstract We have extended the experimental system of Mugnier and Tepfer (1985) to study root-pathogen interactions in vitro. This experimental system effectively simulates the interactions of roots and pathogens when roots are attached to whole plants and growing in the soil. The system consists of plant roots genetically transformed by *Agrobacterium rhizogenes* and a bicompartmental root culture system (BRC system). Using the BRC system we have obtained infections, in vitro, of *Plasmodiophora brassicae* on transformed rape and kohlrabi roots and of *Polymyxa betae* on transformed sugar beet roots. The study of the biology of certain host-pathogen interactions and agrichemical screening and research in crop disease resistance have been enhanced by the BRC system.

Introduction

The pioneering work of Morel (1944, 1948) established that plant tissue culture is a useful tool in the study of plant pathogens in vitro. Since Morel's work, researchers have demonstrated that several aspects of host-pathogen interactions can be effectively studied if an appropriate tissue culture system is used. (As an example, see Helgeson et al., 1972). We are interested in applying plant tissue culture techniques to the study of root-pathogen interactions in vitro. As initial candidates for our studies we have chosen two fungal pathogens of roots: *Plasmodiophora brassicae*, a pathogen of species in the Cruciferae and *Polymyxa betae*, a pathogen of species in the Chenopodiaceae.

Both *P. brassicae* and *P. betae* are intracellular, non-mycelial, zoosporic, biotrophic organisms classified within the Plasmodiophorales (Webster, 1980). *P. brassicae* is the causative agent of club root disease in rape and other *Brassica* species and *P. betae* is associated with rhizomania disease in sugar beet. Rhizomania is reported to be caused by the combined effects of *P. betae* and beet necrotic yellow vein virus, which is believed to be vectored by *P. betae* (Langenberg and Giunchedi, 1982).

We have applied the root culture system developed by Mugnier and Tepfer (1985) to study the infection, in vitro, of rape (*Brassica napus*) roots and sugar beet (*Beta vulgaris*) roots by *P. brassicae* and *P. betae*, respectively.

*To whom correspondence should be sent.

Generation of transformed roots

We obtain transformed roots of rape or sugar beet by inoculating sterile rape stems or sugar beet tubers with *Agrobacterium rhizogenes* strain A4. After 10–30 days incubation at 27°C, roots develop from the site of inoculation. These roots are cultured in solidified Murashige and Skoog (MS) media (Murashige and Skoog, 1962) containing carbenicillin (1 mg/ml) to eliminate contaminating *A. rhizogenes*. The putative transformed roots are tested for sterility and for production of agropine and/or mannopine (Tepfer and Tempé, 1981) to confirm their transformed state. The transformed rape and sugar beet roots can be maintained indefinitely by subculturing on liquid or solid MS media at 27°C.

The bicompartmental root culture (BRC) system

The elements of the BRC experimental system are:

1. plant roots genetically transformed by *A. rhizogenes*
2. a two compartment culture system (Fig. 1).

The inner compartment (see Fig. 1) contains media that mimics the materials that would be supplied to plant roots by the aerial part of a plant (for example, sugar and vitamins). The outer compartment (see Fig. 1) contains media that mimics the environment a root encounters

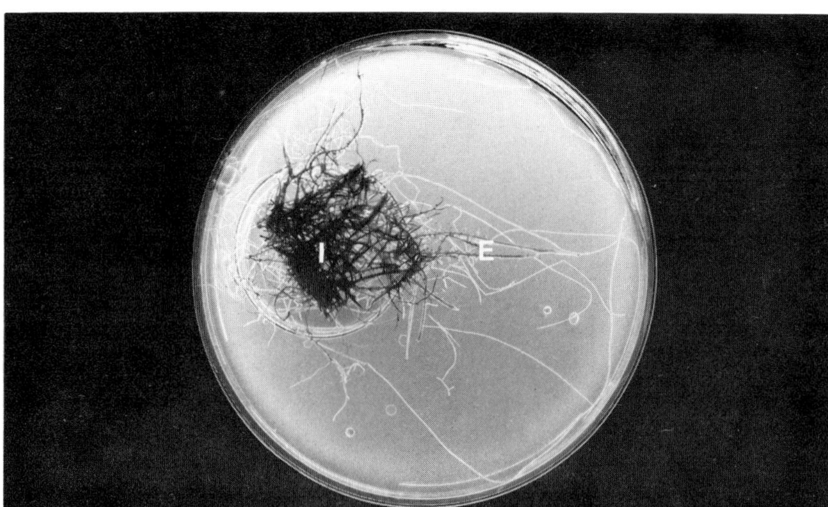

Figure 1. Transformed roots of rape (*Brassica napus*) cultures in a biocompartmental chamber (14 days incubation). I: internal compartment, E: external compartment.

in the soil (for example, nutrient levels, salt concentrations, buffering capacity, pH, and osmoticum).

Root-pathogen studies are conducted by preparing the BRC plates with appropriate media in the two compartments (Ready and Riedel, unpublished observations), inoculating the internal compartment of each BRC plate with genetically transformed roots, incubating the inoculated plates for 7–30 days (depending upon the root species) at 27°C to allow sufficient root growth, and adding fungal spores to roots in the external compartment of each BRC plate. The infected BRC plates are incubated at 27°C for as long as 30 days. During this incubation, fungal infection can be observed by light microscopic examination of the roots in the external compartment of the BRC.

Rape infection with P. brassicae *in vitro*

A fourteen day old BRC plate containing transformed rape roots was pre-layered with 5 ml of sterile water over the roots in the external compartment. One hundred and fifty microliters of a suspension of resting spores (5×10^6/ml) of *P. brassicae* obtained from whole infected plants (a generous gift of Dr. F. M. Humpherson-Jones, National Vegetable Research Station, Warwick) was added to the roots. The infected BRC plate was incubated for 12 hours at 27°C; then 5 ml of sterile H_2O was added to the roots of the external compartment. The infected BRC plates were incubated for an additional 14–28 days at 27°C. BRC plates were checked daily for sufficient moisture in the external compartment. At two day intervals, infected roots were removed aseptically and examined by light microscopy.

Figure 2 shows a typical *P. brassicae* infection obtained in BRC plates with rape roots 21 days after fungal inoculation. We believe that the circular structures seen in the root cortical and hair cells correspond to fungal plasmodia (Tommerup and Ingram, 1971).

Sugar beet infection with P. betae *in vitro*

A root containing resting spores of *P. betae* was excised from a sugar beet plant infected with *P. betae* (generously provided by Dr. W. G. Langenberg, USDA-ARS, University of Nebraska). The excised root was surface sterilized, and ground into a suspension in sterile water. The suspension was used to infect sugar beet roots (pre-layered with 2 ml sterile water) growing in the external compartments of BRC plates that were at least 20 days old. The BRC plates were incubated for 12 hours at 27°C, then a 2 ml sterile water was added to the roots in the external compartments. The infected BRC plates were incubated at 27°C for 30 days. At two day intervals, infected roots were removed from BRC plates and examined by light microscopy.

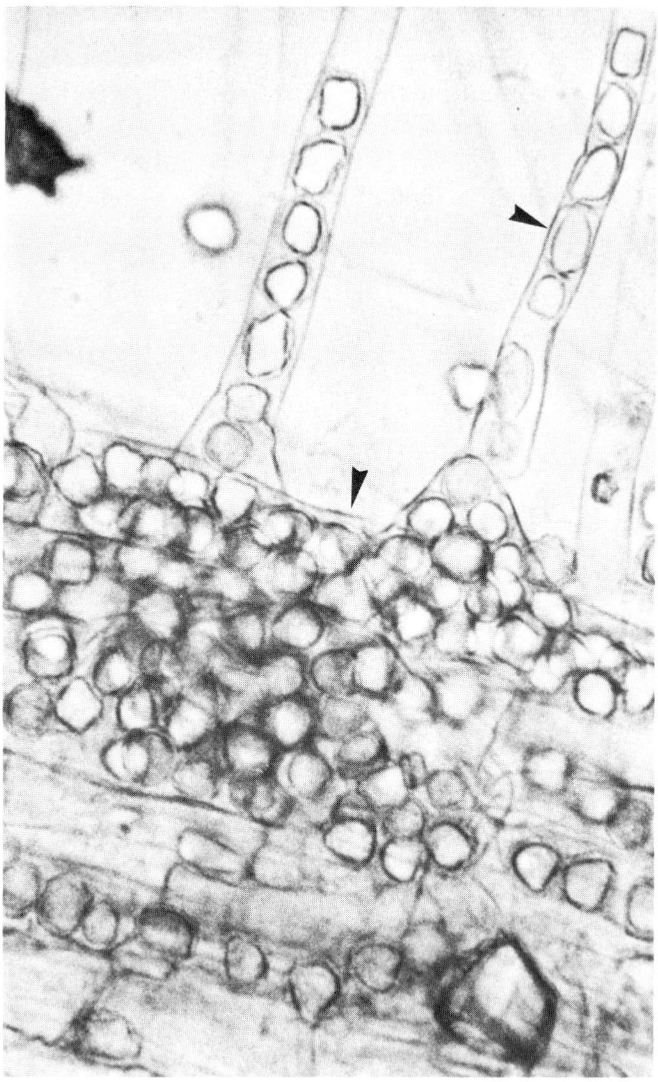

Figure 2. Infection of transformed rape roots with *Plasmodiophora brassicae* (21 days post-infection). Arrows indicate plasmodia (?) in root hair and root cortical cells.

Figure 3 shows a typical *P. betae* infection obtained in BRC plates with sugar beet roots 24 days after fungal inoculation. We believe that the circular structures in the root cortical cells correspond to resting spores of *P. betae* in a typical whole plant infection (Langenberg and Kerr, 1982).

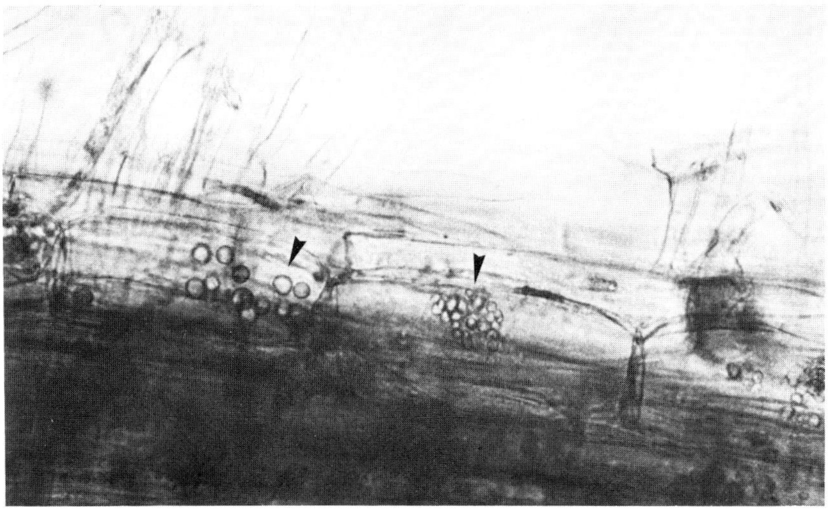

Figure 3. Infection of transformed sugarbeet roots with *Polymyxa betae* (24 days post-infection). Arrows indicate resting spores (?) in root cortical cells.

Experimental advantages of the BRC system

In comparison with whole plant cultures, the BRC system has two major advantages. Unlike whole plant cultures, the BRC system allows root-microorganism interactions to occur in an aseptic, observable, convenient, culture vessel. The BRC system is contained in petri dishes and allows a researcher to conveniently and directly observe root-microbe interactions with a microscope, without perturbing the experimental material. This ease of observation is difficult or impractical to achieve with whole plant studies.

The BRC system allows a researcher to alter root nutrition by varying the media components, in either the inner compartment of the culture or in the outer compartment of the culture (Fig. 1). This dual control of root nutrition is not achieved easily in whole plant culture primarily because of the difficulty of controlling photosynthesis in the aerial part of the whole plant (Feldman, 1984). There is some evidence that changes in photosynthesis alter root development in "whole plant" root cultures (Feldman, 1984).

The BRC system provides root cultures which are physiologically and morphologically similar to roots attached to whole plants. This is not true for callus or cell suspension cultures. It is also generally not true for non-transformed root organ cultures, because:

Non-transformed root organ cultures require phytohormones

(auxins) in their media for growth. An example of the effect of auxin is provided by the case of the biotrophic fungus *Cronartium fusiforme*. The kinetin or 2,4-D needed for maintenance of the non-transformed root organ culture caused significant reductions in fungal growth (Jacobi, 1982).

Non-transformed root organ cultures require high levels of several nutrients in their media for growth. These nutrients are not present in the soil, and alter the physiology of the roots in the organ culture. For example, non-transformed roots require sucrose in the media. Sucrose affects the transport of material within and out of roots. Currently, there is an hypothesis (Hancock and Huisman, 1981) that the long-distance transport of materials in roots has important implications in host-parasite relations.

Experimental uses of the BRC system

The BRC system can be used to establish cultures, in vitro, of soil organisms which require roots for growth. The BRC system offers a researcher the opportunity to culture and study biotrophic fungi and nematodes, root-colonizing bacteria, root-feeding insect pests, and other soil organisms which interact with roots, but which are not necessarily biotrophic. We are continuing to develop the BRC system to use it as a tool in studying the molecular biology of root-pathogen interactions and as a tool in the screening of agrichemicals for useful biological activities.

References

Feldman, L. J. 1984. *Regulation of root development*. Ann. Rev. Plant Physiol. 35:223–242.

Hancock, J. G. and D. C. Huisman. 1981. *Nutrient movement in host-pathogen systems*. Ann. Rev. Phytopathol. 19:309–331.

Helgeson, J. P., J. D. Kemp, G. T. Haberlach and D. P. Maxwell. 1972. *A tissue culture system for studying disease resistance: the black shank disease in tobacco callus cultures*. Phytopathology 62:1439–1443.

Jacobi, W. R. 1982. *Growth medium for dual culture of lobbolly pine callus and Cronartium fusiforms*. Phytopathology 72:136–138.

Langenberg, W. G. and L. Giunchedi. 1982. *Ultrastructure of fungal plant virus vectors: Polymyxa graminis in soilborne wheat mosaic virus-infected wheat and P. betae in beet nucrotic yellow vein virus-infected sugar beet*. Phytopathology 72:1152–1158.

Langenberg, W. G. and E. D. Kerr. 1982. *Polymyxa betae in Nebraska*. Plant Dis. 66:9.

Morel, G. 1944. *Le developement du mildiou sur des tissues de vigne cultures in vitro*. Comptes rendus hebdomadaire des seances de l'Academie des sciences. Paris 218:50–52.

Morel, G. 1948. *Recherches sur la culture associee de parasites obligatoires et de tissues vegetaux*. Annales des Ephiphyties (serie Pathologie Vegetale) 14:1–112.

Murashige, T. and O. Skoog. 1962. *A revised medium for rapid growth and bioassays with tobacco tissue culture*. Physiol. Plant 15:473–497.

Tepfer D. and J. Tempe. 1981. *Production d'agropine far des racines formees sous l'action d'Agrobacterium rhizogenes, souche A*. C. R. Acad. Sc. Paris 292:153–156.

Tommerup, I. C. and D. S. Ingram. 1971. *The life-cycle of Plasmodiophora brassicae Woron. in Brassica tissue cultures and in intact roots*. New Phytologist 70:327–332.

Webster, J. 1980. *Plasmodiophoromycetes*. Pp. 42–53 in *Introduction to Fungi*. Cambridge University Press, Cambridge.

Identification of cis-acting regulatory sequences by P element-mediated germline transformation of *Drosophila* embryos

ERIC HOFFMAN and VICTOR CORCES
Department of Biology, Johns Hopkins University, Baltimore, MD 21218, U.S.A.

Abstract We have used P element-mediated transformation of *Drosophila* embryos as an assay system to determine the role of different DNA sequences located in the 5′ region of the *Drosophila* hsp28 gene on its ability to be induced by heat shock and by the steroid hormone ecdysone. Flies that contain the transformed gene with 1.1 kb of 5′ information express normal levels of RNA after developmental induction by ecdysone but they accumulate lower than wildtype amounts of RNA after heat shock. Transformants which contain 227 bp of 5′ information show ecdysone and temperature-induced expression of the foreign gene, but the RNA accumulates at lower than wildtype levels. Finally, deletion of the sequences located between − 227 and − 124 results in abolishment of transcription induction by heat shock but not by ecdysone. These results suggest that the sequences necessary for heat shock and hormone regulation of transcription are located in different regions of the hsp28 gene.

Introduction

The expression of eucaryotic genes is regulated in a temporal, spatial and quantitative fashion during the development of an organism. Discerning the mechanisms underlying these different levels of control is a first necessary step in the process of understanding the molecular basis of differentiation and the maintenance of the differentiated state. The changes in gene expression that take place in *Drosophila* after heat shock constitute an ideal model system in which to study these problems. After temperature elevation, the transcription of seven major heat shock genes is induced in *Drosophila* cells, whereas transcription of most other previously active genes is repressed. The expression of the heat shock genes in *Drosophila* is also controlled at the translational level; the preexisting mRNAs stay in the cytoplasm but are sequestered from translation, whereas the heat shock mRNAs are efficiently translated (see Schlessinger, Ashburner and Tissieres, 1982 for a review). In addition, some of the heat shock genes are induced by specific developmental cues and in a tissue-specific fashion (Cheney and Shearn, 1983; Ireland et al., 1982). For example, the hsp28 gene is induced in the nurse cells during oogenesis and the RNA is transported to the developing oocyte (Zimmerman, Petri and Meselson, 1983). This gene is then turned off until the prepupae stage when it is induced again by the steroid hormone ecdysone (Hoffman and Corces, 1984). The understanding of the molecular mechanisms governing the control of the expression of these genes will likely involve the isolation of protein factors which interact with specific sequences on the DNA molecule thus altering the transcription of the gene. The fact that seven different genes are coordinately induced by the same stimulus suggests that these control sequences might be common to all of them. Comparison of the DNA sequences of the seven genes has in fact pointed out several conserved regions in the 5′ end of the different *Drosophila* heat shock genes (Holmgren et al., 1981; Ingolia and Craig, 1981). In the past few years several investigators have identified sequences necessary for the temperature inducibility of heat shock genes using both homologous and heterologous transcription assay systems (Corces et al., 1982; Pelham and Bienz, 1982; Mirault, Southgate and Delwart, 1982; Corces and Pellicer, 1984; Dudler and Travers, 1984; Cohen and Meselson, 1984). One of these sequences, the so called "heat shock consensus sequence" is located immediately upstream from the TATA box of the heat shock genes and interacts specifically with a protein factor after temperature induction (Parker and Topol, 1984; Wu, 1984).

The germline transformation protocol (Rubin and Spradling, 1982) affords the possibility of stably transforming *D. melanogaster* with different cloned genes, thus enabling the study of the transcriptional and translational controls operating during heat shock in a homologous system. We have been using this approach to study the *Drosophila* hsp28 gene in order to identify the DNA sequences involved in controlling the temperature and hormone induction of transcription of this gene. We have constructed a size variant of hsp28 (designated hsp18.5) that contains a 207 bp deletion in the protein coding region and can be distinguished from the endogenous gene because it encodes a smaller 0.9 kb RNA. Both the heat shock and the developmental induction of this gene are maintained after transformation into the *Drosophila* germline of a plasmid containing 2.3 kb of 5′ and 0.3 kb of 3′ information. We present evidence indicating that the 5′ sequences of the gene can be dissected into different regions responsible for the temperature and ecdysone control of transcription.

Results

Construction of plasmids used for transformation

In order to distinguish between the endogenous and the transformed hsp28 messages, we made a 207 bp deletion in the protein coding region of the hsp28 gene, immediately 3′ to the ATG protein initiation codon. Plasmid 88.5 (Corces et al., 1980) was linearized with Sma I at nucleotide + 150, 33 bp downstream of the ATG, and then digested with BAL 31 nuclease such that approximately 200 bp were deleted

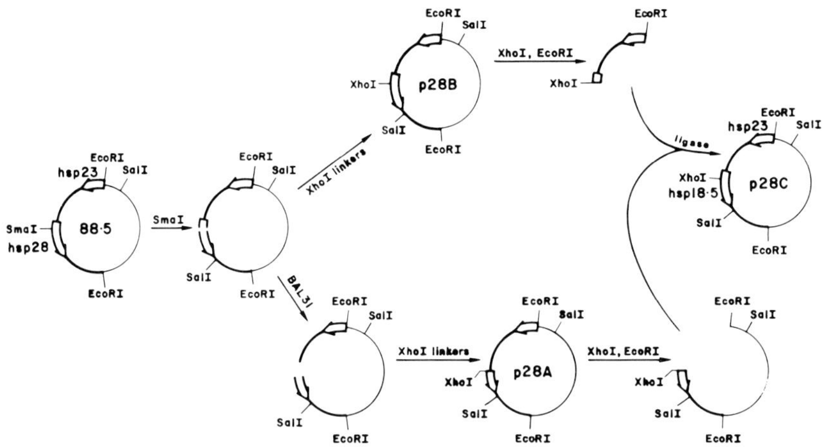

Figure 1. Diagram showing the different steps in the construction of plasmid p28C (see text for a description).

from each end. The digested plasmid was supplied with Xho I linkers, then religated to circularize, and was designated p28A (Fig. 1). Since this construction deleted many of the 5′ regulatory sequences of the gene, an additional plasmid was made (p28B) which contained an Xho I linker at the Sma I site of plasmid 88.5. This plasmid was used to replace the sequences 5′ to the Sma I site that were deleted in p28A. The resulting plasmid contained a deletion of approximately 200 bp downstream of the original Sma I site and will be referred to as p28C. The precise location of the deletion breakpoint in the hsp28 gene was determined by sequence analysis and was found to be from + 150 to + 357. With the addition of hexameric Xho I linker, the resulting plasmid retained the original reading frame and was expected to encode a 950 bp RNA and a 18,500 dalton protein, if properly expressed. This gene will be referred to as hsp18.5. External deletions in the 5′ regions of the hsp18.5 gene were constructed taking advantage of various restriction sites located in this region. Figure 2 shows the organization of part of the 67B locus where the endogenous hsp28 gene is located (Corces et al., 1980) and the various additional plasmids utilized in the transformation experiments. These plasmids contain 1.1 kb, 227 bp and 124 bp of 5′ information and they were made using the Xba I, Dra I and Pst I restriction sites indicated in the map.

Establishment of transformed lines

The different DNA fragments containing the hsp18.5 gene were cloned in the Sal I site of the transformation vector Carnegie 20 (Rubin

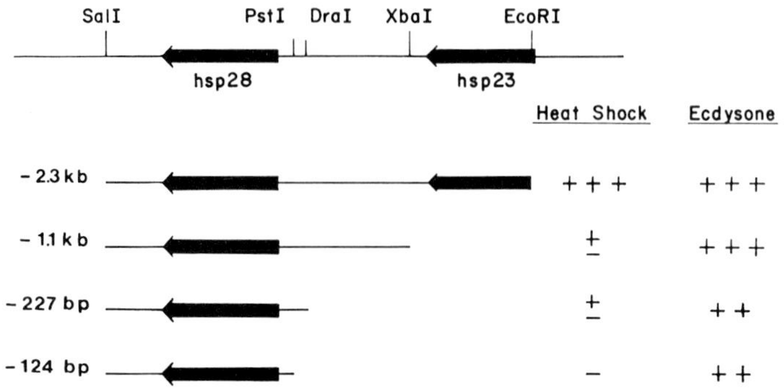

Figure 2. Structure and organization of transforming plasmids. Plasmid p28C was separately digested with Xba I, Dra I, and Pst I, giving rise to various plasmids containing respectively 1.1 kb, 227 bp and 124 bp of 5′ information in the hsp18.5 gene. These DNA fragments were then supplied with Sma I linkers and cloned into the Sal I site of the transformation vector Carnegie 20 (Rubin and Spradling, 1983).

and Spradling, 1983) via Sal I linkers. The orientation of the inserted fragment in the vector was such that the 5′ region of the hsp18.5 gene was adjacent to the P element and the 3′ end was next to the Xanthine dehydrogenase gene. Plasmids containing the hsp18.5 gene with different amounts of 5′ information were then injected into r^{506} preblastoderm embryos using pπ25.1 or pπ25.7 wc as helper plasmids to provide the transposase activity. Adult flies were then backcrossed with the ry^{506} parental stock and transformants were identified in the progeny of this cross by the wildtype eye color due to the expression of the Xanthine dehydrogenase gene (Rubin and Spradling, 1982). The chromosomal location of the transformed gene was determined by in situ hybridization (Pardue and Gall, 1975) of biotinylated hsp18.5 DNA to salivary gland polytene chromosomes. All the insertions were single copy as deduced from the intensity of the hybridization band in the transformed locus as compared with the endogenous one (data not shown). The transformed flies were then rendered homozygous by crosses with appropriate stocks. Transformants in the second chromosome were crossed to a *Sco*/*CyO*;ry^{506}/ry^{506} stock and heterozygous flies containing the *CyO* chromosome and wildtype for the *ry* gene were pairmated. The only survivors of this cross would be homozygous at the transformed locus. A similar approach was followed to obtain homozygous stocks of transformants at the third chromosome, crossing them to a ry^{506}/TM3 ry^{506} stock and following the segregation of the *Sb* marker present in the TM3 chromosome. The transformed flies were maintained as homozygous stocks except those which had a female sterile or homozygous lethal phenotype; these were maintained as heterozygous stocks using the above mentioned balancer chromosomes (see Table 1).

Table 1. Summary analysis of the different transformed stocks. This table shows the characteristics and properties of the different transformed flies analyzed in these studies, their phenotypes and the chromosomal location of the transformed DNA. Also shown is the amount of hsp18.5 RNA that accumulates in the different transformants expressed as a percentage of the amount of hsp28 RNA expressed in the same fly. These numbers were obtained by scanning the corresponding autoradiograms and normalizing the intensity of the RNA bands relative to the number of copies of the transformed gene.

Fly strain		Base pairs 5′ of gene	Chromosome	Position	Genotype	Phenotype	Transcription Heat shock	Transcription Ecdysone
28C	A	2.3 kb	3L	68BC	Hom.	None	100%	100%
	B		2R	30A	Bal.	Fem. Ster.	100%	100%
28X	A	1.1 kb	3L	68C	Hom.	None	22%	135%
	B		3R	84D	Bal.	Hom. Leth.	20%	62%
	C		2R	42F	Hom.	None	21%	
	D		3L	66A	Bal.	Hom. Leth.		98%
	E		3R	82C	Hom.	None		
28A	A	227 bp	3R	95F	Bal.	Fem. Ster.	16%	65%
	B		2 + 3L	65D	Bal.	Hom. Leth.	6%	88%
	C		3R	87AB	Hom.	None	9%	42%
28P	A	124 bp	3L	61D	Hom.	None	0%	34%
	B		3L	97B	Hom.	None	0%	23%
	C		2L	39B	Hom.	None	0%	19%
	D		3R	94B	Hom.	None	0%	44%

Analysis of sequences required for heat shock-induced transcription

In order to identify the putative role of different DNA sequences located in the 5′ region of the hsp28 gene as cis-acting regulatory elements of temperature-induced transcription, we studied the effect of deletion of various sequences on the transcription of the gene after heat shock. We analyzed four different constructions containing 2.3 kb, 1.1 kb, 227 bp and 124 bp of 5′ information respectively. At least three independent transformants containing each plasmid were obtained and total RNA was isolated from adult flies kept at 21°C and heat shocked at 36.5°C for 45 min. The RNA was then subjected to electrophoresis on a 1% agarose-formaldehyde gel (Corces et al., 1981), transferred to BioTrans (ICN Corp.) and hybridized with ^{32}P-labeled hsp18.5 DNA. The results of these experiments are shown in Table 1. The amount of hsp18.5 RNA that accumulates in the different transformants is expressed as a percentage of the amount of hsp28 RNA present in the same fly. Transformants containing 2.3 kb of 5′ information accumulate two different RNA species. The larger RNA, approximately 1.15 kb long is the result of transcription of the endogenous hsp28 locus. The smaller transcript, 0.95 kb in size, is not present in the untransformed parental flies and originates from the transformed hsp18.5 gene (Hoffman and Corces, 1984). The levels of hsp28 and hsp18.5 RNA are approximately the same in these transformants. Table 1 shows that deletion of the sequences located between − 2.3 kb and − 1.1 kb have an effect on the accumulation of the hsp18.5 RNA after temperature elevation. The transcription of the transformed gene containing 1.1 kb of DNA sequences at the 5′ end is still under heat shock control in the sense that the gene is not transcribed in flies kept at 21°C but is induced when the flies are heat shocked at 36.5°C. Nevertheless, the level of hsp18.5 RNA in these transformants is approximately one fifth of that of hsp28 RNA. Deletion of the sequences located between − 1.1 kb and − 227 bp causes a further decrease in the transcription levels of the hsp18.5 gene after temperature elevation but the transformed gene still maintains the heat shock control of transcription. Finally, deletion of the sequences located between − 227 bp and − 124 bp completely abolishes temperature induced transcription of the hsp18.5 gene.

Effect of 5′ deletions on the ecdysone-induced expression of the hsp28 gene

We have previously shown that a transformed hsp18.5 gene containing 2.5 kb of 5′ information shows a completely wildtype developmental expression (Hoffman and Corces, 1984). Both the levels of RNA and the time of development at which the transcription of the gene is induced are the same as those of the endogenous hsp28 gene. In order to study the role of different sequences located in the 5′ region of this gene on the temporal and quantitative regulation of transcription, we used Northern blots to analyze the developmental expression of the transformed hsp18.5 gene in the different transformants obtained in the experiments described above. All transformants containing 1.1 kb, 227 bp or 124 bp of 5′ information in the hsp18.5 gene are able to express the hsp18.5 RNA with the wildtype temporal pattern of transcription (Hoffman and Corces, 1984). The RNA is present in embryos probably due to its expression in the nurse cells of the mother during oogenesis. The gene is then turned off in the larval stages and is expressed in prepupae and early pupae to be turned off again in late pupae. A small amount of RNA visible in the adults is probably due to its expression in the female ovaries. This pattern of expression is the same as that of the endogenous hsp28 gene, and it was maintained in transformants containing 227 bp or 124 bp or 5′ sequences (data not shown). Even though the deletion of these sequences did not have an effect on the temporal pattern of expression of the transformed gene, it had an effect on the total levels of RNA that accumulated during the development of the fly. Table 1 summarizes the results of the Northern blot analysis of the hsp28 and hsp18.5 RNAs that are expressed during the prepupal stages of flies transformed with plasmids containing different amounts of DNA at the 5′ end of the hsp18.5 gene. Transformants containig 1.1 kb of 5′ information show a position effect in the quantitative expression of the hsp18.5 RNA; one of the transformants accumulates the same amount of hsp18.5 and hsp28 RNAs whereas the other two accumulate higher and lower amounts of the transformed RNA respectively as compared with the endogenous one. In average, these transformants accumulate approximately equal amounts of hsp28 and hsp18.5-encoded RNAs. Transformants containing 227 bp of 5′ information accumulate the hsp18.5 RNA at levels between one half and three quarters of those expressed by the hsp28 gene. Finally, deletion of the sequences located between −227 and −124 cause a further decrease in the transcription of the transformed gene, which is now expressed at levels between one fifth and half of those of the endogenous gene.

Heat shock induction of the transformed gene in prepupae

The results described above indicate that the hsp18.5 gene containing 124 bp of 5′ information can be induced by ecdysone during puparium formation but cannot be transcribed after heat shock in adult flies. The lack of temperature inducibility could be due to the deletion of specific sequences that interact with positive heat shock regulatory factors and activate transcription of the gene. Conversely, these sequences could be involved in maintaining an "open" chromatin conformation that is needed for gene activity. This active state should, *a priori*, be independent of the stimulus that activates transcription and should then be present in the transformed gene during the prepupae stage. If this is the case, one would expect that the hsp18.5 gene would be induced by heat shock at this specific developmental stage. To test this hypothesis, we prepared total RNA from heat shocked and non heat shocked prepupae from three of the transformed flies described above. These RNAs were then analyzed by Northern blots (data not shown). The degree of inducibility of the hsp18.5 RNA is the same in prepupae and adult animals, and the transformed gene containing 124 bp of 5′ information cannot be induced by heat shock in the prepupae stage. These results suggest that the region located between −2.3 kb and −124 bp contains sequences that are necessary for heat shock induction of transcription.

Discussion

We have identified the location of different cis-acting regulatory sequences in the 5′ region of a *Drosophila* heat shock gene by analyzing the effect of the deletion of these sequences on the transcription of the gene. The gene we have studied, hsp28, is not only induced by heat shock but it is also turned on by ecdysone, affording the possibility of simultaneously studying the molecular basis for both types of control. Our results indicate the existence of various DNA regions necessary to obtain normal induction of the transformed gene after temperature induction, suggesting that the organization of the heat shock control sequences in the hsp28 gene is very different from that in the hsp70 gene. Dudler and Travers (1984) and Cohen and Meselson (1984) have shown the existence of two different regions located close to the TATA box which are necessary for normal expression of the hsp70 gene. Deletion of the sequences located between − 97 bp and − 68 bp from the start of transcription causes a decrease in the accumulation of RNA after temperature induction, but the gene still maintains its heat shock inducibility; further deletion of the heat shock consensus sequence completely abolished heat shock induced-transcription (Dudler and Travers, 1984; Cohen and Meselson, 1984). In the case of the hsp28 gene, deletion of the region located between − 2.3 kb and − 1.1 kb causes an effect similar to that observed by Dudler and Travers (1984) when the sequences located between − 97 bp and − 68 bp were deleted: the accumulation of the message decreases considerably. We have looked for sequence homologies between these two regions using computer programs, and we have found two sequences in the 5′ region of the hsp28 gene, approximately 1.4 kb from the start of transcription, which are homologous to sequences located in the functionally equivalent region of the hsp70 gene, between − 97 and − 68 bp (Fig. 3A). These two sequences are within a few bases of each other, but their relative distance is different in the hsp70 and hsp28 genes. Deletion of

A

```
hsp70   ACAGTA-AACGGCA....N13....GGTTCGAGAGA
        |||| | |||||||           ||| |||||||
hsp28   ACAGAACAACGGCA....N18....GGTCCGAGAGA
```

B

```
hsp70   CTNGAANNTTCNAG
        ||| |||||| |||
hsp28   CTNAAANNTTA-AG
        ||| ||||||||||
        CT-TAANNTTANAG
          | ||||||||
        GGNAAANNTTCNCT
```

Figure 3. Sequence homologies between the hsp70 and hsp28 *Drosophila* genes. The DNA sequences in the 5′ region of the hsp28 gene concluded to be important for temperature induced transcription were compared to the 5′ sequences of the hsp70 gene. (A) shows the homology of the − 2.3 kb − 1.1 kb region of hsp28 to the − 97 bp − 68 bp region of hsp70. (B) shows the homology to the hsp70 heat shock consensus sequence.

the sequences located between − 1.1 kb and − 227 bp has a very small effect on the heat shock induced transcription of the hsp28 gene (see Table 1) and the observed slight decrease in RNA accumulation is probably due to position effects. On the other hand, deletion of the sequences located between − 227 bp and − 124 bp causes a complete abolishment of transcription of the hsp28 gene after heat shock. This region contains two sequences structurally related to the heat shock consensus sequence (Fig. 3B). This could explain the parallel effects on transcription activation obtained when these sequences are deleted from the hsp28 gene and when the heat shock consensus sequence is deleted from the hsp70 gene (Dudler and Travers, 1984).

The effect of these different deletions on the transcription of the hsp28 gene is very different when one considers the developmental expression of this gene during pupation. Deletion of the sequences located between − 2.3 kb and − 1.1 kb does not appreciably affect the induction of the gene by ecdysone; in addition, flies transformed with a gene containing 124 bp of 5′ information accumulate the hsp18.5 RNA at levels approximately one third of those of the endogenous hsp28 gene. These flies are thus unable to turn on the transformed hsp18.5 gene after temperature induction but they can express the same gene under ecdysone control at levels close to those of the endogenous gene. These results indicate that sequences responsible for these two types of transcriptional control reside in different regions of the gene. The fact that a transformed gene containing 124 bp of 5′ information is able to be induced by ecdysone but not by heat shock suggests that the region located upstream from − 124 contains sequences that do not play a general role in transcription (i.e. entry of DNA polymerase) but rather must interact with regulatory factors specifically involved in heat shock induction.

On the other hand, the sequences involved in ecdysone regulation of transcription must be located close to the gene, between − 124 bp and + 1.4 kb. Sequences responsible for induction of transcription by steroid hormones have been identified in other systems and have been shown to behave as enhancer elements in the sense that they can act in an orientation and distance independent manner (Chandler, Maler and Yamamoto, 1983; Payvar et al., 1983; Scheidereit et al., 1983). Assuming that the sequences responsible for ecdysone induction would possess these properties, we have analyzed the hsp28 gene for homologies with other ecdysone-induced genes for which DNA sequence data was available in the literature. Figure 4 shows the result of such analysis. We have found a consensus sequence conserved among these various genes which is very rich in AT. The location of this sequence

GENE		POSITION	HOMOLOGY	STRAND
Consensus	A A A N A A A A A A A A A A A A			
HSP28	T A A A G A A A A A A T C A A A A A T G	+100	85%	+
HSP26	A A A A C A A A A A A T A A A A A A A A	-230	100%	-
HSP23	C A A G C A A T T A A A A A A C A A A A	+100	80%	+
HSP22	A A A A T A A A A G G T A A A T A A A A	-230	85%	+
SGS4	T A A A T A A A T A A T A A A C A A A A	-450	95%	+
SGS8	A A A A T A A A T A A T T A A A A A A C	+430	90%	-
SGS7	C A A A A A A A A A A A A A A A A A A T	+400	95%	+
SGS3	C A A A A A A A A A A A A A A A A A A T	-1850		+
	T A A T A A A A A T A T G A A A A A A T	-300	85%	-
YP1	T A A A C A A A A A A A A T C A A T A A	-400	85%	+
	A T T A C A A A T T A A A A T A A T C A	-150	70%	+
	A A A A A A A A A A A A A A A A A C A T	+300	95%	+
YP2	C A A A A A A A A A A A A A A A A A A A	-200	95%	+
DDC	A A A A A A A A A A A T A T A A A G C A	+520	85%	+
RANDOM	N N N N N N N N N N N N N N N N N N N N		30%	

Figure 4. Sequence homologies among different ecdysone-induced genes. Sequences of the hsp28 gene shown to be necessary for hormonal induction were compared to other ecdysone-regulated genes. SGS represents genes encoding *Drosophila* salivary gland proteins; YP indicates genes encoding yolk proteins of *Drosophila* and DDC represents the *Drosophila* dopa decarboxylase gene.

with respect to the gene varies in the different examples considered and in three cases is located in the noncoding strand. Its significance in the process of ecdysone induction of transcription will require further study, but it is interesting to note that this sequence overlaps the region described by McGinnis et al. (1983) to be necessary for normal transcription of the Sgs4 gene.

Acknowledgments

This work was supported by Public Health Service Grant GM32036 and Biomedical Research Support Grant 2507RR07041-16 from the National Institutes of Health. E. H. was funded by training grant GM07231 from the National Institutes of Health.

References

Chandler, V. L., B. A. Maler and K. R. Yamamoto. 1983. *DNA sequences bound specifically by glucocorticoid receptor in vitro render a heterologous promoter hormone responsive in vivo.* Cell 33:489–499.

Cheney, C. M. and A. Shearn. 1983. *Developmental regulation of Drosophila imaginal disc proteins: synthesis of a heat shock protein under non-heat shock conditions.* Dev. Biol. 95:325–330.

Cohen, R. S. and M. Meselson. 1984. *Inducible transcription and puffing in Drosophila melanogaster transformed with hsp70-phage lambda hybrid heat shock genes.* Proc. Nat. Acad. Sci. USA 81:5509–5513.

Corces, V., R. Holmgren, R. Morimoto and M. Meselson. 1980. *Four heat shock proteins in Drosophila melanogaster encoded within a 12 kilobase region in chromosome subdivision 67B*. Proc. Nat. Acad. Sci. USA 77:5390–5393.

Corces, V., A. Pellicer, R. Axel and M. Meselson. 1981. *Integration, transcription, and control of a Drosophila heat shock gene in mouse cells.* Proc. Nat. Acad. Sci. USA 78:7038–7042.

Corces, V., A. Pellicer, R. Axel, S-Y. Mei and M. Meselson. 1982. Pp. 27–34 in M. J. Schlessinger, M. Ashburner, and A. Tissieres, eds., *Heat Shock. From Bacteria to Man.* Cold Spring Harbor Laboratory Press, Cold Spring Harbor, NY.

Corces, V. and A. Pellicer. 1984. *Identification of sequences involved in the transcriptional control of a Drosophila heat shock gene.* J. Biol. Chem. 14812–14817.

Dudler, R. and A. A. Travers. 1984. *Upstream elements necessary for optimal function of the hsp70 promoter in transformed flies.* Cell 38:391–398.

Hoffman, E. P. and V. G. Corces. 1984. *Correct temperature induction and developmental regulation of a cloned heat shock gene transformed into the Drosophila germline.* Mol. Cell Biol. 4:2883–2889.

Holmgreen, R., V. Corces, R. Morimoto, R. Blackman and M. Meselson. 1981. *Sequence homologies in the 5' regions of four Drosophila heat shock genes.* Proc. Nat. Acad. Sci. USA 78:3775–3778.

Ingolia, T. D. and E. A. Craig. 1981. *Primary sequence of the 5' flanking region of the Drosophila heat shock genes in chromosome subdivision 67B.* Nuc. Acids Res. 9:1627–1652.

Ireland, R. C., E. Berger, K. Sirotkin, M. A. Yund, D. Osterbur and J. Fristrom. 1982. *Ecdysterone induces the transcription of four heat shock genes in Drosophila S3 cells and imaginal discs.* Dev. Biol. 93:498–507.

McGinnis, W., A. W. Shermoen, J. Heemskerk and S. K. Beckendorf. 1983. *DNA sequence changes in an upstream DNAse I-hypersensitive region are correlated with reduced gene expression.* Proc. Nat. Acad. Sci. USA 80:1061–1067.

Mirault, M-E., R. Southgate and E. Delwart. 1982. *Regulation of heat shock genes: a DNA sequence upstream of Drosophila hsp70 genes is essential for their induction in monkey cells.* EMBO J. 1:1279–1285.

Pardue, M. L. and J. G. Gall. 1975. *Nucleic acid hybridization to the DNA of cytological preparations.* Meth. Cell. Biol. 101:1–17.

Parker, C. S. and J. Topol. 1984. *A Drosophila RNA polymerase II transcription factor binds to the regulatory site of an hsp70 gene.* Cell 37:273–283.

Payvar, F., D. DeFranco, G. L. Firestone, B. Edgar, O. Wrange, S. Okret, J. A. Gustafsson and K. R. Yamamoto. 1983. *Sequence-specific binding of glucocorticoid receptor to MTV DNA at sites within and upstream of the transcribed region.* Cell 35:381–392.

Pelham, H. R. B. and M. Bienz. 1982. *A synthetic heat shock promoter element confers heat inducibility on the herpes simplex virus thymidine kinase gene.* EMBO J. 1:1473–1477.

Rubin, G. M. and A. C. Spradling. 1982. *Genetic transformation of Drosophila with transposable element vectors.* Science 218:348–353.

Rubin, G. M. and A. C. Spradling. 1983. *Vector for P element mediated gene transfer in Drosophila.* Nucl. Acids Res. 11:6341–6351.

Scheidereit, C., S. Geisse, H. M. Westphal and M. Beato. 1983. *The glucocorticoid receptor binds to defined nucleotide sequences near the promoter of mouse mammary tumor virus.* Nature 304:749–752.

Schlessinger, M. J., M. Ashburner and A. Tissieres. 1982. *Heat Shock. From Bacteria to Man.* Cold Spring Harbor Laboratory Press, New York.

Wu, C. 1984. *Two-protein binding sites in chromatin implicated in the activation of heat-shock genes.* Nature 309:229–243.

Zimmerman, J. L., W. Petri and M. Meselson. 1983. *Accumulation of a specific subset of D. melanogaster heat shock mRNAs in normal development without heat shock.* Cell 32:1161–1170.

Part 3

Nutrition

Insulin and insulin receptors in early chick embryo development

FLORA DE PABLO,* LLUIS BASSAS and JESSE ROTH
Diabetes Branch, NIADDK, National Institutes of Health, Bethesda, Maryland 20205, U.S.A.

Abstract The physiological function of insulin in early embryonic life is unknown. Pancreatic insulin synthesis and secretion do not appear in chick embryos until $3\frac{1}{2}$ to 4 days of development. Nevertheless, immuno-active and bioactive insulin is present in eggs and in chick embryos at day 2 of development. Insulin receptors as well as insulin-like growth factor receptors are widespread in embryo tissues during the first week of embryogenesis. We suggest that endogenous insulin plays a physiological role in early embryo growth and differentiation even before emergence of the endocrine pancreas.

Introduction

Development of higher organisms involves differentiation and growth which take place, preferentially, in prenatal life. Both processes are regulated by hormone-like messenger molecules, although the precise mechanism of control is largely unknown. The prevailing feeling among developmental biologists seems to be that the influence of traditional hormones on embryonic development starts late (Tata, 1984; Ham and Veomett, 1980) in part because the usual focus of hormone production is the endocrine gland and these glands first appear at relatively advanced states of differentiation.

In recent studies we found that materials closely related to vertebrate hormones are native to microbes (Le Roith et al., 1981; Le Roith, Shiloach and Roth, 1982), i.e. these messengers are evolutionarily ancient and antedate both the endocrine glands and the nervous system. Therefore, we anticipated that the same classic hormones used for intercellular communication in older vertebrates, e.g. insulin, might possibly arise in early embryos even before a circulatory system and an endocrine system are established.

In an attempt to begin to define a possible role for insulin in early embryos we have shown that (1) material closely related to insulin is present in embryos before the emergence of the endocrine pancreas (De Pablo et al., 1982), and (2) receptors capable of mediating effects of insulin are present (Hendricks, De Pablo and Roth, 1984). These

*Present address: Laboratory of Cellular and Developmental Biology, NIADDK, National Institutes of Health and Endocrinology Division, Department of Medicine, Georgetown University, Washington DC, U.S.A.

aspects are the major subjects of this paper. In addition, we have shown that modest doses of insulin accelerate morphological and biochemical development (De Pablo et al., 1985a). Further, high doses of insulin produce teratogenesis, and this effect is receptor mediated (De Pablo et al., 1985b). More recently we have shown that neutralization of endogenous insulin in vivo increases mortality, and, in the survivors, retards growth and differentiation (De Pablo et al., 1985a).

Broad scope of insulin effects

The broad spectrum of insulin effects makes insulin a strong candidate to play a possible role in early embryogenesis (Fig. 1). The metabolic action of insulin regulating carbohydrate and lipids utilization and storage in adult organisms and late fetuses is well characterized (Cahill, 1971; Angervall, Karlsoon and Martinsson, 1981). The growth stimulatory effects of insulin are quite clear, as reflected by increases in DNA, RNA and protein synthesis in a wide range of in vitro systems (King and Kahn, 1984), and by effects in the more controversial in vivo fetal models (Picon, 1967; Susa et al., 1984; Cooke and Nicoll, 1984).

In the last few years, in addition to its role as a metabolic regulator and growth stimulator, insulin has been suggested as a key hormone in tissue-specific differentiation. Addition of insulin or closely related substances induce morphological and biochemical differentiation in cultured embryo tissues from chicken (muscle and lens epithelium), (De la Haba, Cooper and Elting, 1966; Ridpath et al., 1984; Milstone and Piatagorsky, 1977), rat (muscle, retina neurons and somatotrops) (Ewton and Florini, 1981; Puro and Agardh, 1984; Hemming et al., 1984), mouse (mammary gland) (Bolander et al., 1981), and amphibians (oocytes) (El-Etr, Schorderet-Slatkine and Baulieu, 1979; Maller and Koontz, 1981).

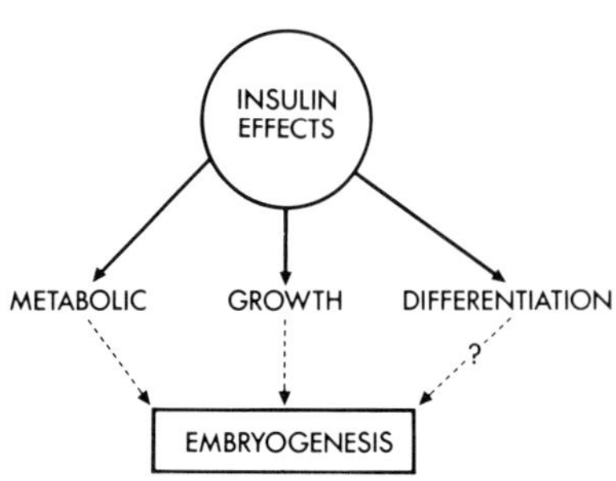

Figure 1. Broad scope of insulin effects.

ACID ETHANOL EXTRACTS FILTERED ON G-50 SEPHADEX

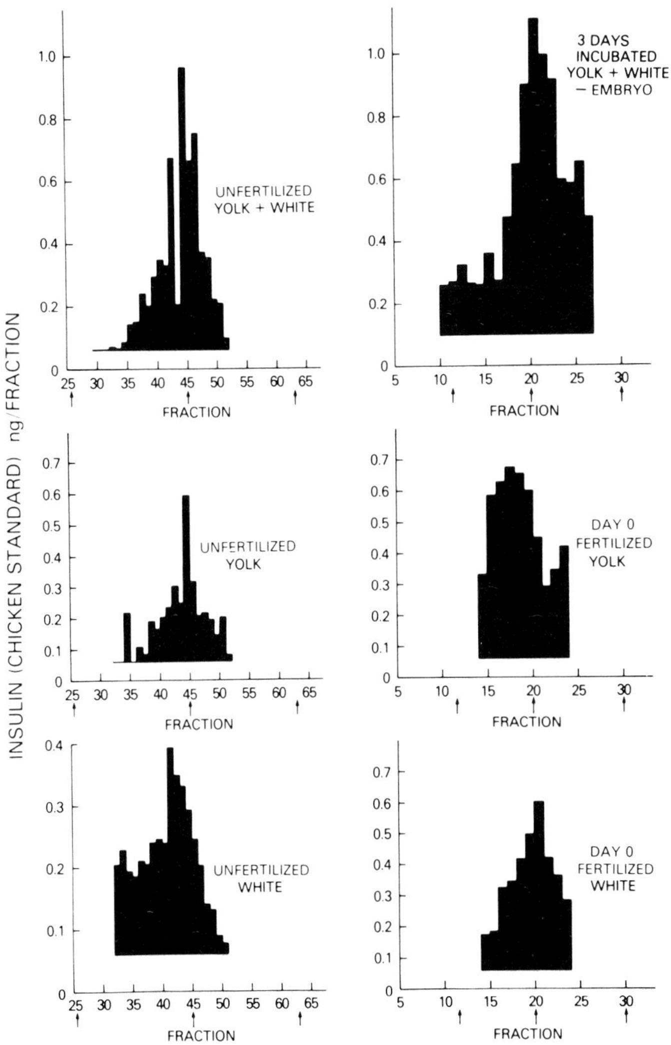

Figure 2. Insulin in egg yolk and white. Acid-ethanol extracts of yolk, white or a mixture were filtered on Sephadex G-50 columns (1.5 × 80 cm left panel samples and 0.9 × 50 cm right panel samples) at 4 C and 2 ml or 1 ml fractions were collected. Each effluent fraction was lyophilized and reconstituted in water from which an aliquot was tested in the radioimmunoassay. The arrows indicate the void peak (left), chicken insulin (middle) and salt peak (right). The horizontal baseline represents the lower limit of detection. The fractions corresponding to the insulin peak were pooled, lyophilized, reconstituted in water, retested in the RIA and assayed for bioactivity.

Insulin in eggs and chick embryos at pre-pancreatic stages

We used a standard acid-ethanol procedure to extract insulin from egg yolk and white, and embryos at different ages. When extracts were partially purified by gel filtration on G-50 Sephadex insulin immunoreactivity eluted in the region of standard insulin (Fig. 2). The specificity of the radioimmunoassay for an avian insulin was assessed by using three different antibodies against porcine insulin or chicken insulin (anti-chicken insulin antibody was a kind gift of Dr. John P. McMurtry, USDA-ARS, Beltsville, Md.). As expected, the material reacted more strongly in the specific chicken insulin immunoassay.

Embryos from days 2 to 8 of development were homogenized and the extracts chromatographed. The content of immunoreactive insulin recovered in the insulin peak, compared to the yolk and white insulin content, revealed an interesting pattern (Fig. 3). With increasing age

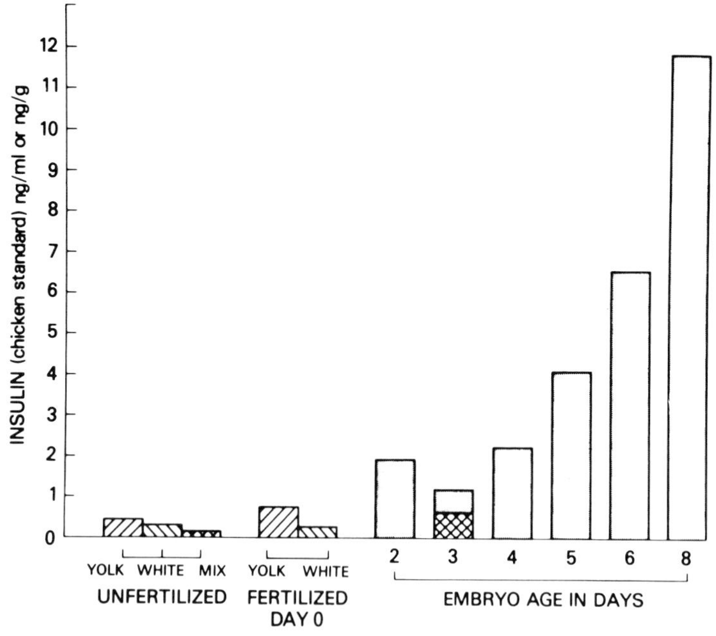

Figure 3. Insulin content of eggs and embryos at different ages. A number of embryos (n = 13 to n = 67 depending on the age) were pooled to prepare the extract. The insulin content is expressed as ng of chicken insulin per g wet weight. For comparison, the insulin content of the yolk and white are shown for both unfertilized and fertilized non-incubated eggs (day 0), expressed as ng/ml. The cross-hatched area wtihin the day 3 embryo bar shows the concentration of insulin in the yolk plus white minus that in the embryo (sample in upper right panel in Fig. 2). Although in concentration the insulin content of yolk plus white appears lower than that of day 2 embryos, the total volume of the egg (50 ml) stores a large amount of insulin (modified from De Pablo et al., 1982).

(and size) of the embryos the content of insulin per g wet weight did not change over days 2 to 4, while there was a clear increase by days 5 to 8. This last period represents the major period of pancreatic islet cell development. All embryo and egg extracts purified by gel filtration stimulated lipogenesis in isolated rat adipocytes. The relative activity in the bioassay when compared to the immunoreactive potency was typical of avian insulin.

We cannot be certain yet of the actual site of insulin synthesis in early embryos. We would suspect that the insulin gene is expressed in tissues other than the differentiated endocrine pancreas at primitive stages of development, but other possibilities can not yet be ruled out. Insulin could be synthesized by the embryonic tissues using mRNA transferred from the mother hen via the egg, or insulin present in embryos before day 4 could have been taken up from the egg yolk (which has low concentration of insulin but in its large volume stores a significant total amount). In situ hybridization studies may in the future help us to solve this question. The relevant conclusion is that the presence of insulin in both embryo and egg makes possible an influence of insulin in early developmental processes.

Receptors for insulin in early embryos

For insulin to act biologically there must be cell receptors to recognize, bind, and transmit the hormonal signal inside the cell, eventually leading to a wide range of biological effects (Roth et al., 1975). It is now recognized that the biological function of insulin is to activate a receptor which contains the full program to activate characteristic intracellular pathways in insulin target tissues (Schechter, 1983).

Initially we studied tissues of embryos day 8 to day 18 (Hendricks et al., 1984) when there is well established pancreatic insulin secretion (Benzo and Green, 1974; Swenne and Lundqvist, 1980) and the hormone has typical metabolic effects (Leibson, Bondareva and Soltitskaya, 1976). Liver and brain of these embryos had receptors with temperature, pH and ligand specificity of insulin receptors (Fig. 4). The relative affinity of different insulins and insulin analogs for binding to the insulin receptor parallels the biological potency of these peptides when they act through the insulin receptor.

An overview of insulin binding to chick tissues throughout ontogeny is shown in Fig. 5. Binding to liver and brain membranes increased with development between days 8 to 18 but the patterns of increase were distinct in the two tissues. Binding to brain preparations was higher than to liver at the end of embryogenesis. The differences in insulin

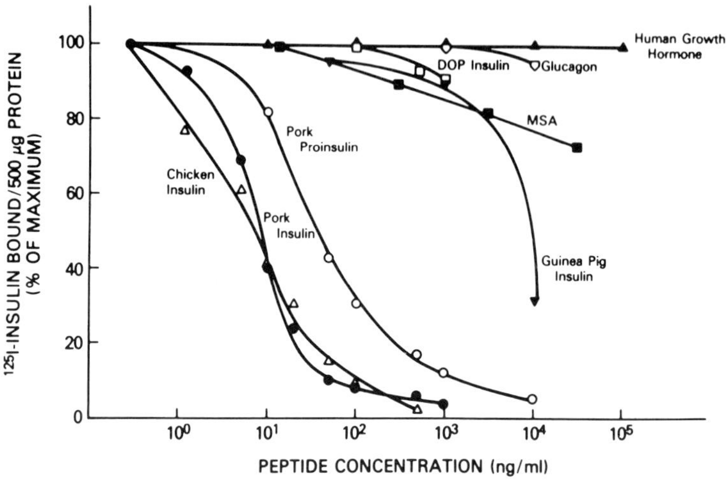

Figure 4. Analog specificity of insulin receptors in 18-day-old chick embryo brain. Membranes were incubated at a protein concentration of 500 µg/ml with 125-I-insulin (0.3 ng/ml) and concentrations of polypeptide hormones ranging from 0–105 ng/ml, for 90 min at 15 C. Specific binding of labeled porcine insulin to triplicate samples is expressed as a percentage of the specific binding in the absence of unlabeled hormone. Maximum specific binding (100%) represented 4.7% of the total radioactivity added. MSA is multiplication-stimulating activity, the rat equivalent of insulin-like growth factor II (modified from Hendricks et al., 1984).

binding in liver and brain during mid to late development suggest tissue-specific mechanisms of regulation and may reflect insulin's diverse role in different tissues.

Pancreatic insulin secretion appears to start in chick embryos between days 4 and 5 (Przybylski, 1967; Dieterlen-Lievre and Beaupain, 1976). However, we found insulin in younger embryos, egg yolk, and white, and looked for insulin receptors at these early stages. Specific insulin binding was present but low at day 2 (whole embryo) and more clearly demonstrable on membranes from the day 3 embryos. By day 4 of development insulin binding was present in the head and body preparations, indicating that receptors for insulin are widely distributed even at this early age.

Subsequently, we examined more extensively the receptors of brain membranes during the first week of chick embryogenesis using 125-I-insulin, 125-I-IGF I and 125-I-IGF II (Bassas et al., 1984). Based on the relative binding of labeled ligands and the competition with unlabeled peptides, we concluded that the dominant receptor at early stages is an IGF-type 1 receptor which binds IGF-I $>$ IGF-II $>$ insulin. We raise the possibility that, at least in some embryo tissues, insulin action may be mediated via an IGF-type 1 receptor. This has been suggested

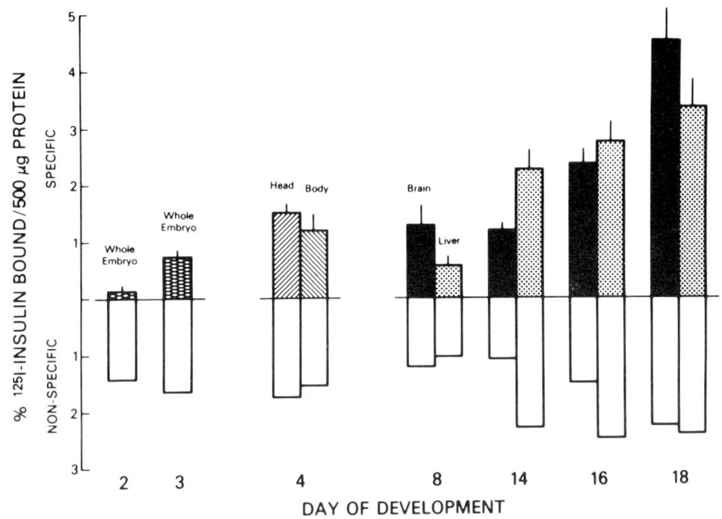

Figure 5. Insulin binding to chick embryo membranes during ontogeny. Membranes prepared from chick embryo on designated days of embryogenesis were incubated with 125-I-insulin in the presence and absence of excess unlabeled insulin. Non-specific binding (empty bars) in the presence of 10 μg/ml insulin was subtracted from total binding to give specific binding (shaded bars). On days 2 and 3 of development membranes were prepared from whole embryos (modified from Hendricks et al., 1984).

in many in vitro studies when insulin was shown to stimulate growth and differentiation (King and Kahn 1984; Maller and Koontz, 1981).

Effects of insulin on embryo growth

The predominant developmental events in chick embryo between days 2 and 8 of embryogenesis are tissue differentiation and organogenesis (Fig. 6). To test our hypothesis that insulin might have a role in the early stages of development, we studied embryos after 2 days of development, i.e. stages 11–14 of Hamburger and Hamilton (1951) classification. The appearance of extraembryonic blood vessels and the changes in head, limb buds and tail are the main criteria to assess morphological development up to day 4 (Fig. 7). From the biochemical point of view we have looked at individual creatine-kinase isozymes during development, as a handle to evaluate biochemical maturation of the embryos (Fig. 8). By using DEAE chromatography, the three isozymes BB, MB and MM can be separated (Yasmineh and Hanson, 1975) and the transitional MB form can be used as a sensitive marker of differentiation between day 2 and day 4 of embryo

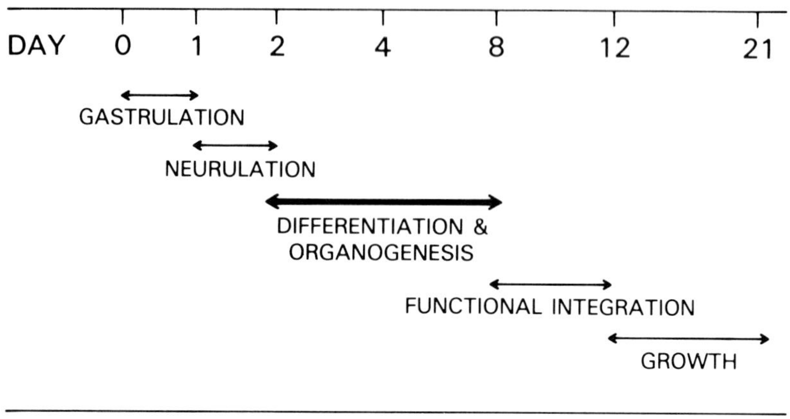

Figure 6. Predominant developmental events in chick embryogenesis.

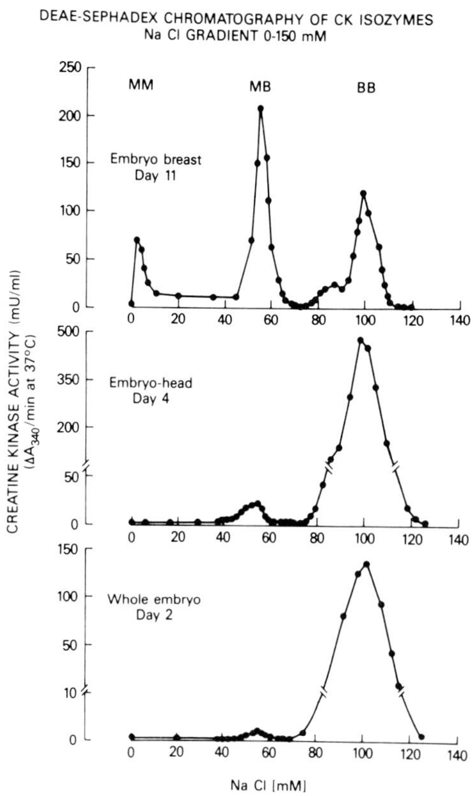

Figure 7. Stage 14 of chick embryogenesis by Hamburger and Hamilton (1952) classification. Approximately 50 hours incubation. The embryo's head has turned, there is distinct enlargement of telencephalon and the primitive heart is S-shaped.

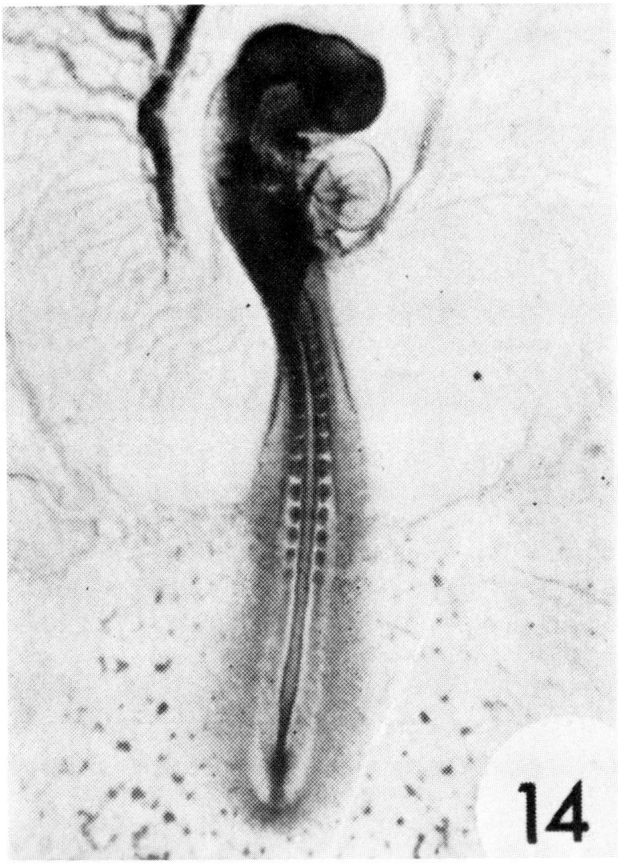

Figure 8. DEAE – Sephadex chromatography of creatine-kinase (CK) isozymes. Fractions were eluted by continuous NaCl gradient (0–150 mM) from a DEAE Sepharose CL-6B column (buffer Tris-HCl 50 mM, pH 7.5). Creatine-kinase activity was measured in each fraction by spectrophotometry at 340 nm, after addition of CK-NAC activated reagent (Boehringer Manheim) and 0.12 nM diadenosine pentaphosphate (myokinase inhibitor). Creatine-kinase catalyzes the reversible reaction ATP + creatine → ADP + p-creatine. Although specially characteristic of muscle, it also occurs in other tissues like heart and brain, where is involved in maintaining proper intracellular ATP/ADP ratios. The homodimer BB appears first in development and, therefore, is called embryonic isozyme. The heterodimer MB form appears later (transitional isozyme) and, finally, the homodimer MM will appear in tissues with terminal muscle differentiation.

development. When we added small doses of insulin (ng amounts/embryo) at day 2 (Table 1) the embryos at day 4 had abnormally advanced development as indicated by an increase in weight, total protein, total creatine-kinase and, more important, the isozyme creatine-kinase-MB (De Pablo et al., 1985a). (Survival rates among the insulin treated embryos were as high as in control injected groups.)

Table 1. Effects of adding and subtracting insulin from early (day 2) chick embryo

	Exogenous insulin added	Endogenous insulin neutralized by anti-insulin antibodies
Death rate	Unchanged	Increased
Morphological stage	Accelerated	Retarded
Weight	Increased	Decreased
Total creatine-kinase	Increased	Decreased
Isozyme creatine-kinase MB	Increased	Decreased

It remains to be elucidated through which receptor this effect of insulin is mediated.

We have been able to demonstrate that insulin does have receptor-mediated effects when it is applied to the embryos in large doses (μg amounts/embryo). It was known that insulin in very high doses caused malformations and death in early chick embryos (Landauer, 1972). By using different analogs of insulin (proinsulin, desoctapeptide-insulin) and insulin-like growth factor II we have shown (De Pablo et al., 1985b) that the teratogenic effect of insulin is specific, probably mediated through an IGF-type receptor and not directly caused by hypoglycemia. In the survivors by day 4, after treatment at day 2, there was a decrease in DNA, RNA and protein content whether glucose was added together with insulin or not (Fig. 9).

The addition of exogenous hormone to the developing embryo, even if given in low doses, might represent only a pharmacological effect. To prove that early embryos need their endogenous insulin physiologically we administered specific insulin antibodies and, therefore, "removed" immunologically their own insulin. Embryos treated with anti-insulin antibodies at day 2 had morphological growth retardation (more evident at day 3) and an increased death rate by days 4 and 5 of development (Table 1). In the survivors the weight, total protein and total creatine-kinase content were decreased and, more remarkable, the creatine-kinase isozyme MB was much lower than corresponded to their age.

Taken together, these experiments suggest to us that during the period of early differentiation, the chick embryo has insulin, receptors for insulin and insulin-like growth factors and biological response pathways for insulin. Further, the endogenous insulin is required for the embryo to grow normally. Insulin, the classic pancreatic hormone that regulates carbohydrate and lipid metabolism, possibly has also a growth or differentiation role in embryogenesis. Agricultural researchers

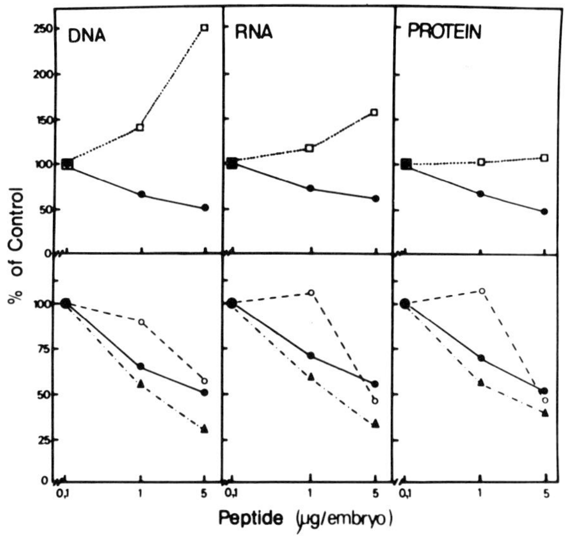

Figure 9. DNA, RNA and protein content in surviving embryos at 4 days of development. Treatment at day 2 was insulin (●——●), insulin + 10 mg glucose (▲-·-·-▲), proinsulin (○----○) or rat IGF-II (□·····□). In each case the value is expressed as a percentage of the content in embryos treated with the low dose (0.1 µg/embryo of the respective peptide). The inhibitory effects of high doses of insulin in protein synthesis and ultimately cell growth are opposite to the well known stimulatory effects at lower concentrations. A hypothetical mechanism by which insulin, proinsulin and DOP in vast doses could reduce macromolecules synthesis could be the competitive inhibition of binding of endogenous growth factors in the embryo.

may eventually use this physiological action of insulin on development to improve commercial species.

References

Angervall, L., K. Karlsoon and A. Martinsson. 1981. *Effects on the rat fetuses of intrauterine injections of insulin.* Diabetologia 20:558–562.

Bassas, L. L., F. De Pablo, M. A. Lesniak and J. Roth. 1984. *Receptors that prefer insulin-like growth factors antedate the insulin receptor during organogenesis of the chick.* J. Steroid Biochem. 20:1423.

Benzo, Z. A. and T. D. Green. 1974. *Functional differentiation of the chick endocrine pancreas: insulin storage and secretion.* Ant. Rec. 180:491–496.

Bolander, F. F., K. R. Nicholas, J. J. Van Wyk and Y. J. Topper. 1981. *Insulin is essential for accumulation of casein mRNA in mouse mammary epithelial cells.* Proc. Nat. Acad. Sci. USA 78:5682–5684.

Cahill, G. F. 1971. *Physiology of insulin in man.* Diabetes 20:785–787.

Cooke, P. S. and C. S. Nicoll. 1984. *Role of insulin in the growth of fetal rat tissues.* Endocrinology 114:638–643.

De la Haba, G., G. W. Cooper and V. Elting. 1966. *Hormonal requirements for myogenesis of striated muscle in vitro: insulin and somatotropin.* Proc. Nat. Acad. Sci. USA 56:1719–1723.

De Pablo, F., J. Roth, E. Hernandez and R. M. Pruss. 1982. *Insulin is present in chicken eggs and early chick embryos.* Endocrinology 111:1909–1916.

De Pablo, F., M. Girbau, J. A. Gomez, E. Hernandez and J. Roth. 1985a. *Does endogenous insulin regulate growth and differentiation of early embryos*? Endocrinology 116:361.

De Pablo, F., E. Hernandez, F. Collia and J. A. Gomez, 1985b. *Untoward effects of pharmacological doses of insulin in early chick embryos: through which receptors are they mediated.* Diabetologia (in press).

Dieterlen-Lievre, F. and D. Beaupain. 1976. *Immunocytological study of endocrine pancreas ontogeny in the chick embryo: normal development and pancreatic potentialities in the early splachnopleure.* P. 37 in T. Adenasanya, I. Grillo, L. Leibson and A. Epple, eds., *The Evolution of Pancreatic Islets.* Pergamon Press, Oxford.

El-Etr, M., S. Schorderet-Slatkine and E. E. Baulieu. 1979. *Meiotic maturation in "Xenopus Laevis" oocytes initiated by insulin.* Science 205:1397–1399.

Ewton, D. C. and J. R. Florini. 1981. *Effects of the somatomedins and insulin on myoblasts differentiation in vitro.* Develop. Biol. 86:31–39.

Ham, R. G. and M. J. Veomett. P. 428 in *Mechanisms of Development.* The CV Mosby Company, St. Louis.

Hamburger, V. and H. L. Hamilton. 1951. *A series of normal stages in the development of the chick embryo.* J. Morphol. 88:49–92.

Hemming, F. J., M. Begeot and M. P. Dubois. 1984. *Fetal rat somatotropes in vitro: effects of insulin, cortisol, and growth hormone-releasing factor on their differentiation. A light and electron microscopic study.* Endocrinology 114:2107–2113.

Hendricks, S. A., F. De Pablo and J. Roth. 1984. *Early development and tissue-specific patterns of insulin binding in chick embryo.* Endocrinology 115:1315–1323.

King, G. L. and C. R. Kahn. *The growth-promoting effects of insulin.* Pp. 223–265 in G. Guroff, ed., *Growth and Maturation Factors,* Vol. 2. John Wiley and Sons. Inc., New York.

Landauer, W. 1972. *Is insulin a teratogen*? Teratology 5:129–135.

Leibson, L., V. Bondareva and L. Soltitskaya. 1976. *The secretion and the role of insulin in chick embryos and chickens.* P. 69 in *The Evolution of Pancreatic Islets.* Pergamon Press, Oxford.

Le Roith, D., J. Shiloach, J. Roth and M. A. Lesniak. 1981. *Insulin or a closely related molecule is native to Escherichia coli.* J. Biol. Chem. 256:6533–6536.

Le Roith, D., J. Shiloach and J. Roth. 1982. *Is there an earlier phylogenetic precursor that is common to both the nervous system and the endocrine systems*? Peptides 3:211–215.

Maller, J. L. and J. W. Koontz. 1981. *A study of the induction of cell division in amphibian oocytes by insulin.* Develop. Biol. 85:309–316.

Milstone, L. M. and J. Piatagorsky. 1977. *Delta-cristallin gene expression in embryonic chick lens epithelia cultured in the presence of insulin.* Expl. Cell Res. 105:9–14.

Picon, L. 1967. *Effect of insulin on growth and biochemical composition of the rat fetus.* Endocrinology 81:1419–1421.

Przybylsky, R. J. 1967. *Cytodifferentiation of the chick pancreas. I. Ultrastructure of the islet cells and the initiation of granule formation.* Gen. Comp. Endocrinol. 8:115–128.

Puro, D. G. and E. Agardh. 1984. *Insulin mediated regulation of neuronal maturation.* Science 225:1170–1172.

Ridpath, J. F., T. W. Huiatt, A. H. Trenkle, R. M. Robson and P. J. Bechtel. 1984. *Growth and differentiation of chicken embryo muscle cell cultures derived from fast- and slow-growing lines.* Differentiation 26:121–126.

Roth, J., C. R. Kahn, M. A. Lesniak, P. Gorden. P. De Meyts, K. Megyesi, D. M. Neville, J. R. Gavin III, A. M. Soll, P. Freychet, I. D. Goldfine, R. S. Bar and J. A. Archer. *Receptors for insulin, NSILA-s and growth hormone: application to disease states in man.* Rec. Prog. Horm. Res. 31:95–139.

Schechter, Y. 1983. *Bound lectins that mimic insulin produce persistent insulin-like activities.* Endocrinology 113:1921–1926.

Susa, J. B., C. Neave, P. Sehgal, D. B. Singer, W. P. Zeller and R. Schwartz. 1984. *Chronic hyperinsulinemia in the fetal rhesus monkey: effects of physiologic hyperinsulinemia on fetal growth and composition.* Diabetes 33:656–660.

Swenne, I., and G. Lundquist. 1980. *Islets structure and pancreatic hormone content of the developing chick embryo.* Gen. Comp. Endocrinol. 41:190.

Tata, J. R. 1984. *The action of growth and development hormones. Evolutionary aspects.* Pp. 1–58 in R. F. Goldberger and K. R. Yamamoto, eds., *Biological Regulation and Development*, Vol. 38. Plenum Press, New York.

Yasmineh, W. G. and N. Q. Hanson. 1975. *Electrophoresis on cellulose acetate and chromatography on DEAE-Sephadex A-50 compared in the estimation of creatine-kinase isoenzymes.* Clin. Chem. 21:381–386.

Agricultural opportunities from nitrogen fixation research

WINSTON J. BRILL
Agracetus, 8520 University Green, Middleton, Wisconsin 53562, U.S.A.

Abstract Most applied nitrogen fixation research currently conducted focuses on the important legume-*Rhizobium* symbiosis. Progress is being made towards increasing the level of nitrogenase activity in the nodule, increasing the flow of carbon to the nodule, decreasing the waste of electrons through hydrogen evolution and increasing the competitiveness of useful *Rhizobium* strains. Because of the intimate association between *Rhizobium* and the root, *Rhizobium* strains with certain foreign genes may be useful to protect the plant from certain pests. Opportunities from research with other nitrogen-fixing bacteria include several approaches aimed towards obtaining cereal plants with the capability to fix nitrogen. Direct genetic engineering of *nif* genes as well as optimizing bacteria-root associations are being examined.

Introduction

The level of research in all aspects of nitrogen (N_2) fixation has been steadily increasing over the past fifteen years. Much of the funding for this work has been driven by the hope that results will eventually be applied to benefit agriculture. The most common benefits mentioned are: (1) increased legume yields, (2) N_2-fixing cereals, and (3) more efficient catalysts for industrial N_2 fixation. If any one of these applications will be realized, the benefit to developed as well as developing countries could be enormous. Granting agencies and researchers certainly have assumed that some of the basic N_2 fixation research will impact areas other than fertilizer production and use. Historically, unpredicted benefits arise from basic research. This article will focus on aspects of the above three predicted benefits and will also discuss an unexpected potential benefit from N_2 fixation research–specific termiticides.

Rhizobium-legume symbiosis

While there is a tremendous amount of biochemical, genetic and molecular data on this symbiosis, applied approaches have focused on hydrogen (H_2)-uptake hydrogenase and on overcoming competition in the soil by indigenous *Rhizobium* strains.

Hydrogen evolution and uptake hydrogenase. All N_2-fixing bacteria have nitrogenases that, besides reducing N_2 to ammonium, also reduce protons to H_2. Nitrogenase-dependent H_2 evolution occurs at the expense of ATP. Approximately half of the electrons that reach nitrogenase are lost as H_2. Thus H_2 evolution seems to be a wasteful reaction for a N_2-fixing bacterium (Simpson and Burris, 1984). Selection for mutants that more efficiently fix N_2 through loss of H_2-evolving capacity has been unsuccessful (Brill, unpublished observations). It seems likely, therefore, that H_2 production is a consequence of the way the complex active site(s) of nitrogenase has evolved.

Some bacteria are able to recover a fraction of the electrons and ATP lost through nitrogenase-dependent H_2 evolution. This recovery occurs through a H_2-uptake hydrogenase, by which H_2 is oxidized and the electrons can either be used directly by nitrogenase or be used to synthesize ATP through oxidative phosphorylation (Dixon, 1978; Emerich et al., 1979). Thus, a *Rhizobium* strain lacking uptake hydrogenase (Hup^-) will evolve H_2. A strain with active uptake hydrogenase (Hup^+) will not evolve H_2. Because of the recycling of electrons and ATP, it is possible that strains with uptake hydrogenase activity will fix N_2 most efficiently.

No *R. meliloti* (nodulates alfalfa) strains are Hup^+ and only one *R. trifolii* (nodulates clover) has been shown to be Hup^+ (Tilak, Scheider and Schlegel, 1984). Of the commercial *R. japonicum* (nodulates soybean) strains less than 25% are Hup^+ (Keyser, Weber and Uratsu, 1984). Therefore, there has been incentive to convert Hup^- strains to Hup^+. The saving of energy by Hup^+ strains was expected to allow *Rhizobium* in the nodules to fix more N_2 and, thereby, increase crop yield. If the *hup* genes can all be cloned, then it should be possible to mobilize them to a Hup^- strain that is desirable for other characteristics (e.g. ease of growth, competitiveness).

So far, no laboratory has yet cloned all of the DNA required for the Hup^+ phenotype, although several genes have been shown to be clustered together in *R. japonicum* (Haugland et al., 1984; Hom, Graham and Maier, 1985) and all *hup* genes may naturally be on a single plasmid (Brewin et al., 1980) in *R. leguminosarum* (nodulates pea). Uptake hydrogenase is a nickel-containing, membrane-bound protein containing two different polypeptides (reviewed in Evans et al., 1985). There probably are at least five different polypeptides essential for uptake hydrogenase activity (Merberg and Maier, 1983), several of them probably are required for electron transport (Maier, 1981). It is not yet known whether all the *hup* genes are clustered together.

Even if all of the *hup* genes are introduced to a Hup^- strain, it is not clear what level of increased N_2 fixation or yield will result. Experiments relevant to this question need to utilize Hup^+ strains and Hup^- point mutants derived from those strains. Small-scale experiments (Evans et al., 1985) showed that the uptake hydrogenase activity increased N content in leaves, seeds, and whole plant material. Larger-scale field experiments are necessary to properly demonstrate the value of *hup*, especially if yield data are included.

If uptake hydrogenase activity allows *Rhizobium* to fix N_2 more efficiently, then it is important to ask the question of why most strains are Hup^-. Perhaps the fact that *hup* is required for *Rhizobium* to grow autotrophically (utilizing H_2 for energy and CO_2 for carbon) may be the reason why certain strains contain uptake hydrogenase (Hanus, Maier and Evans, 1979).

Hup mutants may be selected by taking advantage of the autotrophic growth of Hup^+ *Rhizobium* (Maier, 1981). One class of *R. japonicum* mutants (Merberg and Maier, 1983) produces up to five times more uptake hydrogenase than the wild type. It will be interesting to compare N_2 and yield with such mutants.

Competition. Most commercial *R. japonicum* strains applied to soybean-growing areas do not enter soybean root nodules (Kamicker and Brill, unpublished observations). It seems that indigenous *R. japonicum* in the soil are more competitive and these are the strains that enter the nodules and fix N_2 for the plant. For this reason, molecular improvements of *Rhizobium*, such as introduction of *hup*, must take into account the competitive ability of the strain. If the competition problem can be overcome, this will be a major breakthrough for application of biotechnological improvements of *Rhizobium*.

How many different indigenous strains does an improved inoculant have to compete against? In a recent study (Kamicker and Brill, unpublished results), *R. japonicum* strains from soybean nodules in 19 southeastern Wisconsin farms were differentiated from one another on the basis of polyacrylamide gel electrophoresis of protein patterns (Roberts et al., 1980; Noel and Brill, 1980). The diversity of strains was greater than expected. Thirty-one different strains (representing 543 isolates) were identified. None of the strains represented strains in current commercial inocula. The most predominant strain was only 14% of all the isolates. Even with two farms that used the same soybean cultivar and inoculant and had the same soil type and farm location, there were significant differences among the types of strains isolated. Predominant strains did not correlate with soil type, cultivars used or culture practices during the growing season. These data point to the difficulty in attempts to design (or isolate) competitive strains.

Surface properties of *Rhizobium* probably play a major role in competitiveness. For instance, plant lectins bind to *Rhizobium* surfaces (see Dazzo, 1982). Surface properties also may effect competition factors such as protection from predators, resistance to dessication or salt accumulation, or microcolony formation. In the case of several strains of *R.meliloti*, there is a reversible high frequency (1 in 10^5) conversion between two different types of cell surfaces (Handelsman, Ugalde and Brill, 1984). One type (HA) is defined as those that are highly agglutinable by an alfalfa agglutinin (Paau, Leps and Brill, 1981). The other type (LA) is far less agglutinable. By growth chamber experiments, LA strains nodulate alfalfa earlier than HA strains. LA strains, when mixed with HA strains, outcompete the HA strains. This has also been confirmed in soil experiments (Handelsman and Brill, unpublished observations). The genetic mechanism of the HA-LA strain interconversion is not at all understood, but is reminiscent of phase conversion in *Salmonella typhimurium* (Silverman and Simon, 1980).

Some insight into the biochemical basis of HA-LA differences was gained from analysis of the cell surface polysaccharide (Ugalde, Handelsman and Brill, unpublished observations) which is enriched for galactose in LA strains. A membrane-bound galactosyl transferase was found only in LA strains and not in the HA strains (Ugalde and Brill, unpublished observations).

My laboratory had studied the *R. meliloti*-alfalfa symbiosis for almost a decade and routinely surface-sterilized alfalfa seed with such standard techniques as soaking them in sulfuric acid or Clorox. Seeds treated in this way do not produce bacterial colonies in nutrient media, and were thus judged to be sterile. However, when seedlings from surface-sterilized seeds were placed on rich medium, bacterial growth routinely appeared around the roots (Handelsman and Brill, 1985). Almost every seedling from surface-sterilized alfalfa seeds yielded bacteria. Alfalfa varieties from a wide range of sources all showed the same phenomenon. Most striking was the fact that the bacteria on the roots were the same species – *Erwinia herbicola*. Thus, it might be suspected that *E.herbicola* could play a role in the root-surface ecology of young seedlings and that *E. herbicola* may influence binding or infection by *Rhizobium*.

It was possible to eliminate *E. herbicola* by washing one-day old seedlings with a bacteriocide (Handelsman and Brill, 1985). When bacteriocide-treated alfalfa was inoculated with either HA or LA *R. meliloti* strains, a very different result was observed that when *E. herbicola* was naturally present. *Erwinia herbicola*-free plants nodulated at the rapid rate when either HA or LA strains were added.

The rate of nodulation was the same as that observed with LA-inoculated plants not treated with the bacteriocide. When a strain of *E. herbicola* isolated from an alfalfa root was added to bacteriocide-treated alfalfa and the plants were inoculated with LA or HA *R. meliloti* strains, the rate of nodulation was inhibited only in plants inoculated with HA, but not LA, strains. Thus, the original observation of delayed (and less competitive) nodulation by HA strains seems to be dependent on the 'natural' presence of *E. herbicola*.

A model to explain the results (Handelsman, Ugalde and Brill, 1984; Handelsman and Brill, 1985) assumes that LA strains contain the galactosyl polysaccharide that covers sites that bind the alfalfa agglutinin. When the agglutinin-binding sites are covered, nodulation can occur rapidly. When the galactosyl polysaccharide is missing, as in HA strains, the agglutinin-binding sites are available on the surface to inhibit the rate of nodulation. *Erwinia herbicola* also may cover the agglutinin-binding sites; consequently, *E. herbicola* does not effect LA rate of nodulation but does increase HA nodulation rate. Possibly, the high rate of genetic interconversion between HA and LA represents surface characteristics that are important to *R. meliloti* at different times in its life cycle. For example, LA for nodulation and HA for plant-independent growth. Preliminary work has indicated that LA strains also are more competitive in field situations. Perhaps studies such as these will finally yield insight into methods by which biotechnology can be applied to overcome the problems of competition by indigenous strains.

Catalysis

Even though the active center of nitrogenase has been isolated (Shah and Brill, 1977) and characterized by spectroscopic studies (reviewed by Shah et al., 1984), very little is known about its structure. This center, iron-molybdenum cofactor, can catalyze acetylene reduction to ethylene (Shah, Chisnell and Brill, 1978). However, it does not reduce N_2. So far, work on nitrogenase has not led to the development of new approaches for the design of more efficient catalysts for industrial N_2 fixation. One idea that was pursued and has now been dropped, was the possibility of utilizing derepressed N_2-fixing bacteria to produce ammonium (Gordon and Brill, 1972) in fermention tanks, ponds or in columns with immobilized cells.

Nitrogen-fixing cereals

Transformation of plants with nif. When Dixon and Postgate (1972) transposed the cluster of nitrogen-fixation (*nif*) genes from *Klebsiella pneumoniae* to *Escherichia coli* by plasmid transfer it seemed reasonable to believe that *nif* could be introduced into the plant genome. Now, with the rapid progress being made with plant genetic engineering technology, this achievement will shortly be realized. However, mere incorporation of the 15 *nif* genes, arranged in 7 operons (MacNeil et al., 1978), into the plant is not expected to be sufficient to obtain N_2-fixing plants. Since *nif* is expressed in prokaryotes, the transcription-translation mechanisms in higher organisms are not expected to correctly read the regulatory signals in *nif* genes incorporated in a plant chromosome. Plant plastids, on the other hand, are much more similar to bacteria with regard to transcription-translation mechanisms. For instance, chloroplast genes can be expressed in *E. coli* (Gatenby, Castleton and Saul, 1981). With the use of a transit peptide, it has been possible to genetically engineer a bacterial protein into plant chloroplasts (den Broeck et al., 1985).

In *K. pneumoniae*, the *nif* genes required to synthesize a functional nitrogenase require the products of *nifA* (already in the *nif* cluster of genes) and *ntrA*, which is not in the *nif* cluster (see discussion by Merrick and Dixon, 1984). If the chloroplast does not contain the equivalent of *ntrA*, then *ntrA* may have to be introduced to the chloroplast. However if chloroplast RNA polymerase does not interact properly with the *ntrA* or *nifA* genes, then it may be necessary to express the *nif* genes by removing each operon's natural promoters and replacing them with chloroplast promoters.

Even if all of the *nif* genes are expressed in a ceral chloroplast, there are a multitude of problems that have yet to be overcome. For instance, oxygen (O_2) production by chloroplasts will irreversibly inactivate nitrogenase, which is extremely labile to O_2. One way around the O_2-nitrogenase incompatibility is expression of nitrogenase in plastids that do not produce O_2 – perhaps those in the roots. Promoters need to be designed to express *nif*, therefore, only in roots. A great deal of research is now aimed at isolating tissue-specific promoters.

Two genes in the *nif* cluster produce proteins that transfer electrons from pyruvate to nitrogenase (Roberts et al., 1978; Nieva-Gomez et al., 1980; Shah, Stacey and Brill, 1983). Pyruvate produced in plastids may then be the electron source for N_2 fixation. Reduced ferredoxin also may reduce nitrogenase itself (Yoch et al., 1969). Even if N_2 is fixed in the plant, a big question remaining is whether or not electrons diverted to nitrogenase will be at the expense of important metabolic reactions required for normal growth and yield. The same question is relevant to the high ATP requirement for N_2 fixation.

Nitrogenase contains high iron, sulfur and molybdenum (Eady and

Postgate, 1974). In *K. pneumoniae*, there is an efficient molybdenum (Mo)-uptake system (Imperial, Ugalde and Brill, unpublished results) and it is questionable whether or not a plastid will be able to accumulate sufficient Mo to support N_2 fixation. Even though nitrate reductase in plastids is a Mo-containing enzyme, far less nitrate reductase (and, therefore, Mo) than nitrogenase is required for a unit of ammonium formed from substrate. N_2-fixing plastids would require significantly more iron and sulfur. Will these elements be available to the plastid in sufficient quantities?

It seems, therefore, that there are strategies to follow for designing N_2-fixing cereal plants. The barriers include more than introducing the 15 *nif* genes into cereals. Regulation of protein synthesis, protection from O_2, sequestration of iron and molybdenum, and nitrogenase activity are all major barriers that have yet to be overcome.

Nitrogen-fixing associations. There has been a tremendous effort to utilize N_2-fixing bacteria on cereal roots so that the plant will obtain some of the fixed N. Unlike the *Rhizobium*-legume symbiosis, these associations are less intimate. The N_2-fixing bacteria in the association are not protected from the environment as are *Rhizobium*. In spite of several reports indicating that cereals inoculated with N_2-fixing bacteria such as *Azotobacter*, *Azospirillum*, or *Bacillus* can obtain significant amounts of N through N_2 fixation, none of these experiments have been routinely confirmed (van Berkum and Bohlool, 1980). There are a number of reasons to expect why these types of experiments have not succeeded. For instance, the inoculant bacterium would have to compete with many other bacteria in the soil for root-surface sites. In order for a bound bacterium to fix N_2, it needs an energy source – presumably carbon compounds liberated by the plant to the root surface. Through genetic alteration of both the plant and bacterium, it may someday be possible to design a useful association.

It has been possible to introduce legume-root binding genes from a *Rhizobium* to *Azotobacter* (Bishop et al., 1977; Maier, Bishop and Brill, 1978), thereby allowing *Azotobacter* to bind to legume roots. When we understand more about the surface structure of cereal roots, it should be possible to genetically engineer N_2-fixing bacteria to specifically bind to the cereals. This should give the associated bacteria an advantage in utilizing carbon compounds produced by the plant. In this way a high population of the N_2-fixing bacteria should be maintained on the roots. Through mutation or genetic engineering, the bacteria can be modified to utilize root material more efficiently. This type of association depends on the N_2-fixing bacteria having sufficient root-produced carbon compounds to generate the required energy for N_2 fixation. One would not expect that a cereal bred for high yield under fertilized conditions would retain a high level of metabolite-excreting activity. When corn lines were assayed for the ability to support N_2 fixation by *Azotobacter*, no modern lines exhibited activity; but, several lines of tropical origin did support N_2 fixation (Ela, Anderson and Brill, 1982). These latter lines were bred to increase N_2 fixation even further and this increased fixation was then bred into midwestern U.S. lines. Through further breeding, it should be possible to increase associative N_2 fixation to levels that could have a significant impact on fertilizer use. So far, this type of association has only been tested under growth chamber conditions and needs to be confirmed by field tests once levels of N_2 fixation have been determined to be sufficiently high.

Specific termiticides

Termites cause major damage, worldwide. Besides destroying lumber, they interfere with agricultural practices, especially in tropical countries. Organochlorinated compounds and formulations of copper–chrome–arsenic are commonly used to kill termites. These materials are quite toxic to other life forms and this has created a need to develop termiticides specific for termites.

Termites fed filter paper grow quite well. Their intestines carry N_2-fixing bacteria that reduce acetylene (Breznak et al., 1973); thus, it seems that termites may be able to obtain a significant proportion of their N through N_2 fixation (Bentley, 1984). Other insects tested exhibited no acetylene-reduction activity. If N_2 fixation is important to the dietary needs of termites, then Mo must be required since it is an integral part of the active site of nitrogenase. Tungstate is known to compete with molybdate (Shah et al., 1984). Thus, tungstate might be able to prevent N_2 fixation in termites by preventing utilization of molybdate in nitrogenase.

When termites were fed paper dosed with sodium tungstate (1000 ppm of the metal) in the termite's food, no killing was observed until day 12. After 38 days, 50% of the termites were dead (Brill and Von Meyer, 1985). Salts such as sodium sulfate, zinc sulfate, sodium chloride, ferrous sulfate, ferric chloride, cobalt chloride or sodium stannate had no killing effect.

The interesting and unpredicted result was that sodium molybdate also was an effective termiticide. In fact, it was more effective than sodium tungstate (Brill and Von Meyer, 1985). Again, no killing was apparent until day 12; however, the rate of killing was greater than by tungstate. After 28 days of molybdate feeding, 50% of the termites died. All termites were dead 42 days after they began feeding on molybdate. Other insects such as ants, flies and cockroaches were not killed when similar concentrations of tungstate or molybdate were added to their diets. Other soluble forms of Mo that were tested also killed termites. Several different termite genera were all susceptible to tungstate and molybdate.

The mechanism(s) of these killing effects is not yet understood. As mentioned above, Mo compounds are incorporated into N_2-fixing bacteria through very efficient uptake systems. The termite is not normally expected to feed on a diet with high Mo concentrations. Perhaps the high Mo in the food is rapidly concentrated in the termite intestine through action of N_2-fixing bacteria. This Mo may react in the strongly reducing environment of the gut to form toxic materials or insoluble products that eventually kill the termite.

Unlike termiticides in current use, the Mo and tungsten (W) materials do not immediately kill the insect. This may be a major advantage for the use of termite baits. Termites recognize a toxic environment, perhaps by observing dead bodies, and they learn to avoid the toxic site (Su et al., 1982). Thus, poisons that kill rapidly are not useful as baits. On the other hand, a toxic material that takes days to kill should be preferable as a bait. The lag before killing by tungstate or molybdate can be taken advantage of as a bait. The lag before killing by tungstate or molybdate can be taken advantage of for termite bait use. Another possible application of this finding involves the development of Mo or W compounds that can be inpregnated into wood and chemically rendered insoluble, only to be solubilized in the environment of the termite intestine. In this way, a treated wood can stand up to the weather.

Conclusion

As research progresses, we will determine more precisely how close we are to actual applications of N_2 fixation research. New and unpredicted barriers should invite exciting research challenges. Ideas coming from many disciplines will certainly help to develop important applications that should be relevant to agriculture in developing as well as developed countries.

References

Bentley, B. 1984. *Nitrogen fixation in termites: fate of newly fixed nitrogen.* J. Insect Physiol. 30:653–655.

Bishop, P. E., F. B. Dazzo, E. R. Applebaum, R. J. Maier and W. J. Brill. 1977. *Intergeneric transfer of genes involved in the Rhizobium-legume symbiosis.* Science 198:138–140.

Brewin, N. J., T. M. DeJong, D. A. Phillips and A. W. B. Johnston. 1980. *Contransfer of determinants for hydrogenase activity and nodulation ability in Rhizobium leguminosarum.* Nature (London) 288:77–79.

Breznak, J. A., W. J. Brill, J. W. Mertins and H. C. Coppel. 1973. *Nitrogen fixation in termites.* Nature (London) 244:577–580.

Brill, W. J. and W. C. von Meyer. 1985. *Method and composition for control of termite and shipworms.* U.S. Patent Number 4,504,468.

Dazzo, F. B. 1982. *Leguminous root nodules.* Pp. 431–466 in R. G. Burns and J. H. Slater, eds., *Experimental Microbial Ecology.* Blackwell Scientific Publications, Oxford.

den Broeck, G.V., M. P. Timko, A. P. Kausch, A. R. Cashmore, M. V. Montagu and L. Herrera-Estrella. 1985. *Targeting of a foreign protein to chloroplasts by fusion to the transit peptide from the small subunit of ribulose 1,5-bisphosphate carboxylase.* Nature (London) 313: 358–363.

Dixon, R. A. and J. R. Postgate. 1972. *Genetic transfer of nitrogen fixation from Klebsiella pneumoniae to Escherichia coli.* Nature (London) 237:102–103.

Dixon, R. O. D. 1978. *Nitrogenase-hydrogenase interrelationships in rhizobia.* Biochimie 60:233–236.

Eady, R. R. and J. R. Postgate. 1974. *Nitrogenase.* Nature (London) 249:805–810.

Ela, S. W., M. A. Anderson and W. J. Brill. 1982. *Screening and selection of maize to enhance associative bacterial nitrogen fixation.* Plant Physiol. 70:1564–1567.

Emerich, D. W., T. Ruiz-Arqueso, T. M. Ching and H. J. Evans. 1979. *Hydrogen-dependent nitrogenase activity and ATP formation in Rhizobium japonicum bacteroids.* J. Bacteriol. 173:153–160.

Evans, H. J., F. J. Hanus, S. A. Russell, A. R. Harker, G. R. Lambert and D. A. Dalton. 1985. *Biochemical characterization, evaluation, and genetics of H_2 recycling in Rhizobium.* Pp. 3–11 in P. W. Ludden and J. E. Burris, eds., *Nitrogen Fixation and CO_2 Metabolism.* Elsevier Science Publishing Co., Inc., New York.

Gatenby, A. A., J. A. Castleton and M. W. Saul. 1981. *Expression in E. coli of maize and wheat chloroplast genes for large subunit of ribulose bisphosphate carboxylase.* Nature (London) 291:117–121.

Gordon, J. K and W. J. Brill. 1972. *Mutants that produce nitrogenase in the presence of ammonia.* Proc. Nat. Acad. Sci. USA 69: 3501–3503.

Handelsman, J. and W. J. Brill. 1985. *Erwinia herbicola isolates from alfalfa plants may play a role in nodulation of alfalfa by Rhizobium meliloti.* Appl. and Environ. Microbiol. 49:818–821.

Handelsman, J., R. A. Ugalde and W. J. Brill. 1984. *Rhizobium meliloti competitiveness and the alfalfa agglutinin.* J. Bacteriol. 157:703–707.

Haugland, R. A., M. A. Cantrell, J. S. Beaty, F. J. Hanus, S. A. Russell and H. J. Evans. 1984. *Characterization of Rhizobium japonicum hydrogen uptake genes.* J. Bacteriol. 159:1006–1012.

Hanus, F. J., R. J. Maier and H. J. Evans. 1979. *Autotropic growth of H_2-uptake positive strains of Rhizobium japonicum in an atmosphere supplied with H_2 gas.* Proc. Nat. Acad. Sci. USA 76:1788–1792.

Hom, S. S. M., L. A. Graham and R. J. Maier. 1985. *Isolation of genes (nif/hup cosmids) involved in hydrogenase and nitrogenase activities in Rhizobium japonicum.* J. Bacteriol. 161:882–887.

Keyser, H., D. F. Weber and S. L. Uratsu. 1984. *Rhizobium japonicum serogroup and hydrogenase phenotype distribution in 12 states.* Appl. Environ. Microbiol. 47:613–615.

MacNeil, T., D. MacNeil, G. P. Roberts, M. A. Supiano and W. J. Brill. 1978. *Fine-structure mapping and complementation analysis of nif (nitrogen fixation) genes in Klebsiella pneumoniae.* J. Bacteriol. 136:253–266.

Maier, R. J., P. E. Bishop and W. J. Brill. 1978. *Transfer of genes required for nodulation from Rhizobium japonicum to Azotobacter vinelandii*. J. Bacteriol. 134:1199–1201.

Maier, R. J. 1981. *Rhizobium japonicum mutant strains unable to grow chemoautotrophically with H_2*. J. Bacteriol. 145:533–540.

Merberg, D. and R. J. Maier. 1983. *Mutants of Rhizobium japonicum with increased hydrogenase activity*. Science 220:1064–1065.

Merrick, M. and R. Dixon. 1984. *Why don't plants fix nitrogen*? Trends in Biotechnol. 2:162–166.

Nivea-Gomez, D., G. P. Roberts, S. Klevickis and W. J. Brill, 1980. *Electron transport to nitrogenase in Klebsiella pneumoniae*. Proc. Nat. Acad. Sci. USA 77:2555–2558.

Noel, K. D. and W. J. Brill. 1980. *Diversity and dynamics of indigenous Rhizobium japonicum populations*. Appl. Environ. Microbiol. 40:931–938.

Paau, A. S., W. T. Leps and W. J. Brill. 1981. *Agglutinin from alfalfa necessary for binding and nodulation by Rhizobium meliloti*. Science 213:1513–1514.

Roberts, G. P., W. T. Leps, L. E. Silver and W. J. Brill. 1980. *Use of two-dimensional polyacrylamide gel electrophoresis to identify and classify Rhizobium strains*. Appl. Environ. Microbiol. 39:414–422.

Roberts, G. P., T. MacNeil, D. MacNeil and W. J. Brill. 1978. *Regulation and characterization of protein products coded by the nif (nitrogen fixation) genes of Klebsiella pneumoniae*. J. Bacteriol. 136:267–279.

Shah, V. K. and W. J. Brill. 1977. *Isolation of an iron-molybdenum cofactor from nitrogenase*. Proc. Nat. Acad. Sci. USA 74:3249–3253.

Shah, V. K., J. R. Chisnell and W. J. Brill. 1978. *Acetylene reduction by the iron-molybdenum cofactor from nitrogenase*. Biochem. Biophys. Res. Commun. 81:232–236.

Shah, V. K., G. Stacey and W. J. Brill. 1983. *Electron transport to nitrogenase: purification and characterization of pyruvate: flavodoxin oxidoreductase, the nifJ gene product*. J. Biol. Chem. 258:12064–12068.

Shah, V. K., R. A. Ugalde, J. Imperial and W. J. Brill. 1984. *Molybdenum in nitrogenase*. Ann. Rev. Biochem. 53:231–257.

Silverman, M. and M. Simon. 1980. *Phase variation: genetic switching mutants*. Cell 19: 845–854.

Simpson, F. B. and R. H. Burris. 1984. *A nitrogen pressure of 50 atmospheres does not prevent evolution of hydrogen by nitrogenase*. Science 224:1095–1097.

Su, N-Y., M. Tamashiro, J. R. Yates, and M. I. Haverty. 1982. *Effect of behavior on the evaluation of insecticides for prevention of or remedial control of the Formosan subterranean termite*. J. Econ. Entom. 75:188–193.

Tilak, K. V. B. R., K. Schneider and H. G. Schlegel. 1984. *Autotrophic growth of strains of Rhizobium and properties of isolated hydrogenase*. Curr. Microbiol. 10:49–59.

van Berkum, P. and B. B. Bohlool. 1980. *Evaluation of nitrogen fixation by bacteria in association with roots of tropical grasses*. Microbiol. Rev. 44:491–517.

Yoch, D. C., J. R. Benemann, R. C. Valentine and D. I. Arnon. 1969. *The electron transport system in nitrogen fixation by Azotobacter. II. Isolation and function of a new type of ferredoxin*. Proc. Nat. Acad. Sci. USA 64:1404–1410.

Biochemistry and genetics of microbial iron assimilation

J. B. NEILANDS
Department of Biochemistry, University of California, Berkeley, CA 94720, U.S.A.

Abstract Iron, although essential, is toxic in excess and for this reason the uptake of the element is regulated at the membrane level in all living species. A study of the molecular genetics of the regulatory process may be initiated via use of microorganisms as a model, and the enteric bacterium *Escherichia coli* K-12 is the species of choice. A particular high affinity iron assimilation complex coded on pCol VK-30 of *E. coli* has been demonstrated by cloning techniques to consist of five contiguous genes which perform the biosynthesis and transport of the siderophore aerobactin. The operon is preceded upstream by a regulatory element containing major and minor promoters. Quantitative protection experiments with Sl nuclease used in conjunction with *lac*Z operon and protein fusions proved that gene expression is controlled by iron acting at the transcriptional level. A *lac*Z operon fusion was used to select mutants, generated with Tn5, displaying constitutive expression of all high affinity iron assimilation components in *E. coli*. The mutation was mapped at 15.7 min on the chromosome. The mutation was then cloned on pBR322 and the wild-type gene recovered by homologous recombination. The results suggest that in the presence of iron a chromosomal product acting as a repressor shuts down expression of all components of the siderophore pathway which, in turn, restricts further uptake of iron into the cell.

Introduction

Iron is probably a universal nutrient for all living species. One must write "probably" because, although it is a simple matter to demonstrate growth without added iron, one is always haunted by the presence of contaminating, adventitious iron. The question is most easily addressed in microorganisms, which can be cultivated on defined, minimal media. The results show that even in lactic acid bacteria, notorious for their independence of iron, it has not been possible to come up with less than one iron atom per cell (Archibald, 1983).

The reason for this virtually universal requirement for iron is readily appreciated when we consider some of the biological roles of the element. Depending on the particular coordination, iron enzymes may shuttle electrons over a redox scale spanning the better part of one volt, i.e., from the ferredoxins to cytochrome *c* oxidase. Other essential functions of iron include reduction of dinitrogen and ribonucleotides, the latter reaction affording obligatory precursors to the synthesis of DNA in species as disparate as *Escherichia coli* and man. The work of Trowbridge and Omary (1981) demonstrating the identity of the "tumor antigen" with the transferrin receptor underlines the importance of iron for growth in all animal cells.

In the past decade we have become aware of a counterforce to the broad, positive nutritional role of iron, namely, the potential toxicity of the metal. The precise mechanism of the noxious action of iron remains to be elucidated, but it is known to be related to generation of oxy radicals (Halliwell and Gutteridge, 1984).

When the two roles of iron just described are considered together, it becomes apparent why cells have deemed it necessary to regulate uptake of the metal at the membrane level. Once inside the cell bulk iron is coordinated and recycled, there being no process for its excretion. This holds the intracellular level of this somewhat treacherous metal to just those concentrations necessary to perform vital biological functions.

A survey of the various means whereby iron is transported across membranes reveals considerable variation in the process as it occurs in microorganisms, plants and animals. The one unifying feature, however, is the negative repressive mechanism common to all of these species: cells which are replete with respect to iron stores do not accumulate more of the element from the environment. For technical reasons it is easier to approach an understanding of this problem at the molecular level in microorganisms, and this has been the focus of research in our laboratory for more than three decades. The finding that bacteria and fungi excrete iron-specific ligands as a response to low iron levels was the critical experiment that laid the foundation of our present understanding of microbial iron assimilation (Garibaldi and Neilands, 1955; Garibaldi and Neilands, 1956).

The present article will record progress to date on the application of recombinant DNA techniques to the study of regulation of iron absorption in *E. coli*, a genetically accessible microorganism. In the years ahead, it is certain that similar experiments can be applied to other bacteria, to fungi and to plants and animals. One can only speculate regarding the practical applications of these data. It could include enhanced rates of growth of animal and plant species, as well as protection of these species from infections and neoplastic diseases. On the other hand, one should be cautious about any large scale intervention with a regulatory process that has been built up over a period of 3–4 billion years of evolutionary time.

High and low affinity iron assimilation

When subjected to low-iron stress, aerobic and facultative anaerobic bacteria as well as fungi excrete low molecular weight, high affinity, virtually iron^{III} (Fe III) specific ligands generically termed siderophores

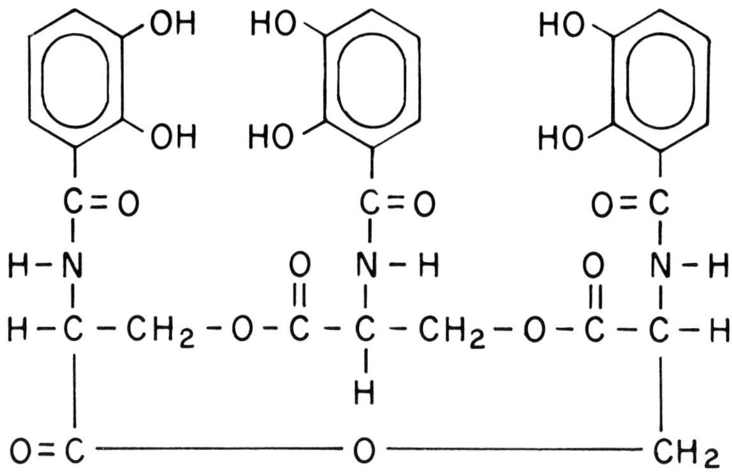

Figure 1. Enterobactin, cyclic trimer of 2,3-dihydroxy-N-benzoyl-L-serine, a siderophore widely distributed in enteric bacteria.

(Neilands, 1984) (Gr. "iron bearers"). *Escherichia coli* and other enteric bacteria elaborate enterobactin (Fig. 1), the cyclic trimer of 2,3-dihydroxy-N-benzoyl-L-serine. Mutants blocked after chorismate in the biosynthetic pathway leading to enterobactin are severely iron-deprived and can only be cultivated in the presence of relatively high levels of the element. Under these conditions iron sufficient for survival and growth is incorporated by a comparatively inefficient, low affinity mechanism. Variations in the catechol type siderophores, of which enterobactin is the prototype, occur in *Agrobacterium tumefaciens* (agrobactin), *Paracoccus denitrificans* (parabactin) and *Vibrio cholerae* (vibriobactin).

A second line of siderophores, widely distributed in both bacteria and fungi, contain hydroxamic [R-CO-N(OH)R′] groups as ligands for Fe^{III}. An example is aerobactin (Fig. 2), first isolated from the enteric bacterium *Aerobacter aerogenes*. Other hydroxamate type siderophores occur in streptomyces (ferrioxamines) and fungi (ferrichromes).

In a few instances both catechol and hydroxamate functions are contained within the same siderophore molecule. Specific examples are the mycobactins from various mycobacterial species and the pseudobactins from *Pseudomonas* species. The recent discovery of a structurally novel ligand in *Rhizobium meliloti* suggests that yet a third family of siderophores may exist. These are best described as complexones as they are derivatives of ethylenediamine (Smith et al., 1985).

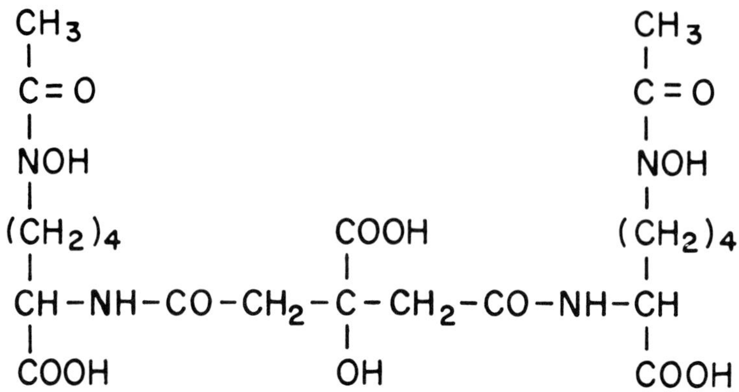

Figure 2. Aerobactin, a dihydroxamate siderophore formed by substitution of each of the two distal carboxyl groups of citrate with residues of N^{e}-acetyl-N^{e}-hydroxyl-L-lysine.

The hallmark of the siderophore molecule is its very high affinity, and hence specificity, for Fe^{III}. Formation constants for Fe^{III} are characteristically in the range of 10^{30}; the high spin, chelated iron is only efficiently released and made available to the cell by reduction to Fe^{II} (Raymond and Carrano, 1979).

A decade ago it was found that *E. coli*, although incapable of synthesis of ferrichrome, maintains an outer membrane receptor for this hydroxamate type siderophore which is elaborated by all species of *Penicillia* and by many other fungi (Wayne and Neilands, 1975). The ferrichrome receptor acts as a binding locus for an array of lethal agents, including bacteriophages T1, T5, Φ80, UC-1, as well as for colicin M and albomycin. Siderophores, in the iron complex form, appear too large to penetrate the small, water-filled pores of the outer membrane of enteric bacteria. This has required the elaboration of specialized receptors and, over evolutionary time, such surface proteins have become the target for lethal agents.

Additional examples of ferri-siderophore receptors acting as attachment sites for lethal agents are the colicin B-ferric enterobactin receptor and the cloacin-ferric aerobactin receptor, both in the outer membrane of *E. coli*. This coincidence renders cloning of the siderophore systems quite straightforward since the presence of a recombinant plasmid in a transformed cell renders that cell susceptible to the specific lethal agent by virtue of the presence of the common receptor.

Molecular genetics of aerobactin

Isolation of aerobactin from Aerobacter aerogenes. Gibson and Magrath (1969) showed that low iron grown cultures of the enteric bacterium *Aerobacter aerogenes* contained two siderophores, enterobactin and aerobactin. The latter siderophore is one of three known derivatives of citric acid in which the two distal carboxyl groups of the tricarboxylic acid are substituted by side chains terminating in hydroxamic acid functions. The original member of the series was discovered by Lochhead, Burton and Thexton (1952) as a product of *Arthrobacter pascens* active for *Arthrobacter terregens*; it was named "terregens factor". The siderophore was subsequently chemically characterized as the decarboxylation product of aerobactin and re-named arthrobactin. The final member, schizokinen, was characterized from *Bacillus megaterium* by Mullis, Pollack and Neilands (1971). This siderophore is unique in containing a three-carbon hydroxamic acid sidechain. It has been isolated from *Anabaena* species (Simpson and Neilands, 1976) and other cyanobacteria. It may be significant that aerobactin is apparently quite widely distributed in enteric species and has now been reported from *Shigella flexneri*, *Salmonella austin*, *S. memphis* and other *Salmonella* species, from *Enterobacter cloacae* and from various clinical isolates of *E. coli* (McDougall and Neilands, 1984; Bindereif and Neilands, 1985a).

Aerobactin as E. coli *virulence factor.* Williams (1979) and colleagues described a novel, hydroxamate type siderophore present in invasive strains of *E. coli* bearing the V ("virulence") plasmid. Although the plasmid encodes synthesis of a colicin, also labelled V, Williams showed by mutational analysis that it is the siderophore, subsequently identified as aerobactin (Warner et al., 1981), which accounts for the virulent character of the clinical isolates. Enterobactin determinants are present on the chromosome of these strains, but the catechol type siderophore is apparently inactivated by adsorption to serum albumins and other proteins (Konopka and Neilands, 1984). Thus, while synthesis of enterobactin may be better than complete siderophore auxotrophy, aerobactin is clearly a virulence factor in enteric bacteria.

Cloning the aerobactin system. The enterobactin system chromosomally encoded at min. 14 on the *E. coli* chromosome has been cloned on bacteriophage μ and shown to be organized into several transcriptional units spread across some 26 kb of DNA (Laird et al., 1980). With the demonstration that the pColV-K30 plasmid carries the determinants of aerobactin, we immediately seized upon this system as a desirable one for cloning (Bindereif and Neilands, 1983). Aerobactin is structurally simpler than enterobactin, the biosynthesis of the latter siderophore having been shown to require a large number of genes.

The presence of aerobactin genes on a plasmid, albeit at 70 Mdal. a large one, meant that the siderophore had already been "cloned" in Nature and it only remained to sub-clone the iron transport activity. This could be achieved by using susceptibility to cloacin, a bacteriocin from *Enterobacter cloacae*, which uses the ferric aerobactin receptor.

As vector for cloning aerobactin we selected pPlac, a small, multicopy derivative of pBR322 bearing the origin of replication and the β-lactamase gene on 2 kb of DNA. A *Hin*dIII total digest of pColV-K30 DNA was ligated into pPlac and the ligation mixture used to transform strain 294 of *E. coli*, which was then plated on ampicillin. Resistant colonies were replica plated to media containing both ampicillin and cloacin. A single transformant resistant to ampicillin and sensitive to cloacin was isolated and the 18.3 kb plasmid contained therein designated pABN1. Complete digestion of pABN1 with *Hin*dIII and *Eco*R1 yielded an 8.7 kb plasmid, designated pABN5. The larger plasmic synthesized the complete aerobactin system, including the 74 K outer membrane receptor of the ferri-aerobactin complex. The smaller plasmid formed only aerobactin. Growth of cells bearing these plasmids at various iron levels indicated that the regulatory elements had been captured on the cloned sequences of DNA. However, the system was relatively insensitive to repression by iron when compared to the low copy number pColV-K30. This has been attributed to a gene dosage effect, i.e., a deficit of repressor.

Aerobactin operon organization and expression. Having in hand the cloned sequences of the complete aerobactin system the first task was to locate the promoter region (Bindereif and Neilands, 1985b). For this purpose subclones of pABN1, pABN5 and pABN15 were constructed. These contained, respectively, 16.3, 6.7 and 2.1 kb of DNA rightward of the presumed start site. Since pABN5 contained all of the biosynthetic genes for aerobactin, it could be presumed that the internal *Eco*R1 site in pABN1 cut into the gene for the 74 K outer membrane receptor for ferric aerobactin. In vitro "runoff" transcripts from templates of digests of these plasmids were analyzed on gels and shown to afford smaller and smaller segments of RNA as the DNA was truncated leftward. It was concluded that the aerobactin gene cluster was driven by a single strong promoter and organized into one operon. In particular, it enabled the conclusion that the regulatory region is situated at the far left side of the map of pABN5.

To obtain a more precise idea of the location of the promoter, a 645 bp segment of the sense strand, 5′ labeled at one end, was used as probe for RNA protected against digestion by S1 nuclease. The RNA was either transcribed in vitro from a ca. 1.3 kb template or total RNA from *E. coli* carrying the pColV-K30 plasmid and grown at low iron. The result showed a major and a minor start site separated by about 50 bp with in vitro RNA; in vivo, only the major start site was evident. RNA sequencing was performed in order to assign the major start site to position + 1 shown in Fig. 3.

The DNA sequence of the double stranded upstream array across the promoter region is shown in Fig. 3. At position + 31 and + 34 are the codons for methionine and isoleucine, respectively. The DNA stretches rightward into the long open reading frame of a 63 K protein. At the juncture to the next protein, there is a new Shine-Dalgarno site but no new promoter. Thus it is very possible that the promoter(s) shown in Fig. 3 drive the entire aerobactin gene cluster, although a second minor promoter further downstream cannot be excluded. Such secondary promoters might enable the cell to form transport proteins for environmentally encountered aerobactin without need for synthesis of the siderophore.

Upstream of the first structural gene we find the common 7 bp spacing to a strong ribosome binding sequence, AGGAGcTG, which

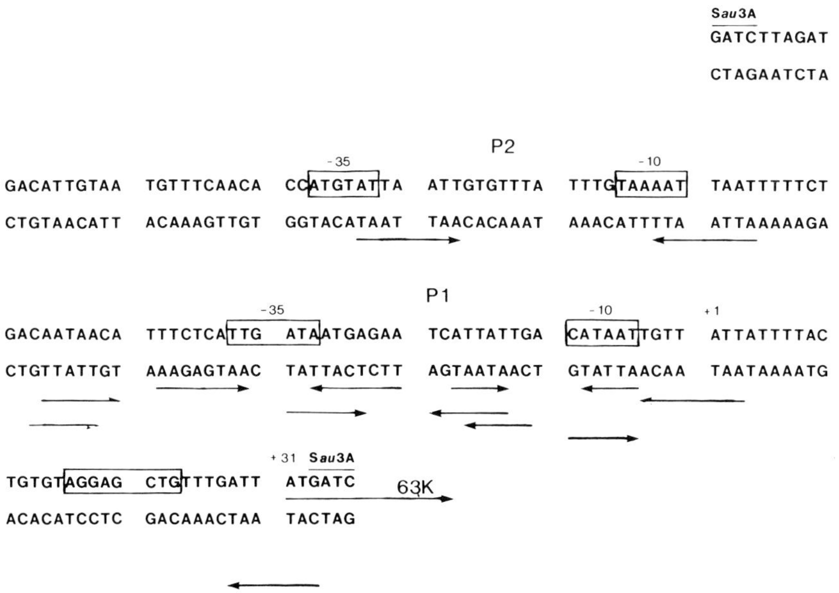

Figure 3. A 156 base pair *Sau*3A fragment across the main (P1) and minor (P2) promoter regions upstream of the aerobactin gene complex. Sequences in P1 and P2 at the − 10 and − 35 regions, as well as the Shine-Dalgarno ribosome binding site, are boxed. Inverted repeat sequences are shown by arrow pairs. The original publication (Bindereif and Neilands, 1985b) must be consulted for identification of the full complement of palindromes, symmetry centers and homologies between P1 and P2 present in this A–T rich base sequence.

spans the region + 16 to + 23. Around the major promoter, P1, the − 10 and − 35 sites are in the expected locations for a strong promoter and are separated by the optimum 17 bp. The sequences at − 10 and − 35 each differ from the consensus sequence by a single pyrimidine residue. At P2, some 50 bp upstream from P1, the sequences at − 10 and − 35 differ substantially from the consensus and the spacing is reduced to 16 bp. Although P2 appears to be a weak promoter from the sequence and from the fact that it does not appear to generate a transcript in vivo, we cannot rule out the possibility that it may play some role in guiding the RNA polymerase to P1.

A notable feature of this sequence is the ca. 80% AT content, some 30% higher than the average AT composition of the *E. coli* chromosome. In addition, the sequence reveals a marked propensity for the formation of secondary structures in the DNA or transcribed RNA. As seen in Fig. 3, the region around P1 is particularly rich in dyad symmetries. Thus there would appear to be abundant opportunity for perturbation of the DNA structure by binding of specific proteins, such as a repressor, in the P1 zone of the aerobactin promoter.

The aerobactin gene complex is flanked by IS1 elements and hence has the general features of a transposable element (McDougall and Neilands, 1984).

Regulation of *E. coli* siderophore systems

Quantitative assay of specific RNA. It has already been noted that expression of the aerobactin system from the low copy number pColV-K30 plasmid is very sensitive to iron repression while the same gene complex on high copy number plasmids, such as pBR322 derivatives, is scarcely regulated by iron. In order to demonstrate that iron acts directly at the transcriptional level, the 16.3 kb insert of pABN1 was cloned into pRK2501 (Bindereif and Neilands, 1985b). The latter, a derivative of pRK2, is maintained at a copy number of eight per chromosome of *E. coli*. When *E. coli* 294 was transformed with the resulting plasmid, pABN100, and grown at high, low and very low levels of iron, there was a quantitative increase in the content of specific RNA as measured by protection of a labeled probe against S1 nuclease digestion.

Iron regulation of a protein fusion. To rule out the possibility that iron was acting at the RNA level a protein fusion was constructed in which a 152 bp fragment spanning P1 and P2 (Fig. 3) was ligated, via a three codon linker, to *lac*Z. As may be seen in Fig. 3, this construct carried only the first two codons of the 63 K protein of the first structural gene of the aerobactin operon. Upon insertion into *E. coli* SE5000 the protein fusion plasmid, pABN40, was found to form β-galactosidase in high levels upon treatment with bipyridyl, an iron chelator. Although the levels of enzyme formed were considerably lower than when the *lac*Z gene carrying its own Shine-Dalgarno sequence was inserted in the middle of the operon, the experiment again indicates that control by iron is exerted primarily at the operator level. The lower level of expression of β-galactosidase in the protein fusion may be explained by less efficient binding of the transcript to the ribosome.

The fur *mutation.* Having localized the site of action of iron, the next question to pose is the form of iron. As a regulator of gene expression, ferric or ferrous ions are hardly semantic enough to act as an effector of a macromolecule such as DNA. It was therefore not surprising when Ernst, Bennett and Rothfield (1978) reported the isolation from *Salmonella typhimurium* of a mutant, designated *fur* (*f*erric *u*ptake *r*egulation), constitutive in the expression of all high affinity iron systems. The mutation was described in *E. coli* and provisionally mapped near *lac*, subsequently revised to 15.7 on the *E. coli* chromosome (Hantke, 1984; Bagg and Neilands, 1985).

The following procedure was used in this laboratory for isolation of *fur* mutants. The *lac*Z gene, minus its promoter, was retrieved from pJA1 and ligated into a *Bam*H1 site of pABN5. Since the latter is high copy and not tighly regulated by iron, the fusion was transferred to the low copy number pColV-K30 by homologous recombination. Adding a non-utilized iron chelator, deferriferrichrome A, to *E. coli* cells harboring pColV-K30:*lac*Z resulted in an order of magnitude increase in synthesis of β-galactosidase. Tn5 was selected as the mutagenic agent since, among other desirable features, it confers resistance to kanamycin. The mutagenized culture was plated on iron-rich MacConkey medium containing 20 mM citrate. The latter serves to enhance iron supply and increase the incidence of red, derepressed colonies that are not merely defective in the enterobactin pathway (T. Fleming and M. McIntosh, personal communication). The *fur*:Tn5 mutants, all of which map at the same locus, exhibit the desired phenotypic characters. They produce high levels of catechol and hydroxamate, as well as β-galactosidase, when grown in complex media such as nutrient broth. Examination of the outer membranes shows the constitutive presence of the brace of ca. 80 K iron regulated proteins typical of *E. coli* K12.

Using kanamycin resistance as a marker, it was possible to clone the *fur* mutation and then recover the *fur* gene from wild-type DNA by homologous recombination. The gene product has been identified with both radiographic techniques (Hantke, 1984) and with Coomassie stain (Bagg and Neilands, unpublished) as a ca. 20 K polypeptide.

Conclusions

For the first time there is now available a fully cloned iron assimilation pathway comprised of a siderophore biosynthesis-transport system and the matching regulatory element, a putative repressor protein.

Before considering the various mechanisms whereby expression of this system can be regulated, it is pertinent to review what is known about the amount and form of iron in *E. coli*. The total iron content of the organism depends on growth conditions and may range from 11 to 890 μM, according to one report (Archibald, 1983). How much of this is merely hydrated or weakly coordinated ferrous ion can only be speculated; assays with bleomycin indicate that even low iron grown cells contain a substantial level of "loose", non-protein bound iron. The latter cannot be in the ferric form *and* uncoordinated at biological pH. In fact, it is very unlikely that the regulatory iron could be ferric as this would be immediately bound by siderophores; we know that these can be eliminated by mutation without perturbing the regulatory process. Similarly, mutants blocked in heme synthesis are still iron regulated (Klebba, McIntosh and Neilands, 1983). Whatever the form of iron, we know that it can be complexed by either ferric or ferrous specific ligands, as exemplified by deferriferrichrome A and bipyridyl, respectively. As the ferrous ion level of cells may, depending on the redox level, be of the order of μM (Williams, 1982), we may assume tentatively that at least part of the control iron is the hydrated ferrous ion.

There would appear to be at least three different mechanisms whereby iron could function in concert with the Fur protein to regulate expression of siderophore systems in *E. coli*. Iron could bind to Fur as a co-repressor, the holorepressor then in turn binding to the operator site so as to compete with attachment of RNA polymerase at the promoter. In a modification of this mechanism, *Fur* may form a transcriptionally competent complex with the operator and only become inhibitory when iron binds with the DNA-*Fur* complex. This model would have the advantage of very rapid response to iron repletion, which is observed (Klebba, McIntosh and Neilands, 1983). A third model would have *Fur* processed to a derivative, *Fur'*, which then acts as a repressor.

Determination of the sequence of the *fur* gene, which is in progress, may help to define the potential metal binding activity of its product. Additional questions to be posed are the number and types of metal ions bound by *Fur* as well as the subunit structure of the active repressor. With the *E. coli* work as a solid foundation, it will then be possible to extend this type of investigation into the eukaryotic world of fungi, plants and animals.

References

Archibald, F. 1983. *Lactobacillus plantarum, an organism not requiring iron.* FEMS Microbiol. Lett. 19:29–32.

Bagg, A. and J. B. Neilands. 1985. *Mapping of a mutation affecting regulation of iron uptake systems in Escherichia coli K-12.* J. Bacteriol. 161:450–453.

Bindereif, A. and J. B. Neilands. 1983. *Cloning of the aerobactin-mediated iron assimilation system of plasmid ColV.* J. Bacteriol. 153:1111–1113.

Bindereif, A. and J. B. Neilands. 1985a. *Aerobactin genes in clinical isolates of Escherichia coli.* J. Bacteriol. 161:727–735.

Bindereif, A. and J. B. Neilands. 1985b. *Promoter mapping and transcriptional regulation of the iron assimilation system of plasmid ColV-K30 in Escherichia coli K-12.* J. Bacteriol. 162: 1039–1046.

Ernst, J. F., R. L. Bennett and L. I. Rothfield. 1978. *Constitutive expression of the iron enterochelin and ferrichrome uptake systems in a mutant strain of Salmonella typhimurium.* J. Bacteriol. 135:928–934.

Garibaldi, J. A. and J. B. Neilands. 1955. *Isolation and properties of ferrichrome A.* J. Am. Chem. Soc. 77:2429–2430.

Garibaldi, J. A. and J. B. Neilands. 1956. *Formation of iron binding compounds by microorganisms.* Nature 177:526–528.

Gibson, F. and D. I. Magrath. 1969. *The isolation and characterization of a hydroxamic acid (aerobactin) formed by Aerobacter aerogenes 62-I.* Biochim. Biophys. Acta 192:175–184.

Halliwell, B. and J. M. C. Gutteridge. 1984. *Oxygen toxicity, oxygen radicals, transition metals, and disease.* Biochem. J. 219:1–14.

Hantke, K. 1984. *Cloning of the repressor protein gene of iron regulated systems of Escherichia coli K12.* Mol. Gen. Genet. 197:337–341.

Klebba, P. E., M. A. McIntosh and J. B. Neilands. 1982. *Kinetics of biosynthesis of membrane proteins in Escherichia coli.* J. Bacteriol. 149:880–888.

Konopka, K. and J. B. Neilands. 1984. *Aerobactin mediated utilization of transferrin iron.* Biochemistry 23:2122–2127.

Laird, A. J., D. W. Ribbons, G. C. Woodrow and I. G. Young. 1980. *Bacteriophage u-mediated gene transposition and in vitro cloning of the enterochelin gene cluster of Escherichia coli.* Gene 11:347–357.

Lochhead, A. G., M. O. Burton and R. H. Thexton. 1952. *A bacterial growth factor synthesized by a soil bacterium.* Nature 170:282–283.

McDougall, S. and J. B. Neilands. 1984. *Plasmid and chromosome code aerobactin synthesis. Insertion sequences flank operon in plasmid coded systems.* J. Bacteriol. 159:300–305.

Mullis, K. B., J. R. Pollack and J. B. Neilands. 1971. *Structure of schizokinen, an iron transport compound from Bacillus megaterium.* Biochemistry 10:4894–4898.

Neilands, J. B. 1984. *Siderophores of bacteria and fungi.* Microbiol. Sci. 1:9–14.

Raymond, K. N. and C. J. Carrano. 1979. *Coordination chemistry and microbial iron transport.* Accts. Chem. Res. 12:183–189.

Simpson, F. and J. B. Neilands. 1976. *Siderochromes in Cyanophyceae: isolation and characterization of schizokinen from Anabaena sp.* J. Phycol. 12:44–48.

Smith, M. H., J. N. Shoolery, B. Schwyn, I. Holden and J. B. Neilands. 1985. *Rhizobactin, a structurally novel siderophore from Rhizobium meliloti.* J. Amer. Chem. Soc. 107:1739–1743.

Trowbridge, I. S. and M. B. Omary. 1981. *Human cell surface glycoprotein related to cell proliferation is the receptor for transferrin.* Proc. Nat. Acad. Sci. USA 78:3039–3043.

Wayne, R. and J. B. Neilands. 1975. *Evidence for common binding sites for ferrichrome compounds and bacteriophage 80 in the cell envelope of Escherichia coli.* J. Bacteriol. 121:497–501.

Warner, P. J., P. H. Williams, A. Bindereif and J. B. Neilands. 1981. *ColV plasmid specified aerobactin synthesis by invasive strains of Escherichia coli.* Infect. Immun. 33:540–545.

Williams, P. H. 1979. *Novel iron uptake system specified by ColV plasmids: an important component in the virulence of invasive strains of Escherichia coli.* Infect. Immun. 26:925–932.

Williams, R. J. P. 1982. *Free manganese (II) and iron (II) cations can act as intracellular controls.* FEBS Lett. 140:3–10.

Molecular biology of zinc

ROBERT J. COUSINS
Department of Food Science and Human Nutrition, Institute of Food and Agricultural Sciences, University of Florida, Gainesville, Florida 32611, U.S.A.

Abstract The ubiquitous distribution of zinc (Zn) in microbial, plant and animal cells suggests a wide range of potential functions for this transition element. The physiological role of Zn has centered on its interaction with enzymes. For example, the cessation of growth associated with nutritional Zn deficiency involves a reduction in DNA polymerase activity and concomitant reduction in DNA synthesis. Similarly, RNA polymerases I, II, and III are all Zn metalloenzymes and in appropriate model systems a reduction in the synthesis r-RNA, mRNA and t-RNA, respectively, is observed with cellular Zn depletion. The effect of Zn at the genetic level is most dramatically shown through induction of metallothionein synthesis. This protein influences the kinetics of Zn uptake by cells. It has been clearly demonstrated that metallothionein synthesis is under transcriptional control. Zinc and glucocorticoids, as well as glucagon and epinephrine (via elevation of cellular cAMP levels) all induce transcription of the metallothionein gene. Recently it has been shown that a 12 bp fragment of the metallothionein gene promoter sequence regulates expression of heterologous genes in vivo, which demonstrates that Zn can have a direct effect on expression of specific genes. These results illustrate the potential of this relatively nontoxic metal in biotechnology applications where Zn responsive fusion genes could be utilized.

Introduction

Zinc (Zn) has a particularly broad range of functions related to gene expression. This diversity includes structural, enzymatic and gene promoter activation roles. Therefore, this plethora of effects places Zn in a special category with respect to other metals that have biological significance. Early nutrition experiments with plants (Raulin, 1869) and animals (Todd et al., 1934) demonstrated its essentiality as well as its integral relationship to growth and development, which has been subsequently placed within the perspective of DNA replication and protein synthesis. In this review, the various levels of Zn action in gene expression will be discussed and integrated into an overview of the physiological importance of this essential nutrient.

Zinc complexes of polynucleotides

Nucleotides bind metal ions through electron donation of oxygen and nitrogen ligands as well as the phosphate groups. The latter exhibits the strongest stability constants with transition metals. For Zn^{2+} stability constants for 1:1 complexes of 5′-AMP and 5′-ATP are 2.7 and 4.9, respectively (Eichhorn, 1973). These stabilities are similar to those calculated for other transition metals (Mn, Co, Ni and Cu), with stability increasing proportionately with phosphate content. $Zinc^{2+}$ binding to individual purine or pyrimidine bases is minimal. The significant binding of Zn^{2+} with ATP presents an obstacle for in vitro experiments where the effect of ATP on a specific process related to Zn is being measured.

Binding of metal ions to polynucleotides affects both the chemical and physical properties of the macromolecules. Cations, particularly Mg^{2+}, neutralize the negative charges of the phosphates in DNA molecules, thus decreasing the tendency of adjacent nucleotides to repel each other. In this way cations stabilize the DNA helix. DNA stabilization is of the order of $Mg^{2+} > Co^{2+} > Ni^{2+} > Mn^{2+} > Zn^{2+} > Cd^{2+} > Cu^{2+}$.

Ribonucleic acid is also stabilized and destabilized by metal ions in roughly the same order as is observed for DNA. However, hydroxides of some metals, particularly Zn^{2+}, increases the rate of RNA hydrolysis. This is most prevalent for phosphodiester bonds associated with uracil and least prevalent with guanine. The entire area of the metallobiochemistry of the nucleic acids has been thoroughly reviewed by Eichhorn (1973).

Fazakerly (1984) has provided evidence that Zn^{2+} complexes promote transition of B form DNA (right handed helix form) in high moisture conditions to the Z form (left handed helix form) without dehydration. Using shifts in circular dichroism spectra the $B \rightarrow Z$ transition was detectable at 1:1, Zn-ligand complex concentrations as low as 3 μM. It has been proposed that Z helix regions of DNA may have regulatory functions related to gene expression. Therefore, the $B \rightarrow Z$ helix transition caused by zinc may also have physiological significance since it occurs within physiological concentrations of Zn^{2+}.

Cellular distribution of zinc

From the distribution of Zn within cells, it might be deduced that this metal could have many functions related to replication and protein synthesis. Smeyers-Verbeke et al. (1977) determined the distribution of Zn within subcellular fractions of rat liver. When expressed as μg Zn per g of protein the subcellular fractions had the following distribution of Zn: nuclei (80), mitochondria (97), rough microsomes (218), smooth microsomes (192) and cytosol (246). Together the nuclei and cytosol account for most of the cellular Zn (22 and 60%, respectively). Alfaro and Heaton (1974) found the nuclear pellet, microsomes, and cytosol from rat liver contained 26, 17 and 41 μg Zn per g dry matter, respectively. While these values may vary between studies due to the fractionation conditions used, Zn is clearly not excluded from any major cellular fraction. Probably the situation is roughly the same in situ, but the effect of tissue disruption and lysis of the plasma membrane could distort the true intracellular distribution considerably.

In contrast to total cellular Zn, the nuclei of liver cells contained the greatest amount of radioactivity when rats were administered ^{65}Zn (Weser and Bischoff, 1970). This suggests that early after cellular uptake, a substantial portion of newly acquired Zn enters the nucleus. If this nuclear uptake phenomenon is found in other cell types, the nucleus must be viewed as an organelle that is interactive with extracellular Zn pools. Kinetic analysis of Zn^{2+} uptake by liver parenchymal cells in primary monolayer culture suggest that a major portion of uptake is accounted for as exchange with existing Zn pools within the cells (Pattison and Cousins, 1985). Estimates of Zn turnover by hepatocytes suggest a $t_{1/2}$ of about 15 hr. A labile pool which displays fast uptake kinetics accounts for cellular Zn accumulation.

Bryan et al. (1981) have examined the association of Zn with nucleoprotein fractions of DNA following nuclease treatment. About 0.1 μg Zn per mg DNA was tightly bound to chromatin. Zinc was associated with higher molecular weight oligonucleosomes. When Zn is removed by chelating agents, an apparent dissociation occurs, and the metal is associated with low molecular weight products following nuclease digestion. These chromatin localization studies are constant with the functions of Zn in replication and protein synthesis.

Chromatin, DNA polymerase and replication

Arrested growth is a universal characteristic of nutritional Zn deficiency in microorganisms, plants and animals including man (Underwood, 1977; Prasad et al., 1978).

Zinc does not exhibit any direct redox chemistry as does other abundant metals, e.g. iron or copper. Therefore, it is well suited for a catalytic role in nucleic acid metabolism since free radicals and reactive oxygen species would be minimized. Nevertheless, Zn^{2+} acts as a Lewis acid and can catalyze nucleotide phosphate ester formation. The unique properties of Zn in catalysis in biological systems has been reviewed in detail by Williams (1984).

Evidence that *Escherichia coli* DNA polymerase contains Zn as a functional constituent was provided by Slater et al. (1971). $^{65}Zinc$ migrated with the enzyme during sucrose density gradient centrifugation and the Zn content amounted to 2 gm atoms/mole. Removal of Zn^{2+} with appropriate chelators inhibited the polymerase activity

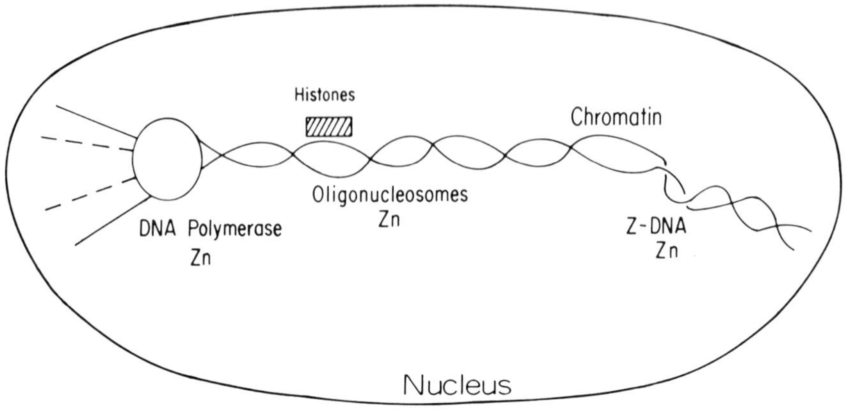

Figure 1. Effect of zinc on chromatin structure and function. In various model systems, zinc has been shown to influence DNA polymerase activity, histone protein synthesis and conversion of B form DNA into Z form DNA.

which was subsequently restored by readdition of Zn^{2+}. A detailed analysis of *E. coli* DNA polymerase I demonstrated that Zn^{2+} is necessary for activity and has a function in coordinating addition of the deoxyribonucleotides at the 3-OH terminus of the DNA (Springgate et al., 1973). The "zinc-free" apoenzyme had only 7% of the polymerase activity found in the holoenzyme. It is of particular interest that $^{65}Zn^{2+}$ exchanged with constituent Zn^{2+} ions suggesting a similar exchange could occur in situ with Zn^{2+} entering the nucleus. More recent data suggest that *E. coli* DNA polymerase I is not a metalloenzyme (Ferrin et al., 1983). However, the Zn content did not correlate with enzymatic activity. This conclusion is supported by experiments with Phage T7 DNA polymerase which showed that the enzyme did not contain appreciable amounts of Zn, did not require Zn^{2+} for activity, and Zn^{2+} was actually an inhibitor (Slaby et al., 1984). In order to resolve this discrepancy, it was suggested that chelation-related inhibition was caused by denaturation or production of inhibitory fragments of DNA. Using the classic approach of metal analysis of the enzyme and appropriate changes in enzyme activity during chelator treatment, Auld et al. (1974) provided strong evidence that RNA-dependent DNA polymerase (reverse transcriptase) from avian myeloblastosis virus is a Zn metalloenzyme.

A variety of model systems have been utilized for the purpose of establishing a relationship between Zn nutrition and growth as related to cell division. Initial experiments by Lieberman et al. (1963) clearly showed that DNA synthesis in cultured rabbit kidney cortex cells was dependent upon the presence of Zn^{2+} in the growth medium. Both

DNA polymerase and thymidine kinase activities were found to be similarly Zn dependent whereas other enzyme activities were not affected. Thymidine kinase activity was also found to be decreased in Zn deficiency (Swenerton et al., 1969; Prasad and Oberleas, 1974). Convincing evidence for a Zn requirement for DNA synthesis in chick embryo cells was presented by Rubin (1972). EDTA prevented synthesis which was specifically restored by Zn^{2+}, but not by other selected cations. It was suggested that there is a continual requirement for Zn to sustain DNA synthesis in cultures of replicating cells. With a similar approach, Chesters (1972) and Williams and Loeb (1973) detected a Zn requirement for DNA synthesis by phytohemagglutinin-stimulated lymphocytes.

Sandstead and Rinaldi (1969) observed that liver DNA synthesis, based on ^{3}H-thymidine incorporation, was reduced in Zn deficient rats compared to pair-fed and ad libitum-fed controls. Extensive studies of nucleic acid metabolism as related to Zn deficiency have used the growth characteristics of *Euglena gracilis* as a model system (Price and Vallee, 1962). Evidence indicates that DNA synthesis and replication in these eukaryotic cells is Zn-dependent (Falchuk et al., 1975). Chromatin from Zn adequate (+ Zn) and Zn deficient (− Zn) *Euglena* cells exhibit markedly different patterns of histone composition based on electrophoretic and immunologic evidence (Mazus et al., 1984). Histones H1, H3, H2A, H2B and H4 were found in + Zn cells while only H1 and H3 were identified in − Zn cells. Since nucleosome function requires an interaction between DNA and histones, these observations imply that Zn is necessary for normal chromatin function. Collectively these results suggest that if the Zn concentration of the nucleus is decreased, the reduction in DNA synthesis, is most likely accounted for by changes in chromatin, e.g. histones, or by a concomitant decrease in DNA polymerase activity.

RNA polymerases and protein synthesis

Cellular RNA levels were shown to decrease when cells are cultured in Zn deficient growth medium. These include *Rhizopus* (Wegener and Romano, 1963) and *Euglena* (Wacker, 1962; Schneider and Price, 1962). Similarly, Macapinlac et al. (1968) found Zn deficiency reduced the RNA content of rat testis. Since ^{14}C-adenine incorporation into the RNA was not altered they suggested that this nutritional deficiency increased RNA catabolism. These observations are consistent with the findings of others (Prasad et al., 1973; Somers and Underwood, 1969). Incorporation of ^{3}H-uridine into rat brain was not altered by Zn deficiency (O'Neal et al., 1970). RNA synthesis has also been shown to

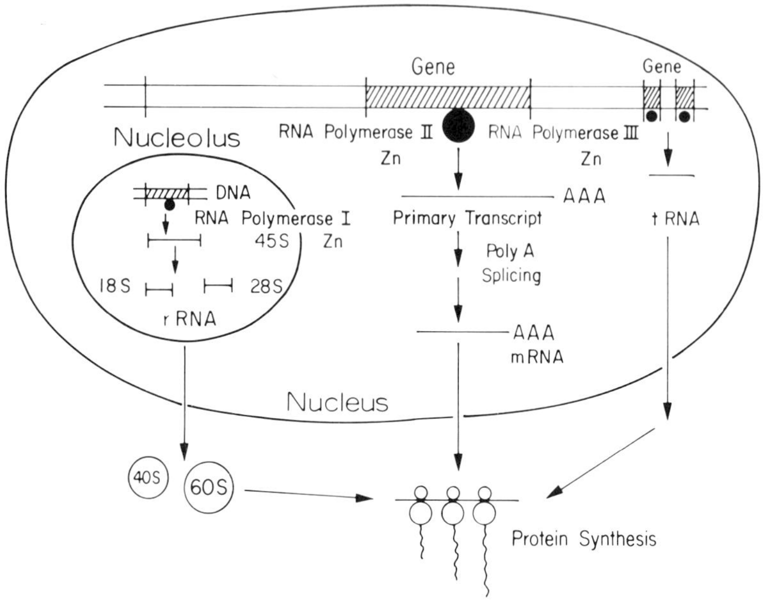

Figure 2. Effect of zinc on RNA and protein synthesis in general. In various model systems, zinc has been shown to influence the activities of RNA polymerases I, II and III. Zinc may also stabilize polysomes and help maintain protein synthesis through interaction with initiation and elongation factors.

decrease as a result of Zn deficiency in rat liver (Terhune and Sandstead, 1972) and brain (Fosmire et al., 1974).

Despite the equivocal nature of the effect of dietary Zn deficiency on RNA synthesis there is considerable evidence that RNA polymerases I, II, and III are Zn metalloenzymes. Scrutton et al. (1971) used classical chelator-inhibition and metal co-migration chromatographic techniques to show an RNA polymerase from *E. coli* was a Zn metalloenzyme. These methods plus restoration of polymerase activity by Zn^{2+} were used to establish that polymerase I (Falchuk, 1977), polymerase II (Falchuk, 1976; Petranyi et al., 1977), and polymerase III (Wandzilak and Benson, 1977) contain zinc as a catalytically essential element. The mRNA's from Zn deficient *Euglena* were not grossly altered, based upon translational activity, suggesting the deficiency did not influence RNA polymerase II directly. However, the proteins that are synthesized varied, which is suggestive of some defect in protein synthesis. One would expect that during a nutritional stress, such as Zn deficiency, changes in the expression of specific genes would occur. Furthermore, there is some evidence that polyribosome formation and stability is related to Zn (Tal, 1969; Sandstead et al., 1971).

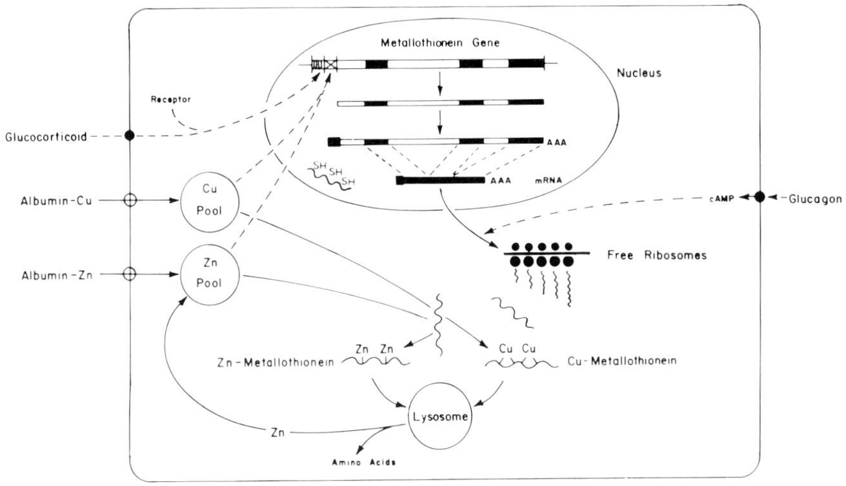

Figure 3. Schematic representation of factors involved in the turnover of metallothionein in hepatocytes is shown. The effect of zinc in initiating transcription of the metallothionein gene by interacting with the promoter is shown. From Cousins, 1985, with permission.

Zinc and metallothionein

There is no single phenomenon that demonstrates the molecular biology of Zn more vividly than its induction of metallothionein. Although initial experiments dealt with induction by cadmium, a toxic heavy metal, and showed that transcriptional regulation was involved (Squibb and Cousins, 1974), the experiments dealing with metallothionein induction by relatively non-toxic metals are of more significance from a biomedical and biotechnological perspective. The physiological basis of metallothionein synthesis, degradation and function have been reviewed in detail elsewhere (Cousins, 1985).

Early nutritional studies where Zn was injected, showed that hepatic, intestinal, and renal metallothionein synthesis in rats was augmented by Zn. The levels of dietary Zn needed to increase synthesis were within normal levels ca. 25 mg/kg (Richards and Cousins, 1975, 1976a, b; Squibb et al., 1977). Induction was blocked by actinomycin D suggesting increased transcription accounted for the change. Actinomycin D and cordycepin also blocked the ability of the liver to take up the extra supply of Zn needed for induction, which as a result, remained circulating in the plasma. These studies were the first to suggest that Zn^{2+} uptake or retention by hepatocytes may require ongoing metallothionein synthesis. Subsequent experiments with isolated rat liver parenchymal cells demonstrated that Zn^{2+} uptake/exchange above basal levels required hormonal stimulation (Failla and Cousins, 1978).

Specifically, glucocorticoids were found to induce metallothionein synthesis in the hepatocytes, and this increase in metallothionein synthesis was directly correlated to cellular Zn^{2+} uptake and accumulation. Blocking metallothionein synthesis similarly blocked hepatocyte Zn^{2+} uptake and accumulation. Glucocorticoid administration to rats similarly increased synthesis of metallothionein in liver (Etzel et al., 1979). Recent experiments with isolated hepatocytes have shown that epinephrine, Bt_2-cAMP and glucagon all lead to increased Zn^{2+} uptake and metallothionein synthesis (Cousins and Burges, 1985). The effect of glucagon is inhibited by insulin. These responses are similar to the induction of gluconeogenic enzymes in hepatocytes. Epinephrine and glucagon also induce metallothionein in intact rats (Brady and Helvig, 1984). We have also found that *E. coli* endotoxin and interleukin-1 will induce metallothionein synthesis which is accompanied by a transient hypozincemia (Etzel et al., 1982; DiSilvestro and Cousins, 1984a, b).

The physiological role of increased cellular Zn^{2+} and metallothionein as regulated by hormones and host defense mediators such as interleukin-1 is not known. Increased cellular Zn^{2+} could affect the synthesis of metalloenzymes either as a direct participant or as a source of Zn^{2+} ions (Li et al., 1980; Udom and Brady, 1980). Recently, Coppen et al. (1985) observed that oxidative damage and free radical formation, induced in isolated hepatocytes by a variety of agents, e.g. t-butyl hydroperoxide, can be reversed by supplemental Zn^{2+} in the culture medium. Electron spin resonance data show the inhibition is inversely correlated with cellular metallothionein levels. These data agree with the observation that metallothionein is an extremely efficient $\cdot$OH radical scavenger (Thornalley and Vasak, 1985). Of possible relevance to the host defense aspects of this protein is our finding that the glucocorticoid analog dexamethasone potentiates the effect of Zn administration on metallothionein synthesis and hepatic Zn uptake (Quinones and Cousins, 1984). A similar combination was recently found more effective than Zn alone in inhibiting the growth of murine melanoma cells (Kreutzfeld et al., 1985), suggesting a relationship between tumor cell growth and metallothionein. The studies on metallothionein function strongly suggest that increased functional copies of this gene within the genome would be beneficial to organisms so endowed. Experiments are currently in progress to evaluate the effects of additional metallothionein genes on cellular function.

Expression of the metallothionein gene is clearly regulated by hormones and the dietary Zn supply. Shapiro et al. (1978) demonstrated that the induction of this protein by Zn involved changes in levels of translatable metallothionein mRNA. This was accounted for totally in the free polysomal mRNA pool (Shapiro and Cousins, 1980). Since these experiments, the metallothionein gene has been cloned from several species (Durnam et al., 1980; Hamer and Walling, 1982; Karin and Richards, 1982; Andersen et al., 1983; Butt et al., 1984).

The coding sequence that regulates the expression of the metallothionein gene by metals has been identified (Stuart et al., 1984; Carter et al., 1984). These were used to construct fusion genes with varying lengths of 5′ flanking sequence including metal regulation sites. A sequence of no more than 12 base pairs is required for regulation of the heterologous gene constructs by cadmium. Future experiments with the 5′ flanking sequence will probably delineate the site of regulation by cAMP, which is probably similar to the cAMP regulatory region for the phosphoenolpyruvate carboxykinase gene (Wynshaw-Boris et al., 1984).

Palmiter, Chin and Brinster (1982a) were successful in constructing fusion genes using the metallothionein gene promoter. Specifically, metallothionein-growth hormone fusion genes were introduced into fertilized mice eggs and were used to develop a strain of transgenic mice (Palmiter et al., 1982b). The transgenic mice responded to a high exogenous level of Zn in the drinking water and additional growth hormone was produced which stimulated growth far beyond what was normally expected. These classic experiments demonstrate the potential of supplying an activator of the promoter of a fusion gene via the diet or drinking water after the desired gene(s) have been introduced into the genome.

Summary

The ubiquitous distribution of Zn in microbial, plant and animal cells suggests a wide range of potential functions for this transition element. Until recently, the physiological role of Zn has centered on its interaction with enzymes. For example, the cessation of growth associated with nutritional Zn deficiency may involve a reduction in DNA polymerase activity and concomitant reduction in DNA synthesis. Similarly, RNA polymerases I, II and III are all Zn metalloenzymes and in appropriate model systems a reduction in the synthesis of r-RNA, mRNA, and t-RNA, respectively, is observed with cellular Zn depletion. The effect of Zn at the genetic level is most dramatically shown through induction of metallothionein synthesis. This protein influences the kinetics of Zn uptake by cells. It has been clearly demonstrated that metallothionein synthesis is under transcriptional control. Zinc and glucocorticoids, as well as glucagon and epinephrine (via elevation of cellular cAMP levels) all induce transcription of the metallothionein gene. Recently, it has been shown that a 12 bp fragment of the metallothionein gene promoter sequence regulates expression of heterologous genes in vivo which demonstrates that Zn can have a direct effect on expression of specific fusion genes. These results illustrate the potential of this relatively nontoxic metal in biotechnology applications where Zn responsive fusion genes could be utilized.

Acknowledgements This review paper is dedicated to Professor F. Aloysius Wood, former Dean for Research at the Institute of Food and Agricultural Sciences of the University of Florida in recognition of his pioneering efforts in stimulating the application of biotechnology to agricultural problems. The author wishes to thank Mrs. Ann C. Coutu for preparing the manuscript and Mr. Walter Jones for drawing two of the figures. Research from the author's laboratory discussed in this review was supported by NIH Grants AM 31127 and AM 31651.

References

Alfaro, B. and F. Heaton. 1974. *The subcellular distribution of copper, zinc, and iron in liver and kidney. Changes during copper dificiency in the rat.* Br. J. Nutr. 32:435–445.

Anderson, R. D., B. W. Birren, T. Gans, J. E. Piletz and H. R. Herschman. 1983. *Molecular cloning of the rat methallothionein-I (MT-1) mRNA sequence.* DNA 2:15–22.

Auld, D. S., H. Kawaguhi, D. M. Livingson and B. L. Vallee. 1974. *RNA-dependent DNA polymerase (reverse transcriptase) from avian myeloblastosis virus: a zinc metalloenzyme.* Proc. Nat. Acad. Sci. 71:2091–2095.

Brady, F. O. and B. Helvig. 1984. *Effect of epinephrine and norepinephrine on zinc thionein levels and induction in rat liver.* Amer. J. Physiol. 247:E318–E322.

Bryan, S. E., D. L. Vizard, D. A. Beary, R. A. LaBiche and K. J. Hardy. 1981. *Partitioning of zinc and copper within subnuclear nucleoprotein particles.* Nucl. Acids Res. 9:5811–5821.

Butt, T. R., E. J. Sternberg, J. A. Gorman, P. Clark, D. Hamer, M. Rosenberg and S. T. Crooke. 1984. *Copper metallothionein of yeast, structure of the gene, and regulation of expression.* Proc. Nat. Acad. Sci. 81:3332–3336.

Carter, A. D., B. K. Felder, M. T. Walling, M. F. Jubier, C. J. Schmidt and D. H. Hamer. 1984. *Duplicated heavy metal control sequences of the mouse metallothionein-I gene.* Proc. Nat. Acad. Sci. 81:7392–7396.

Chesters, J. K. 1972. *The role of zinc ions in the transformation of lymphocytes by phytohaemagglutinin.* Biochem. J. 130:133–319.

Coppen, D. E., R. J. Cousins and D. E. Richardson. 1985. *Effect of zinc on chemically induced peroxidation in rat liver parenchymal cells in primary culture.* Fed. Proc. 44:6404.

Cousins, R. J. 1985. *Absorption, transport, and intracellular metabolism of copper and zinc: special reference to metallothionein and ceruloplasmin.* Physiol. Rev. 65:238–309.

Cousins, R. J. and A. R. Burges. 1984. *Influence of glucagon and epinephrine on metallothionein induction and zinc exchange in rat liver parenchymal cells.* Fed. Proc. 44:3403.

DiSilvestro, R. A. and R. J. Cousins. 1984. *Mediation of endotoxin-induced changes in zinc metabolism in rats.* Amer. J. Physiol; 247:E436–E441.

DiSilvestro, R. A. and R. J. Cousins. 1984. *Glucocorticoid independent mediation of interleukin-1 induced changes in serum zinc and liver metallothionein levels.* Life Sci. 35:2113–2118.

Durnam, D. M., F. Perrin, F. Gannon and R. D. Palmiter. 1980. *Isolation and characterization of the mouse metallothionein-I gene*. Proc. Nat. Acad. Sci. USA 77:6511–6515.

Eichhorn, G. L. 1973. Pp. 1191–1243 in *Inorganic Biochemistry*. Elsevier Scientific Publishing Co., New York.

Etzel, K. R., S. G. Shapiro and R. J. Cousins. 1979. *Regulation of liver metallothionein and plasma zinc by the glucocorticoid dexamethasone*. Biochem. Biophys. Res. Comm. 89:1120–1126.

Etzel, K. R., M. R. Swerdel, J. N. Swerdel and R. J. Cousins. 1982. *Endotoxin-induced changes in copper and zinc metabolism in the syrian hamster*. J. Nutr. 112:2363–2373.

Failla, M. L. and R. J. Cousins. 1978. *Zinc uptake by isolated rat liver parenchymal cells*. Biochem. Biophys. Acta 538:434–444.

Falchuk, K. H., A. Krishan and B. L. Vallee. 1975. *DNA distribution in the cell cycle of Euglena gracilis, cytofluorometry of zinc deficient cells*. Biochemistry 14:3439–3444.

Falchuk, K. H., B. Mazus, L. Ulpino and B. L. Vallee. 1976. *Euglena gracilis DNA dependent RNA polymerase II: a zinc metalloenzyme*. Biochemistry 15:4468–4475.

Falchuk, K. H., L. Ulpino, B. Maxus and B. L. Vallee. 1977. *E. Gracilis RNA polymerase I: a zinc metalloenzyme*. Biochem. Biophys. Res. Comm. 74:1206–1212.

Fazakerly, G. V. 1984. *Zinc Z-DNA*. Nucl. Acids Res. 12:3643–3648.

Ferrin, L. J., A. S. Mildvan and L. A. Leob. 1983. *Metal content of DNA polymerase I purified from overproducing and wild type Escherichia coli*. Biochem. Biophys. Res. Comm. 112:723–727.

Fosmire, G. J., Y. Y. Al-Ubaidi, E. Halas and H. H. Sandstead. 1974. *The effect of zinc deprivation on the brain*. Adv. Expl. Med. Biol. 48:329–345.

Hamer, D. H. and M. Walling. 1982. *Regulation in vivo of a cloned mammalian gene: cadmium induces the transcription of a mouse metallothionein gene in SV 40 vectors*. J. Mol. Appl. Genet. 1:273–288.

Karin, M. and R. I. Richards. 1982. *Human metallothionein genes – primary structure of the metallothionein-II gene and a related processed gene*. Nature (London) 299:797–802.

Kreutzfeld, K. L., K. Y. Lei, M. D. Bergman and F. L. Meyskens, Jr. 1985. *Dexamethasone and zinc in combination inhibit the anchorage-independent growth of S-91 cloudman murine melanoma*. Life Sci. 36:823–827.

Li, T. Y., A. J. Kramer, C. F. Shaw III and D. H. Petering. 1980. *Ligand substitution reactions of metallothionein with EDTA and apo-carbonic anhydrase*. Proc. Nat. Acad. Sci. USA 77:6334–6338.

Lieberman, I., R. Abrams, N. Hun and P. Ove. 1963. *Levels of enzyme activity and deoxyribonucleic acid synthesis in mammalian cells cultured from the animal*. J. Biol. Chem. 238:3935–3962.

Macapinlac, M. P., W. N. Pearson, G. H. Barney and W. J. Darby. 1968. *Protein and nucleic acid metabolism in the testes of zinc-deficient rats*. Nutrition 95:569–577.

Mazus, B., K. H. Falchuk and B. L. Vallee. 1984. *Histone formation, gene expression, and zinc deficiency in Euglena gracilis*. Biochemistry 23:43–47.

O'Neal, R. M., G. W. Pla, M. R. S. Fox, F. S. Gibson and B. E. Fry, Jr. 1970. *Effect of zinc deficiency and restricted feeding on protein and ribonucleic acid metabolism of rat brain*. J. Nutr. 100:491–497.

Palmiter, R. D., H. Y. Chen and R. D. L. Brinster. 1982a. *Differential regulation of metallothionein-thymidine kinase fusion genes in transgenic mice and their offspring*. Cell 29:701–710.

Palmiter, R. D., R. L. Brinster, R. E. Hamer, M. E. Trumkbauer, M. G. Rosenfeld, N. C. Birnberg and R. M. Evans. 1982b. *Dramatic growth of mice that develop from eggs microinjected with metallothionein-growth hormone fusion genes*. Nature 300:611–615.

Pattison, S. E. and R. J. Cousins. 1985. *Zinc transport and metabolism by hepatocytes*. Fed. Proc. 44 (in press).

Petranyi, P., J. J. Jendrisak and R. R. Burgess. 1977. *RNA polymerase II from wheat germ contains tightly bound zinc*. Biochem. Biophys. Res. Comm. 74:1031–1038.

Prasad, A. S., D. Oberleas, D. Koniuch and E. DuMouchelle. 1973. *Ribonuclease and deoxyribonuclease activities in zinc-deficient tissues*. J. Lab. Clin. Med. 82:461–466.

Prasad, A. S. and D. Oberleas. 1974. *Thymidine kinase activity and incorporation of thymidine into DNA in zinc-deficient tissue*. J. Lab. Chin. Med. 83:634.

Prasad, A. S., P. Rabbani, A. Abbasil, E. Bowersox and M. R. S. Fox. 1978. *Experimental zinc deficiency in humans*. Ann. Intern. Med. 89:483–490.

Price, C. A. and B. L. Vallee. 1962. *Euglena gracilus, a test organism for study of zinc*. Plant Physiol. 37:428–433.

Quinones, S. R. and R. J. Cousins. 1984. *Augmentation of dexamethasone induction of rat liver metallothionein by zinc*. Biochem. J. 219:959–963.

Raulin, J. 1969. *Etudes cliniques sur la vegetation*. Ann. Sci. Nat. Bot. Biol. Vegetale 11:93–299.

Richards, M. P. and R. J. Cousins. 1975. *Mammalian zinc homeostasis: requirements for RNA and metallothionein synthesis*. Biochem. Biophys. Res. Comm. 64:1215–1223.

Richards, M. P. and R. J. Cousins. 1976a. *Metallothionein and its relationship to the metabolism of dietary zinc in rats*. J. Nutr. 106:1591–1599.

Richards, M. P. and R. J. Cousins. 1976b. *Zinc-binding protein: relationship to short term changes in zinc metabolism*. Proc. Soc. Expl. Biol. Med. 153:52–56.

Rubin, H. 1972. *Inhibition of DNA synthesis in animal cells by ethylene diamine tetraacetate, and its reversal of zinc*. Proc. Nat. Acad. Sci. 69:712–716.

Sandstead, H. H. and R. A. Rinaldi. 1969. *Impairment of deoxyribonucleic acid synthesis by dietary zinc deficiency in the rat*. J. Cell Physiol. 73:81–84.

Sandstead, H. H., M. Terhune, R. N. Brady, D. Gillespie and W. L. Holloway. 1971. *Zinc deficiency: brain DNA protein and lipids; and liver ribosomes and RNA polymerase*. Clin. Res. 19:83–89.

Schneider, E. and C. A. Price. 1962. *Decreased ribonucleic acid levels: a possible cause of growth inhibition in zinc deficiency*. Biochem. Biophys. Acta 55:406–410.

Scrutton, M. C., C. W. Wu and D. A. Goldthwait. 1971. *The presence and possible role of zinc in RNA polymerase obtained from Escherichia coli*. Proc. Nat. Acad. Sci. 68:2497–2501.

Shapiro, S. G. and R. J. Cousins. 1980. *Induction of rat liver metallothionein mRNA and its distribution between free and membrane-bound polysomes*. Biochem. J. 190:755–764.

Shapiro, S. G., K. S. Aquibb, LK. A. Markowitz and R. J. Cousins. 1978. *Cell-free synthesis of metallothionein directed by rat liver polyadenylated messenger RNA*. Biochem. J. 175:833–840.

Slaby, I., B. Kind and A. Holmgren. 1984. *T7 DNA polymerase is not a zinc-metalloenzyme and the polymerase and exonuclease activities are inhibited by zinc ions*. Biochem. Biophys. Res. Commun. 122:1410–1417.

Slater, J. P., A. S. Mildvan and L. A. Loeb. 1971. *Zinc in DNA polymerase*. Biochem. Biophys. Res. Commun. 44:37–43.

Smeyers-Verbeke, J., C. May, P. Drochmans and D. L. Massart. 1977. *The determination of Cu, Zn, and Mn in subcellular rat liver fractions*. Anal. Biochem. 83:746–753.

Somers, M. and E. J. Underwood. 1969. *Ribonuclease activity and nucleic acid and protein metabolism in the testes of zinc-deficient rats*. Austral. J. Biol. Sci. 22:1277–1282.

Springgate, C. F., A. S. Mildvan, R. Abramson, J. L. Engle and L. A. Leob. 1973. *Escherichia coli deoxyribonucleic acid polymerase I, a zinc metalloenzyme*. J. Biol. Chem. 248:5987–5993.

Squibb, K. S. and R. J. Cousins. 1974. *Control of cadmium binding protein synthesis in rat liver*. Environ. Physiol. Biochem. 4:24–30.

Squibb, K. S., R. J. Cousins and S. L. Feldman. 1977. *Control of zinc-thionein synthesis in rat liver*. Biochem. J. 164:223–228.

Stuart, G. W., P. F. Searle, H. Y. Chen, R. L. Brinster and R. D. Palmiter. 1984. *A 12-base-pair DNA motif that is repeated several times in metallothionein gene promoters confers metal regulation to a heterologous gene*. Proc. Nat. Acad. Sci. 81:7318–7322.

Swenerton, H., R. Shrader and L. S. Hurley. 1969. *Zinc-deficient embryos: reduced thymidine incorporation*. Science 166:1014–1015.

Tal, M. 1969. *Metal ions and ribosomal conformation*. Biochim. Biophys. Acta 195:76–80.

Terhune, M. W. and H. H. Sandstead. 1972. *Decreased RNA polymerase activity in mammalian zinc deficiency*. Science 177:68–69.

Thornalley, P. J. and M. Vasak. 1985. *Possible role for metallothionein in protection against radiation-induced oxidative stress. Kinetics and mechanisms of its reaction with superoxide and hydroxyl radical*. Biochim. Biophys. Acta 827:36–44.

Todd, W. R., C. A. Elvehjem and E. B. Hart. 1934. *Zinc in the nutrition of the rat*. Am. J. Physiol. 197:146.

Udom, A. O. and F. O. Brady. 1980. *Reactivation in vitro of zinc-requiring apo-enzymes by rat liver zinc-thionein*. Biochem. J. 187:326–335.

Underwood, E. J. 1977. *Zinc*. Pp. 196–242 in *Trace Elements in Human and Animal Nutrition*. Academic Press, New York.

Wacker, W. E. C. 1962. *Nucleic acids and metals*, III. *Changes in nucleic acid, protein and metal content as a consequence of zinc deficiency in Euglena gracilis*. Biochemistry 1:859–865.

Wandzilak, T. M. and R. W. Benson. 1977. *Yeast RNA polymerase III: a zinc metalloenzyme*. Biochem. Biophys. Res. Comm. 76:247–252.

Wegener, W. S. and A. H. Romano. 1963. *Zinc stimulation of RNA and protein synthesis in Rhizopus nigrican*. Science 142:1669–1670.

Weser, U. and E. Bichoff. 1970. *Incorporation of* ^{65}*Zn in rat liver nuclei*. Europ. Biochem. J. 12:571–575.

Williams, R. O. and L. A. Loeb. 1973. *Zinc requirement for DNA replication in stimulated human lymphocytes*. J. Cell Biol. 58:594–601.

Williams, R. J. P. 1984. *Zinc: what is its role in biology*? Endeavor 8:65–70.

Wynshaw-Boris, A., T. G. Lugo, J. M. Short, R. E. K. Fournier and R. W. Hanson. 1984. *Identification of a cAMP regulatory region in the gene for rat cytosolic phosphoenolypyruvate carboxykinase (GTP)*. J. Biol. Chem. 259:12161–21269.

Part 4

Health and diseases

Novel approaches to improving disease resistance in plants

N. T. KEEN
Department of Plant Pathology, University of California, Riverside, CA 92521, U.S.A.

Abstract Improved disease and pest resistance in plants is one of the more lucrative possibilities for utilization of recombinant DNA technology. This optimism results from recent increases in our knowledge of how naturally occurring disease resistance functions and to the fact that modulation of resistance expression is often controlled by single dominant alleles in plants. Historically these disease resistance genes could only be transferred between plants that can be intercrossed. Recently, however, technologies have been developed which permit gene transfer between taxonomically diverse plants by (i) protoplast fusion and selection of somatic recombinants or (ii) introduction of specific DNA sequence into plants via Ti plasmid vectors. Disease resistance genes have not yet been isolated to permit full utilization of these technologies, but several rationales are being pursued in attempts to molecularly clone them. These include the use of transposable elements to tag resistance gene sequences, the screening of c-DNA libraries from inoculated plants and the use of specific pathogen metabolites called elicitors as probes to detect the protein products of plant disease resistance genes.

Introduction

Increasing the disease and pest resistance of cultivated plants is an appealing goal for the utilization of biotechnology to increase food and fiber production. This paper will suggest various lucrative experimental approaches, including the molecular cloning of disease resistance genes. I will summarize some of what is known about how naturally occurring disease resistance works but will largely rely on citations to other reviews and recent original papers. Finally, I will describe some current research projects which have applicability to the goal of introducing disease resistance genes into foreign plants. I will use references sparingly, instead relying on review papers where possible. The terminology of plant-pathogen interactions is frequently confused and oppressive, so I will attempt to define terms at the outset.

Disease resistance implies that pathogen damage is reduced relative to that occurring on a susceptible plant, presumably by some heritable plant trait(s). This generally results from a passive or active defense mechanism which impairs pathogen development. As discussed more fully later, the resistance mechanism controlled by single disease resistance genes is usually the hypersensitive reaction (HR). In cases where this or other mechanisms function, the plant is called resistant

Submitted for Beltsville Symposium X, "Biotechnology for Solving Agricultural Problems", May 5–9, 1985.

and the plant-pathogen interaction is considered incompatible; conversely, if resistance does not occur, the plant is susceptible and the interaction is referred to as compatible. Pathogen genes which complement plant resistance genes have conventionally been called avirulence genes because the dominant alleles generally condition incompatibility on the complementary host resistance genotype. Pathogen races are biotypes of a single pathogen species, pathovar or forma speciales which are distinguished by the unique pattern of compatible or incompatible reactions of various host resistance genotypes. Race is an arbitrary designation for those pathogen biotypes which fall into a certain category based on the differentials employed and does not refer to the specific genes involved in determination of specificity. Pathovars and forma speciales are subgroups of a pathogen species which are distinguished by their ability to attack members of one plant species, but generally not other plant species. These groupings therefore represent a higher level of plant-pathogen specificity than race.

The HR is involved in plant resistance against only certain races of a pathogen taxon (race specific resistance) or against entire pathovars or species (general resistance). Although the recognitional elements are clearly different in these two types of resistance, they appear to invoke the same HR expressive mechanisms. Hypersensitive resistance also involves two distinct chronological aspects. The determinative phase denotes the recognitional period shortly after pathogen infection when the pathogen may be detected by plant cells and a signal is generated which initiates the plant HR. The expressive phase involves the resultant complex of biochemical events which are believed to account for the cessation of pathogen development.

Plant biotechnology and the potential of manipulating disease resistance genes

With the development of efficient techniques for the transformation of foreign DNA into plant cells by the use of Ti plasmid vectors, cell fusions and direct transformation, plant biology is entering an exciting new era. Now that it is possible to introduce desired DNA sequences into many plant species, the remaining questions are "what genes do we introduce and where do we get them?" There are several current suggestions in this arena, some of which are summarized in Table 1. The most immediate progress will be made with single dominant genes that give well defined and easily screened phenotypes. Genes for herbicide resistance or new storage proteins that may affect the nutritional status of plant tissue have already been transformed

Table 1. Some approaches to improving crop plants using biotechnology

1. Introduce genes contributing to increased yield.
2. Introduce genes for greater quality and nutritive value of edible plant parts.
   a. Genes coding for storage proteins with high levels of essential amino acids.
   b. Genes such as amylases to alter the sugar/starch ratio of food stuffs.
3. Use engineered microorganisms which interact with crop plants to increase productivity.
   a. *Rhizobium* sp. with improved characteristics and competitiveness.
   b. Microorganisms which counteract stress agents, including diseases and pests, element shortages or excesses, frost damage, etc.
4. Introduce genes into plants that permit tolerance to stresses.
   a. Drought and salinity tolerance.
   b. Herbicide resistance.
   c. Heat tolerance and resistance to cold.
   d. Resistance to toxic chemicals and elements in the soil.
   e. Tolerance to air pollution.
5. Improved resistance to diseases caused by bacteria, fungi, insects, and nematodes.
   a. Transfer plant disease resistance genes to unrelated plants.
   b. Introduce genes to broaden the number of phytoalexins made by a certain plant.
   c. Genes which alter the amount of phytoalexins or induced structural barriers made in response to a pathogen.
   d. Transform engineered genes into plants that code for secreted proteins which are antagonistic to microbial pathogens and insects.
      i. lysozyme production to antagonize bacteria.
      ii. ribonuclease to attack RNA plant viruses.
      iii. chitinase and glucanase to antagonize insects and fungi.
      iv. *Bacillus thuringiensis* and other proteins which antagonize insects.
   e. Clone genes into plants that diminish the plant environment available to a pathogen.
      i. genes coding for efficient microbial iron-binding siderophores, thus out-competing pathogens for iron.
      ii. efficient permeases which transport glucose and amino acids into plant cells, thus leaving low amounts in plant intercellular spaces for pathogen growth.
   f. Introduce a rearranged immunoglobulin gene coding for an IgA which recognizes a conserved cell surface antigen of pathogens or a protein important in its pathogenesis.

experimentally and may be routinely used in crop production in the near future. These examples have limited utility, however, when it comes to the more general questions of increased yield and improved plant quality. It has been widely noted that productivity of several major crop plants has levelled in recent years, and that recombinant DNA technology poses an appealing alternative approach to seeking dramatic increases. While this is a useful goal, the genes controlling plant yield and quality are generally not well defined and, to the contrary, occur as a large number of genes with small but additive effects. There is also the problem that introduced foreign genes may not be regulated properly in the recipient plant and detrimental effects may, in fact, be observed. All of these limitations can be expected to temper the rate at which biotechnology impacts on plant productivity. For the near future, gene exchange using Ti plasmid vectors and other methods will probably be of most use to science itself in the imposing task of further understanding basic plant biology that is required before a systematic approach to plant improvement through biotechnology can be marshalled.

Disease resistance genes and their transfer to unrelated plants

Biotechnological approaches will meet with the most rapid progress through the introduction into foreign plants of single Mendelian genes that exhibit well defined phenotypes and significantly increase plant productivity. At present, this list is short, but approaches which may provide improved disease and pest control are appealing candidates for such work. As noted in Table 1, number 5, disease control may be effected by transferring naturally occurring disease resistance genes to unrelated plants, by introducing foreign genes which make the plant defense system more effective, or by transforming genes that create a hostile plant environment for pathogens and pests. Points 5b and 5c are discussed more fully in Keen (1985) and will not be elaborated on here. The approaches in point 5d (Table 1) are simple, since several candidate enzymes have already been cloned for various sources and their engineering and transformation into plants should be straightforward. This would permit assessment of their potential as disease control agents. Similarly, the approaches in point 5e of Table 1 would appear to be readily testable in attempts to impair the success of pathogens. Since many pathogens develop in the intercellular spaces of plant hosts, any device which reduces the nutrient supply in this location by promoting more efficient nutrient uptake by plant cells would be expected to effect disease control. Mammalian immunoglobulin genes which recognize features of plant pathogens such as cell surface glycoproteins could be cloned (point 5f, Table 1), and the appropriately engineered genes introduced into plants. Alternatively, a gene coding for an antibody clone that recognises a pathogen peptide toxin or enzyme important in pathogenesis could be expected to impede the development of pathogenesis. Appealing as experiments such as these are, plant biologists are currently more interested in the possibility of cloning naturally occurring disease resistance genes and transforming them into unrelated plants for disease control (point 5a of Table 1).

Hundreds of resistance genes have been demonstrated to occur in the germplasm of various cultivated crop plants, and they frequently produce considerable economic benefit by preventing losses due to diseases and insect pests. Disease resistance genes are often, indeed generally, inherited as dominant single gene characters which can readily be transferred between plants of a single plant species by crossing. With the exception of those few cases of permissible interspecific crosses, however, resistance genes from one plant species cannot be transferred to another species by conventional plant breeding due to infertility. Since pathogens generally do not obey such plant taxonomic restrictions and to the contrary often have wide plant host ranges, it is clear that resistance genes from one plant species effective against a certain pathogen would be of potential use if they could be transferred to an unrelated plant species lacking useful resistance genes. To facilitate such an approach requires that we have some understanding about how natural disease resistance functions in higher plants. Fortunately, progress is occurring in both of these areas.

The nature of resistance mechanisms modulated by disease resistance genes

Resistance mechanisms have thus far been investigated almost exclusively by comparative biochemical approaches, but these have often met with considerable ambiguity (see Keen and Staskawicz, 1984). In most of the plant-pathogen systems studied, it appears that the expression of single gene disease resistance results from a complex active defense mechanism called the hypersensitive response (HR). The most obvious symptom of HR is the inducible but relatively rapid necrosis of host cells in the immediate vicinity of the infecting pathogen. This is coupled with series of biochemical events in the dying cells as well as healthy plant cells surrounding the infection site. Some of these are: (i) the inducible synthesis of antibiotic secondary metabolites called phytoalexins by healthy cells and their transport to the dead, hypersensitive plant cells and their intercellular spaces (for recent reviews see Darvill and Albersheim, 1984; Keen, 1985; Kuc and Rush, 1985); (ii) the synthesis in at least some plant HRs of inducible structural barriers such as lignin and glycoproteins around the infection site (Roby, Toppan and Esquerre-Tugaye, 1985; Vance, Kirk and Sherwood, 1980). The presumption is that one or more of these mechanisms inhibit pathogen growth and/or wall off the pathogen from further spread through the plant. An early symptom of the HR in many plants is a rapid and pronounced change in plasma membrane permeability which may result from the production of superoxide anion (Doke, 1983) or singlet oxygen (Salzwedel, Daub and Huang, 1985) by the necrosing host cells.

Most HR can be blocked by inhibitors of plant transcription and translation such that cell necrosis and the subsequent biochemical symptoms do not occur and the pathogen is not inhibited. A good deal is now known about plant gene activation during phytoalexin biosynthesis, as Dr. Lamb describes in this volume. More specific inhibitors of the metabolic pathways leading to phytoalexin biosynthesis have also frequently been shown to block resistance expression, but host cell necrosis still occurs. Such experiments support the idea that phytoalexins or other mechanisms, but not plant cell death per se, are causally related to the cessation of pathogen development. While the full inventory of biochemical events associated with the hypersensitive response of plants is far from complete, it is clear that all of these reactions are inducible. That is, the expressive biochemical mechanisms of HR do not occur until after penetration and initial infection by an incompatible pathogen race. If a genetically compatible pathogen race invades, recognition does not occur and the subsequent events associated with resistance expression, including plant cell necrosis and phytoalexin accumulation, do not occur. These events a priori suggest that resistance involves two distinct stages, the first a recognitional phase involving detection of some feature of the invading pathogen by plant cells and the second a subsequent series of biochemical events associated with containment of the pathogen. A wealth of genetic evidence in plants clearly indicates that most plant disease resistance genes modulate the hypersensitive response (see Keen, 1982). For a resistance gene to initiate the HR also requires that the pathogen contain a complementary dominant allele called an "avirulence gene". If the pathogen contains a recessive "virulence" allele at the avirulence locus, plant recognition of the pathogen does not occur, the HR is not invoked and the pathogen colonizes and damages the plant. In a complex plant-pathogen interaction such as wheat and the stem rust fungus, at least 50 different plant disease resistance genes have been detected. The pathogen, *Puccinia graminis* f. sp. *tritici*, also contains 50 independent avirulence genes, each complementary to a different plant resistance gene.

While we do not know the molecular mechanisms underlying such gene-for-gene complementarity, the simplest model is a direct elicitor-receptor mechanism. Such models propose that the pathogen avirulence gene product itself or a secondary gene product (called an elicitor) is recognized by the plant resistance gene product to initiate the hypersensitive response. Thus, pathogens contain molecules whose specific chemical structure is determined directly or indirectly by dominant avirulence alleles and these molecules are selectively recognized only by plants carrying complementary resistance gene alleles. There are now a few cases in which such race-specific pathogen elicitors have been demonstrated (Anderson, 1980; Bruegger and Keen, 1970; DeWit et al., 1985; Keen and Legrand, 1980; Mayama and Keen, 1984), but none of them have yet been well characterized or subjected to genetic analysis. In none of the cases, therefore, has it yet been experimentally proven that the race specific elicitors are in fact primary or secondary gene products of pathogen avirulence genes.

Recent work with bacterial pathogens has resulted in the molecular cloning of race specific avirulence genes associated with induction of the plant HR. Gabriel (1984) has cloned several different avirulence genes from *Xanthomonas campestris* pv. *malvacearum*, a pathogen of cotton. These clones were shown to each complement a specific single plant disease resistance gene, but the molecular genetics of the avirulence genes has not yet been studied. More extensive work has been done by Staskawicz and collaborators with *Pseudomonas syringae* pv. *glycinea*, a pathogen of soybean. Three different avirulence genes have thus far been isolated. Initially, an avirulence gene was cloned from race 6 of the bacterium by screening a cosmid library and was shown to confer all race 6 incompatibilities on a series of host cultivars (Staskawicz, Dahlbeck and Keen, 1984). This gene has subsequently been sub-cloned as a ca. 3 kb fragment (Dahlbeck and Staskawicz, unpublished observations). Introduction into other *P. syringae* pv. *glycinea* races and, significantly, *Xanthomonas campestris* pv. *glycines* resulted in HR on all soybean cultivars resistant to race 6. Transposon mutagenesis with Tn5 suggested that a single gene of *ca*. 2.5 kb conferred the race 6 phenotype. This was recently confirmed when Napoli and Staskawicz (unpublished observations) used *E. coli* maxicells and DNA sequence data to demonstrate that the race 6 avirulence gene of *P. syringae* pv. *glycinea* appears to code for a single large protein. The remaining unanswered question is whether this protein functions per se as a race specific defence elicitor in soybean. The cloning work is thus far consistent with earlier results by Bruegger and Keen (1979) indicating that the various races of *P. syringae* pv. *glycinea* contain race specific elicitors associated with their outer membranes. If the bacterial avirulence gene products themselves turn out to be the recognitional elements which are detected by plant disease resistance gene products, labelling of the pathogen molecules provides a direct approach to isolation of the complementary plant molecules by affinity techniques. Identification of the putative resistance gene products in turn should permit the identification of DNA clones from soybean that contain disease resistance genes. This approach will be discussed more fully later.

Isolation of a putative plant disease resistance gene by the above rationale may also be done by employing the host specific toxin, victorin. Victorin is a series of structurally related, low molecular weight peptides produced by the fungus *Helminthosporium victoriae*. Extremely low concentrations of the peptides produce necrotic symptoms on oat plants containing a single dominant allele called *Pc-2* (Scheffer and Livingston, 1984). Victorin does not affect oat plants carrying the recessive *Pc-2* allele, any other known oat genotype or any other known plant, but is a potent metabolite on *Pc-2* oats, giving necrotic effects at concentrations as low as 1 pg ml^{-1} (Keen, Midland and Sims, 1983; Mayama and Keen, unpublished observations). Of considerable interest, the Pc-2 allele is a disease resistance gene against the fungus pathogen *Puccinia coronata*, conditioning a HR and phytoalexin production in response to incompatible pathogen races (Mayama et al., 1982). Perhaps not surprising, therefore, the biochemical effects of victorin on *Pc-2* oat tissues [rapid increase in plasma membrane permeability, cell necrosis and phytoalexin production (Mayama and Keen, 1984)] are precisely those expected if victorin were behaving as a specific elicitor of the HR. Significantly, victorin affects the permeability of most cells in *Pc-2* oat plants within minutes (Scheffer and Livingston, 1984), thus indicating that the *Pc-2* gene is constitutively expressed. Victorin active molecules have recently been purified by several groups (Keen et al., 1983; Macko, unpublished observations; Walton and Earle, 1984). While structures have not yet appeared, it is known that the active molecules are related basic peptides containing unusual amino acids. The in vitro labelling of these metabolites to high specific activity and their use as probes is being employed as a route to attempt isolation of the *Pc-2* gene product (Bubrick and Keen, unpublished observations). Use of these probes should facilitate identification of the *Pc-2* gene product on protoplasts, isolated plasma membranes, and possibly on electrophoretic blots of plasma membrane proteins from *Pc-2* oats.

Schemes for the molecular cloning of plant disease resistance genes

While the introduction of disease resistance genes into disparate plants is an appealing idea, cloned resistance genes are not presently available. Obtaining such clones is an imposing task since the identity of resistance gene products is not known and the only current method for their detection is to inoculate plants with the appropriate pathogen. Nevertheless, several avenues are presently being considered to clone plant disease resistance genes (Table 2).

Scheme number one in Table 2 is only feasible in plants that have several genes mapped and in which one or more of them is relatively close to a resistance gene. The approach would be most feasible with a plant having a relatively small genome, such as *Arabidopsis* spp. The

Table 2. Possible rationales for the cloning of plant disease resistance genes

1. Identify a closely linked marker gene with a readily identifiable phenotype; clone this gene and "walk" along the chromosome in order to clone the resistance gene.
2. Attempt to clone resistance genes by function. In this scheme, large plant genomic DNA fragments would be transformed into a different plant species or into different cultivars of the same species which lack the gene, and plants would be screened for resistance by inoculation with an appropriate pathogen strain.
3. Screen a cDNA library in the same way as in number 2.
4. Use transposable elements to mutate a disease resistance gene. Isolate clones containing the transposable element by hybridization analysis and use flanking sequence to identify the native resistance gene in a DNA library prepared from the wild-type plant.
5. Select monoclonal antibodies which are specific for protein encoded by plant disease resistance genes and then use them to select genomic clones from a phage expression library.
6. Electrophorese plasma membrane proteins on two dimensional gels from near-isogenic plant lines, one of which contains a dominant disease resistance gene. Any novel protein from the latter would be a candidate resistance gene product.
7. Employ pathogen avirulence gene products to isolate the receptor proteins which are predicted to be the products of plant disease resistance genes. Once these latter proteins are identified, prepare antibodies to them and screen an expression library of genomic plant DNA or a c-DNA library.

same is true for attempts to clone by function (nos. 2 and 3), since the number of plant transformants to screen would otherwise be prohibitively large. It should be noted that all of the schemes in Table 2 require that the plant be readily transformed and the transformants regenerated for the testing of phenotype by inoculation with an appropriate pathogen race. If disease resistance genes are inducible after pathogen infection, then they might be cloned from a cDNA library prepared after cascade hybridization of inoculated plant DNA with non-inoculated plant DNA. However, if resistance genes are constitutive, as previous work with victorin has indicated, then a cDNA approach may not work. The use of transposable elements to inactivate a resistance gene (scheme 4, Table 2) is an appealing approach, since the mutated gene would then have a molecular tag which could facilitate subsequent cloning of the wild-type gene. The method suffers, however, because many plants do not have useful transposable elements, and those that do usually have many copies of the element. Monoclonal antibodies (scheme 5) have the high specificity necessary to detect the protein products of functional alleles for disease resistance genes. If resistance gene products are localized in the plant plasma membrane, as predicted by many models, then screening antibodies prepared against membrane proteins from near isogenic resistant and susceptible plants may result in the isolation of a resistance-gene specific antibody which could subsequently be employed to select the gene from a phage expression library of plant DNA. Another related approach to the identification of resistance gene products is the electrophoresis of plasma membrane proteins, hoping to find a unique polypeptide (scheme 6). The final scheme is based on the previously discussed prediction that avirulence gene products and plant resistance gene products may physically interact (scheme 7, Table 2). If this in fact occurs, isolation of the resistance gene product should be possible by using a suitable labelled derivative of the pathogen metabolite. Once resistance gene products were identified by schemes 5–7, cloning of the genes should be straightforward by using antibodies directed to the protein products. It is anticipated that one or more of the above rationales will result in the cloning of resistance genes in the near future and will therefore permit testing of whether they are functionally expressed following transformation into other plant species.

Prospects for introducing disease resistance genes into foreign plants via biotechnology

If any of the approaches discussed in the last section result in the cloning of plant disease resistance genes, it should be a straightforward matter to transform them into various other higher plants with a Ti plasmid vector and test whether they in fact confer resistance to the expected pathogens. It is of concern that the known specific resistance genes that are effective against only certain races of a single pathogen species are probably not the genes that will have widespread utility in agriculture. This is because they are too specific, conferring resistance to only specific biotypes of one pathogen species. More lucrative are general resistance genes that are effective against several entire pathogen species or sub-species, such that virulent races would not be expected to readily evolve. If such plant resistance genes were introduced into a foreign plant, they would be expected to function against the same pathogen species or sub-species as they did in the original plant and we would not predict the evolution of virulent races. However, do these plant genes exist? We have little information because unrelated plants are generally not cross fertile and the desired genes can therefore not be detected by conventional crossing. In a few cases where interspecific crosses have been accomplished, however, there are tantalizing examples of resistance genes that have been moved across species limitations and have not led to the rapid appearance of virulent races. Recent work with various pathovars of *Pseudomonas syringae* supports the suggestion that such general resistance genes not only exist but may be cloned by some of the same approaches noted in the last section for race specific resistance genes.

Avirulence genes have recently been cloned from *Pseudomonas syringae* pv. *tomato* which, when introduced into *P. syringae* pv.

glycinea, caused this bacterium to elicit a HR on several tested soybean cultivars (Kobayashi and Keen, 1985). The cloned sequences appear to contain pathovar specific avirulence genes which may account for the general observation that biotypes of pv. tomato invariably elicit a hypersensitive defense reaction on all soybean cultivars. It will accordingly be of considerable interest to attempt isolation of gene products from the cloned pv. *tomato* avirulence genes in order to search for complementary soybean resistance genes in the manner discussed previously. The cloning of such general disease resistance genes against pv. *tomato* would permit their transformation into tomato plants and testing of the resulting transformants for resistance to pv. *tomato*.

Another approach to transferring genes for disease resistance between plants that cannot be intercrossed is the use of protoplast fusions (Maliga, 1984). The method can also be used for genes that have not been cloned. It involves the fusion of cells of resistant and susceptible species followed by screening for somatic recombinants in which a resistance gene has been transferred into the susceptible parent. Dr. Avi Nachmias of the Gilat Experiment Station, Negev, Israel, is pursuing such a project with a well known gene that confers tolerance in tomato plants to the vascular pathogen *Verticillium dahliae*. The gene, called *Ve*, is inherited as a single, dominant, Mendelian character. It confers resistance to disease symptoms and pathogen colonization. Unfortunately, genes analogous to *Ve* have not been detected in other solanaceous plants such as potato and eggplant, and these plants may be seriously damaged by *V. dahliae*. These latter plants also cannot be intercrossed with tomato. However, since protoplasts of potato or eggplant may be fused with those from tomato, it should be possible to attempt introduction of the *Ve* gene into potato or eggplant if a suitable selection existed for detection of somatic recombinants. Nachmias and co-workers have shown that the pathogenicity of race 1 *V. dahliae* is largely accounted for by its production of a peptide toxin of ca. 20 amino acid residues (Buchner, Nachmias and Burtin, 1982; Nachmias, Buchner and Krikun, 1982; Nichmias, Buchner and Burstein, 1985a). Significantly, the peptide from *V. dahliae* race 1 does not significantly affect tomato plants containing the *Ve* gene but does damage plants lacking this allele (Nachmias et al., 1985b). The *Ve* gene thus appears to behave differently from the large number of disease resistance genes which modulate a HR. To the contrary, *Ve* appears to confer tolerance to the damaging effects of the Verticillium toxin. This suspicion was confirmed by Nachmias et al. (1985b) with the demonstration that *V. dahliae* race 2 produces a toxic peptide of altered amino acid composition that overcomes *Ve* gene resistance and gives severe symptoms on tomato plants carrying this gene. While the sequencing and synthesis of the peptides from races 1 and 2 of *V. dahliae* is still underway (Buchner and Burstein, unpublished observations), it is clear that the *Ve* gene confers insensitivity to the toxic peptide of *V. dahliae* race 1 and the gene would therefore be of potential use in other plants attacked by race 1 of the fungus. Since the toxic peptide from race 1 of *V. dahliae* can be used to select intact plants and cultured cells containing the *Ve* gene, Nachmias and collaborators are employing it to detect desired recombinants resulting from fusions of potato or eggplant cells with *Ve* tomato cells (see Fig. 1).

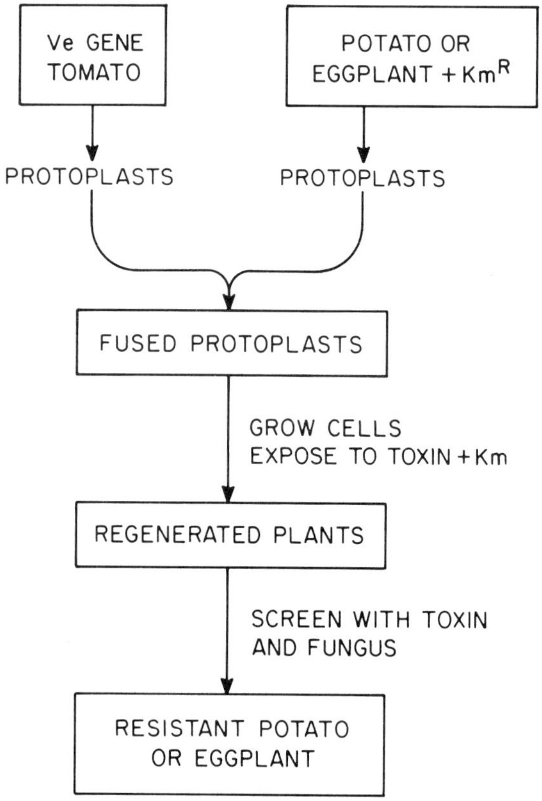

Figure 1. Proposed scheme for transfer of the tomato *Ve* gene to eggplant or potato by cell fusions and selection of somatic recombinants. Recipient eggplant and potato cells carry kanamycin resistance introduced by a Ti plasmid vector (Horsch et al., 1985). This permits counter selection against the unfused *Ve* tomato protoplasts on kanamycin medium. Race 1 *Verticillium dehliae* toxic peptide is used to kill protoplasts lacking the *Ve* gene. In this way, it should be possible to select regenerated plants with eggplant or potato phenotypes, which contain the *Ve* gene.

Conclusions

Plant-pathogen systems have two primary appeals to further molecular biologic investigations of plants. First, the study of these interactions has historically resulted in considerable information on basic plant biology, including the discovery and/or study of most plant growth hormones, effects of pathogen-produced toxins and enzymes on plant metabolism, the physiology of translocation and utilization of the Ti plasmid from *Agrobacterium tumefaciens* as a transformation vector. It is probable that plant-pathogen systems will continue to make heavy contributions in these areas. Second, as I have outlined in this paper, several rationales are available for using the tools of biotechnology to effect improved plant disease and pest control in practical agriculture. Some of them will require considerable effort, such as those involving the cloning of resistance genes. Others, including the introduction of cloned genes coding for proteins that are antagonistic to pathogens, should be more rapidly tested for their effectiveness. The probability is high that one or more of the approaches to increasing disease resistance will succeed. This is not a trivial prospect since losses to pathogens and pests are large on a world basis.

References

Anderson, A. J. 1980. *Differences in the biochemical composition and elicitor activity of extracellular components produced by three races of a fungal plant pathogen, Colletotrichum lindmuthianum*. Can. J. Microbiol. 26:1473–1479.

Bruegger, B. B. and N. T. Keen. 1979. *Specific elicitors of glyceollin accumulation in the Pseudomonas glycinea-soybean host-parasite system*. Physiol. Plant Pathol. 15:43–51.

Buchner, V., A. Nachmias and Y. Burtin. 1982. *Isolation and partial characterization of a phytotoxic glycopeptide from a protein-lipopolysaccharide complex produced by a potato isolate of Verticillium dahliae.* FEBS Lett. 138:261–264.

Darvill, A. and P. Albersheim. 1984. *Phytoalexins and their elicitors – a defense against microbial infection in plants*. Ann. Rev. Plant Physiol. 35:243–275.

DeWit, P. J. G. M., A. E. Hofman, G. C. M. Velthuis and J. A. Kúc. 1985. *Isolation and characterization of an elicitor of necrosis isolated from intercellular fluids of compatible interactions of Cladosporium fulvum (syn. Fulvia fulva) and tomato*. Plant Physiol. 77:642–647.

Doke, N. 1983. *Involvement of superoxide anion generation in the hypersensitive response of potato tuber tissues to infection with an incompatible race of Phytophthora infestans and to the hyphal wall components*. Physiol. Plant Pathol. 23:345–357.

Gabriel, D. W. 1984. *Molecular cloning of specific avirulence genes from Xanthomonas malvacearum*. Proc 2nd Int. Symp. on the Molecular Genetics of the Bacterial-Plant Interaction (in press).

Horsch, R. B., J. E. Fry, N. L. Hoffman, D. Eichholtz, S. G. Rogers and R. T. Fraley. 1985. *A simple and general method for transferring genes into plants*. Science 227:1229–1231.

Keen, N. T. 1982. *Specific recognition in gene-for-gene host-parasite systems*. Adv. Plant Pathol. 1:35–82.

Keen, N. T. 1985. *Phytoalexins and their involvement in plant disease resistance*. Iowa State J. Sci. (in press).

Keen, N. T. and M. Legrand. 1980. *Surface glycoproteins: evidence that they may function as the race specific phytoalexin elicitors of Phytophthora megasperma f. sp. glycinea*. Physiol. Plant Pathol. 17:175–192.

Keen, N. T., S. Midland and J. J. Sims. 1983. *Purification of victorin*. Phytopathology 73:830.

Keen, N. T. and B. J. Staskawicz. 1984. *Gene cloning as an approach to understanding specificity in plant-pathogen systems*. Pp. 344–358, in W. M. Dugger and S. Bartnicki-Garcia, eds., *Structure, function, and biosynthesis of plant cell walls*. Amer. Soc. Plant Physiol. Rockville, MD.

Kobayashi, D. Y. and N. T. Keen. 1985. *Cloning of a factor from Pseudomonas syringae pv. tomato responsible for a hypersensitive response on soybean*. Phytopathology (in press).

Kúc, J. and J. S. Rush. 1985. *Phytoalexins*. Arch. Biochem. Biophys. 236:455–472.

Maliga, P. 1984. *Isolation and characterization of mutants in plant cell culture*. Ann. Rev. Plant Physiol. 35:519–542.

Mayama, S. and N. T. Keen. 1984. *Production of victorin by various isolates of Helminthosporium victoriae and its phytoalexin elicitor activity*. Phytopathology 74:850.

Mayama, S., Y. Matsuura, H. Iida and T. Tani. 1982. *The role of avenalumin in the resistance of oat to crown rust, Puccinia coronata f. sp. avenae*. Physiol. Plant Pathol. 20:189–199.

Nachmias, A., V. Buchner and Y. Burstein. 1985a. *Biological and immunological characterization of a low molecular weight phytotoxin isolated from a protein-lipopolysaccharide (PLP) complex produced by a potato isolate of Verticillium dahliae Kleb*. Physiol. Plant Pathol. (in press).

Nachmias, A., V. Buchner and J. Krikun. 1982. *Comparison of protein-lipopolysaccharide complexes produced by pathogenic and non-pathogenic strains of Verticillium dahliae Kleb. from potato*. Physiol. Plant Pathol. 20:213–221.

Nachmias, A., V. Buchner, L. Tsror, Y. Burstein and N. T. Keen. 1985b. *Differential phytotoxicity of peptides from Verticillium dahliae races 1 and 2*. Phytopathology (in press).

Roby, D., A. Toppan and M-T. Esquerre-Tugaye. 1985. *Cell surfaces in plant-microbial interactions.* V. *Elicitors of fungal and plant origin trigger the synthesis of ethylene and of cell wall hydroxyproline-rich glycoproteins in plants*. Plant Physiol. 77:700–704.

Salzwedel, J. L., M. E. Daub and J-S. Huang. 1985. *Delay of bacterially-induced hypersensitive cell death in suspension cultured tomato cells by a singlet oxygen quencher*. Phytopathology (in press).

Scheffer, R. P. and R. S. Livingston. 1984. *Host-selective toxins and their role in plant diseases*. Science 223:17–21.

Staskawicz, B. J., D. Dahlbeck and N. T. Keen. 1984. *Cloned avirulence gene of Pseudomonas syringae pv. glycinea determines race-specific incompatibility on Glycinea max (L.) Merr.* Proc. Nat. Acad. Sci. USA 81:6024–6028.

Vance, C. P., T. K. Kirk and R. Sherwood. 1980. *Lignification as a mechanism of disease resistance*. Ann. Rev. Phytopathol. 18:259–288.

Walton, J. D. and E. D. Earle. 1984. *Characterization of the host-specific phytotoxin victorin by high-pressure liquid chromatography*. Plant Sci. Lett. 34:231–238.

Molecular response of plants to infection

CHRIS J. LAMB, JOHN N. BELL, CAROLE C. CRAMER, SANDRA L. DILDINE, CLAUDE GRAND, SUSAN A. HEDRICK, THOMAS B. RYDER and ALLAN M. SHOWALTER*
Plant Biology Laboratory, Salk Institute for Biological Studies, P.O. Box 85800, San Diego, CA 92138, U.S.A.

Abstract Plants exhibit resistance to disease involving inducible defenses including phytoalexin and hydroxyproline-rich glycoprotein (HRGP) accumulation, lignin deposition and increased activity of certain hydrolytic enzymes. Treatment of suspension-cultured cells of *Phaseolus vulgaris* L. with fungal elicitors redirects RNA synthesis leading to induction of mRNAs encoding phytoalexin biosynthetic enzymes such as phenylalanine ammonia-lyase and chalcone synthase; the lignin precursor biosynthetic enzyme cinnamyl alcohol dehydrogenase and HRGP. Accumulation of defense mRNAs is also observed during race:cultivar specific interactions between the fungus *Colletotrichum lindemuthianum* and *P. vulgaris* hypocotyls. There are clear temporal and spatial differences in the pattern of mRNA accumulation between incompatible (host resistant) and compatible (host susceptible) interactions. The data suggest that specific activation of plant defense genes is a key early component in the sequence of events leading to expression of defense reponses during race:cultivar specific interactions and that an elicitation signal is transmitted intercellularly to pre-activate defense genes in hitherto uninfected tissue. These observations indicate a number of potential sites for biotechnological manipulation and enhancement of disease resistance.

Introduction

Plants exhibit natural resistance to disease which has been exploited by breeders to reduce crop losses and hence increase yield. Detailed studies of the interaction between *Phytophthora infestans* and *Solanum tuberosum* (potato) by Muller and associates in the 1940s lead to the general concept that plants resist infection by inducible disease resistance mechanisms including specifically the synthesis of low molecular weight antimicrobial compounds termed phytoalexins (Deverall, 1982). Subsequent investigations have shown that accumulation of phytoalexins is a general response of plants to infection, and furthermore that there are a number of other inducible defense mechanisms including: accumulation of hydroxyproline-rich glycoproteins (HRGP), deposition of lignin-like material and other wall-bound phenolics, deposition of callose, stimulation of the activity of certain hydrolytic enzymes such as glucanase and chitinase, as well as the accumulation of proteinase inhibitors (Bell, 1981; Sequeira, 1983).

* Department of Biology, Washington University, St. Louis, MO 63130, U.S.A.

Although the genetics, physiology and cytology of plant:pathogen interactions have been extensively studied, until recently relatively little was known at the biochemical level about how plants respond to infection and resist disease. Over the last few years we have examined the interaction between *Phaseolus vulgaris* L. (dwarf French bean) and *Colletotrichum lindemuthianum*, causal agent of anthracnose, as a model for analysis of the molecular mechanisms underlying expression of plant defense responses. In the present chapter these studies are used to illustrate current concepts of the molecular basis for plant disease resistance and to provide a framework for discussion of possible avenues for biotechnological manipulation and enhancement of this agriculturally important trait.

Interaction between P. vulgaris *and* C. lindemuthianum *as an experimental system*

Advantages of this experimental system for molecular analysis are: (a) the disease resistance profile of *P. vulgaris* has been characterized genetically to some extent and physiologically (Bailey, 1982; Meiners, 1981). In particular, different cultivars (e.g. Kievit, Red Kidney) undergo highly specific interactions with physiological races of *C. lindemuthianum*. (b) The physiology and cytology of this interaction has been extensively studied using hypocotyls, the organ which together with leaves is the natural site of attack by *C. lindemuthianum* (Bailey, 1982; O'Connell, Bailey and Richmond, 1985). There is strong evidence that the differential accumulation of the isoflavonoid derivative phaseollin and other structurally related phytoalexins such as phaseollidin, phaseollinisoflavan and kievitone in incompatible and compatible interactions plays a crucial role in the specificity of host resistance (Bailey, 1982). (c) A sexual stage for *C. lindemuthianum* has recently been observed thereby opening up the possibility of genetic analysis of the pathogen (J. A. Bailey, personal communication). (d) Phytoalexin accumulation can be stimulated by treatment of suspension-cultured cells of *P. vulgaris* with elicitors present in filtrates from cultures of *C. lindemuthianum* or heat-released from mycelial cell walls (Dixon, 1980).

Elicitor activates genes encoding phytoalexin biosynthetic enzymes

Plant cell cultures can provide aseptic, relatively homogenous cell populations growing under precisely defined conditions. Such systems are excellent for isotope-labeling experiments and biochemical studies including isolation and purification of specific metabolites, proteins etc. (Dixon, 1980). Early work with suspension-cultured *P. vulgaris* cells centered on the development of optimal conditions for the induction of phaseollin accumulation by elicitors (Dixon and Fuller, 1976, 1978). Furthermore, [^{14}C]phenylalanine incorporation studies indicated that phaseollin was synthesized *de novo* from phenylalanine during elicitor-induced phytoalexin accumulation and it is probable that control of synthesis is an important regulatory element in phytoalexin accumulation (Moesta and Grisebach, 1980).

Phenylalanine is the precursor of a wide variety of phenylpropanoid compounds including lignin, coumarins, flavonoids and esters of hydroxycinnamic acids in addition to isoflavonoid-derivatives such as phaseollin. A central pathway from phenylalanine via cinnamic acid and *4*-coumaric acid to *4*-coumaroyl-CoA has been elucidated and *4*-coumaroyl-CoA is a key intermediate from which branch pathways originate associated with the biosynthesis of particular groups of phenylpropanoid compounds (Hahlbrock and Grisebach, 1979). The enzymes of the central pathway: phenylalanine ammonia-lyase (PAL), cinnamic acid 4-hydroxylase (C4H) and *4*-coumarate:CoA ligase (4CL) have been well characterized, as have the first two enzymes in a branch pathway specific for flavonoid and isoflavonoid biosynthesis: chalcone synthase (CHS) and chalcone isomerase (CHI). The enzymology of the later stages of isoflavonoid phytoalexin biosynthesis is under study in a number of laboratories and enzymes responsible for aryl migration and dimethylallyl transfer have been detected in cell free extracts (Hagmann and Grisebach, 1984; Zähringer, Ebel and Grisebach, 1978).

In cell cultures of *P. vulgaris* PAL, C4H, 4CL, CHS and CHI activities are present only at relatively low levels. Following treatment with elicitors there are marked and rapid increases in the activities of these enzymes, concomitant with the onset of phytoalexin accumulation (Cramer et al., 1985a; Dixon and Bendall, 1978; Robbins and Dixon, 1984). Although unlike the case of fine control, there is no theoretically rigorous criterion for the identification of specific sites of coarse control, these studies strongly suggest that increases in the activity of phytoalexin biosynthetic enzymes regulate accumulation of phaseollin and other isoflavonoid-derived phytoalexins in *P. vulgaris* cell suspension cultures.

In vivo labeling studies have shown that elicitor causes rapid, marked, concomitant but transient increases in the rates of synthesis of PAL, 4CL, CHS and CHI with maximum rates of synthesis occurring 3–4 hours after elicitation, during the phase of rapid increase in enzyme activity (Cramer et al., 1985a; Lawton et al., 1983, unpublished observations). These data strongly suggest that stimulation of enzyme synthesis is an important element regulating increased enzyme activity and hence phytoalexin accumulation. However, it should be noted that there also appears to be post-translational control mechanisms regulating the activities of at least some of these enzymes in elicitor-treated cells (Dixon et al., 1983; Lawton, Dixon and Lamb, 1980; Lawton et al., 1983; Robbins and Dixon, 1984) and under certain circumstances the enzyme activity levels are related by reciprocal changes in the rate of enzyme production and removal (Lawton et al., 1980). Such a multiple control system provides for rapid, amplified and flexible responses to environmental signals in slowly growing plant cells and may be of widespread biological significance (Lamb, Merritt and Butt, 1979).

In order to investigate further molecular mechanisms underlying elicitor stimulation of the synthesis of phytoalexin biosynthetic enzymes in this system, cloned cDNA libraries containing sequences complementary to mRNA present in elicitor-treated cells have been constructed. From these libraries a number of cDNA clones containing PAL and CHS sequences have been identified (Edwards, unpublished observations; Ryder et al., 1984). Using [^{32}P]-labeled PAL and CHS cDNA sequences as probes in RNA blot hybridization experiments, we have demonstrated rapid, marked but transient coordinate increases in PAL mRNA and CHS mRNA in response to elicitor treatment. There was a close correspondence between the induction kinetics for enzyme synthesis as measured by in vivo labeling techniques, the level of translatable mRNA activity and the amount of hybridizable mRNA for PAL and CHS in both total cellular mRNA and polysomal RNA fractions. Thus it can be concluded that the transient accumulation of PAL and CHS mRNA is the major factor governing the rate of enzyme synthesis throughout the phase of rapid increase in enzyme activity. Furthermore, the rapid accumulation of PAL and CHS mRNA from very low, almost undetectable basal levels in unelicited cells, strongly suggests that elicitor initially causes a transient coordinate increase in the transcription of PAL and CHS genes.

This has been confirmed by in vivo and in vitro labeling of newly-synthesized PAL and CHS mRNA (Cramer et al., 1985b; unpublished observations). Newly-synthesized mRNA was purified by organomercurial affinity chromatography following in vivo labeling with 4-thiouridine: a technique initially used to study developmental regulation of gene expression in yeast (Stetler and Thorner, 1984). During adaptation of this technique to a plant system we demonstrated uptake of 4-thiouridine into suspension cultured cells, linear incorporation into RNA as a function of time and a substantial enrichment for newly-synthesized mRNA following subsequent affinity chromatography

(Cramer et al., 1985b). Blot hybridization of newly-synthesized RNA purified in this way from elicitor-treated and control cells pulse labeled with 4-thiouridine, showed that elicitor markedly stimulated the rates of PAL and CHS mRNA synthesis. Transcriptional activation of PAL and CHS genes was confirmed recently by analysis of in vitro run-off transcripts in nuclei isolated from control and elicitor-treated cells (unpublished observations).

Interestingly, isolated newly-synthesized RNA, in vivo labeled with 4-thiouridine, can serve as a templete for protein synthesis in in vitro translation systems (Cramer et al., 1985b). By a number of criteria this in vitro translation is of high fidelity and hence it is possible to compare the pattern of translation products encoded by newly-synthesized mRNA to the pattern encoded by pre-existing RNA or total cellular mRNA. By two-dimensional gel electrophoretic analysis of corresponding sets of in vitro translation products we have shown that elicitor causes a rapid and pronounced change in the pattern of protein synthesis and that this change is the result of a dramatic switch in the pattern of mRNA synthesis. This switch involves down regulation of the synthesis of a small number of mRNA species and up regulation, often from very low basal levels, of the synthesis of a large number of new mRNA species.

Expression of other disease resistance mechanisms also involves defense gene activation

These observations raise the question of whether expression of other plant disease resistance mechanisms also involves accumulation of specific mRNAs. Hydroxyproline-rich glycoproteins are major structural components of plant cell walls (Cooper, Chen and Varner, 1984). In addition to hydroxyproline (Hyp), cell wall HRGPs are rich in serine, valine, tyrosine and lysine. Moreover, the protein moiety of cell wall HRGPs contains a characteristic repeating pentapeptide sequence, Ser-$(Hyp)_4$. Recently a carrot genomic clone encoding cell wall HRGP has been isolated, sequenced and shown to contain 25 Ser-$(Pro)_4$ repeat units, the unhydroxylated precursors of the Ser-$(Hyp)_4$ repeat units, distributed throughout the 306 amino acid coding sequence (Chen and Varner, 1985). The carbohydrate moiety of cell wall HRGPs is composed largely of short oligoarabinosides attached *O*-glycosidically to most of the hydroxyproline residues and to a much lesser extent of galactose which is *O*-glycosidically linked to some of the serine residues (Cooper et al., 1984). The accumulation of cell wall HRGP in response to infection has been observed in a number of systems and is correlated with expression of disease resistance (Esquerré-Tugayé et al., 1985).

Hydroxyproline-rich glycoproteins HRGPs may function in defense as specific agglutinins of microbial pathogens (Leach, Cantrell and Sequeira, 1982) and/or as structural barriers either directly or by providing sites for lignin deposition (Whitmore, 1978). Cloned genomic HRGP sequences have been used to demonstrate, by RNA blot hybridization, that elicitor treatment of suspension cultured bean cells causes a marked increase in hybridizable HRGP mRNAs (Showalter, unpublished observations). Accumulation of three transcripts, size 1.6, 2.7 and about 5.6 kilobases was observed. Interestingly, the response was less rapid but more prolonged than that observed for mRNAs encoding enzymes of phytoalexin biosynthesis, with a lag of 3–4 hours followed by a phase of rapid accumulation between 6 and 12 hours after elicitation, after which the mRNAs remained at high levels.

In addition, recent studies have demonstrated that elicitor stimulates the translatable activities of mRNAs encoding chitinase and cinnamyl alcohol dehydrogenase (CAD) in suspension-cultured bean cells (unpublished observations). Chitinase has lysozymal activity and may function in defense by degradation of the fungal cell wall polymer chitin (Boller, 1985; Boller et al., 1983; Pegg and Young, 1982). Cinnamyl alcohol dehydrogenase is an enzyme of phenylpropanoid metabolism specific to a branch pathway for synthesis of lignin monomers (Sarni, Grand and Boudet, 1984). Interestingly, elicitor stimulation of the level of CAD mRNA activity is more rapid than that previously observed for mRNAs encoding phytoalexin biosynthetic enzymes, with maximum rates of synthesis occurring between 1.5 and 2.0 hr after elicitation (unpublished observations). More rapid stimulation of CAD compared to the phytoalexin biosynthetic enzymes was also observed at the enzyme activity level. The different kinetics for accumulation of mRNAs encoding (a) CAD, (b) PAL and CHS and (c) HRGP may reflect distinct stimuli or a single stimulus leading to either sequential effects or divergent signal pathways.

Defense gene activation during race:cultivar specific interactions between C. lindemuthianum *and* P. vulgaris *hypocotyls*

The picture beginning to emerge from these studies is that elicitation of disease resistance mechanisms in cultured cells characteristically involves specific activation of defense genes. However, although a simple and convenient experimental system, elicitor-treated cells display neither the complex spatial relationships nor the race:cultivar specificity which characterize many plant:pathogen interactions (Dixon, 1980). These considerations have prompted us to commence study of the regulation of defense gene expression in race:cultivar specific interactions between *C. lindemuthianum* and hypocotyls of *P. vulgaris*. Hypocotyls, a natural site of the infection, can be reproducibly infected, without prior mechanical damage, by surface innoculation with conidia, the natural infective propagule (Bailey, 1982). Detailed physiological and cytological studies have established the biological importance of phytoalexin accumulation in this system (Bailey 1982; O'Connell et al., 1985), which therefore provides a model for molecular analysis of disease resistance under conditions closely resembling natural infection.

In the incompatible interaction following application of spores of race β to the unwounded surface of hypocotyls of the cultivar Kievitsboon Koekoek there is a period of 2–3 days during which the spores germinate and the fungus penetrates the cuticle. Immediately the surveillance mechanisms of the plant cell detect the presence of the fungus and there is a hypersensitive response in the initially infected cells with marked but apparently localised accumulation of phytoalexins and restriction of further fungal growth (Bailey, 1982; O'Connell et al., 1985). In contrast in the compatible interaction with race γ, the surveillance mechanisms of the plant cells apparently do not operate, the infected cells remain alive and the fungus is able to undergo extensive biotrophic intercellular growth. After a further 4–5 days (i.e. 6–7 days from spore innoculation) widespread host cell death and development of watery, spreading anthracnose lesions occurs (Bailey, 1982; O'Connell et al., 1985). There is marked phytoalexin accumulation at this stage, associated with attempted lesion limitation. Depending on environmental factors and the physiological status of the plant, activation of defense responses at this late stage of a compatibie interaction may restrict lesion development and prevent complete rotting of the hypocotyl (Rowell and Bailey, 1983).

Infection of hypocotyls with *C. lindemuthianum* causes marked, co-ordinated increases in the rates of synthesis of PAL, CHS and CHI as measured by in vitro translation of isolated polysomal mRNA, with clear temporal and spatial differences between compatible and incompatible interactions (Bell et al., 1984; Cramer et al., 1985a). More recently we have extended this analysis to demonstate, by RNA blot hybridization, differences between incompatible and compatible interactions in the pattern of accumulation of mRNAs encoding the phytoalexin biosynthetic enzymes PAL and CHS and mRNAs encoding HRGP (Showalter et al., 1985; unpublished observations). Thus in an incompatible interaction with race β, an early accumulation of these mRNAs, localized mainly but not entirely in tissue adjacent to the site of infection was observed prior to onset of phytoalexin accumulation and expression of localized, hypersensitive resistance. In contrast, in a compatible interaction with race γ, there was no early accumulation of mRNA but rather a delayed widespread response associated with phytoalexin accumulation during attempted lesion limitation. As in elicitor-treated suspension-cultured cells, marked accumulation of HRGP mRNA in each type of interaction occurred somewhat after the onset of accumulation of PAL and CHS mRNAs. The data suggests that specific accumulation of plant defense mRNAs is an early component in the sequence of events leading to expression of defense responses during race:cultivar specific interactions and that an elicitation signal is transmitted intercellularly to pre-activate defense genes in hitherto uninfected cells.

Polymorphism of defense genes and proteins

From a cDNA library of about 20,000 clones containing sequences complementary to mRNA in elicitor-treated cells, 48 clones were found to contain CHS sequences (Ryder et al., 1984). By restriction fragment length analysis, these clones were grouped into five classes and representatives of each of these classes sequenced. There was strong conservation of sequence in the coding regions of these five cDNAs, and most of the base substitutions were silent with respect to amino acid sequence (unpublished observations). However, within the 3′ non-coding region of the five transcripts, there was considerable sequence divergence. Southern blotting of genomic DNA from *P. vulgaris* reveals the presence of a number of fragments containing CHS sequences and recently we have isolated and characterized CHS genes from a *P. vulgaris* genomic library (unpublished observations). Overall the data indicate the presence of at least six CHS genes within the genome, some of which may be closely clustered and furthermore that at least five of these genes are activated by elicitor treatment.

Polymorphism of CHS has also been observed at the protein level. Two-dimensional gel electrophoretic analysis of immunoprecipitated CHS synthesized in vivo in elicitor-treated cells or by in vitro translation of mRNA from elicited cells, indicates the presence of about ten or eleven different CHS polypeptides with similar molecular ratio (M_r) but different pI (unpublished observations).

Polymorphism of PAL and HRGP has also been observed. At the protein level, PAL exists in a series of 4 forms of the native tetrameric enzyme which exhibit different kinetic properties and which in cell suspension cultures are differentially induced by elicitor (Bolwell et al., 1985). Two-dimensional gel electrophoretic analysis of PAL polypeptide subunits synthesized in vitro and in vivo also reveals extensive polymorphism (Bolwell et al., 1985). As noted above polymorphism of HRGP has been observed at the mRNA level, with at least three different size classes of transcript (Showalter et al., 1985).

In each of these cases, the relationship between polymorphism at the gene, mRNA and protein levels remain to be established, but it seems probable that this polymorphism is of considerable biological importance. Thus, polymorphism may reflect gene amplification to enhance the ability of the plant to respond effectively to infection following perception of an appropriate signal. Furthermore, genetic polymorphism may allow evolution of a family of genes encoding closely related polypeptides with similar but subtly different functional properties attuned to optimal operation in specific biological circumstances. Similarly, genetic polymorphism allows a set of genes encoding identical or very similar polypeptides to be positioned in different regulatory circuits, thereby allowing a flexible response to a variety of environmental conditions.

Strategies for biotechnological enhancement of plant disease resistance

Analysis of the molecular events underlying expression of disease resistance mechanisms during an incompatible and a compatible interaction between *P. vulgaris* and *C. lindemuthianum*, taken together with previous physiological and cytological studies, provides a framework (depicted in Fig. 1) for the design of rational strategies for biotechnological enhancement of disease resistance.

Incompatibility probably arises following a molecular recognition event between plant and pathogen at the plasma membrane of the first host cell invaded. It should be possible to isolate and manipulate by gene transfer techniques, the genes encoding the plant cell surface proteins responsible for recognition of specific pathogens. These genes represent the disease resistance genes manipulated by plant breeders during conventional breeding programs which confer in many cases resistance as a single gene trait. Strategies for identification and isolation of such disease resistance genes are discussed in detail by Keen in the preceding chapter. This approach has the advantage that one is dealing with a series of single gene traits, but consequently may suffer from the same difficulty as that encountered by breeders, namely that the beneficial effects of transfer of a specific disease resistance gene can be rapidly negated by changes in the natural pathogen population.

Since expression of localized resistance and hypersensitivity following early molecular recognition of a pathogen already represents a very successful defense reponse, the possibilities for enhancement of the local response in an incompatible interaction may be limited. However, observation of the pre-activation of defense genes in hitherto uninfected cells distant from the initial site of infection by an incompatible pathogen provides a clue to the molecular basis for the phenomenon of induced systemic resistance or immunity (Kuć and Caruso, 1977; Sequeira, 1983, 1984). It should be noted that observation of defense gene activation in distant uninfected cells does not imply that the induced mRNAs are translated into active protein leading to complete activation of the actual defense response (phytoalexin synthesis, HRGP accumulation, etc.). Indeed, it is possible to speculate that the immunized state reflects accumulation of mRNAs encoding defense proteins so that following a second challenge infection phytoalexin synthesis HRGP accumulation, etc. can be more rapidly and effectively activated, than in the non-immunized state.

Hence it may be possible to engineer plants such that the relevant genes are placed in different regulatory circuits leading to enhancement of this immunity. For example, it is possible to envision modifying plants such that these genes are pre-activated in response to developmental signals or environmental signals other than prior infection. While activation of these genes in such a way will place an additional energy burden on the plant, this may be minimized for example by expression only in specific organs or even tissues such as epidermis or only at specific developmental stages under circumstances where the plant is particularly vulnerable to disease. An alternate strategy would be to find chemicals that pre-activated defense genes in appropriate ways to achieve the state of induced systemic immunity prior to an initial infection and without placing an undesirably large energy burden on the modified plant. Such chemicals may be discovered by various screening strategies (Cartwright et al., 1977) or alternatively by synthesis of analogs of the putative natural signal factor that transmits the stress response from the infected cells to other cells.

Although in compatible interactions early molecular recognition by genetically determined mechanisms does not occur, the plant responds to the presence of the pathogen at a later stage to activate defense responses as part of an attempt to limit lesion development. Detailed physiological studies and cytological studies suggest that the key step here is the switch from biotrophic to necrotrophic growth by fungal hyphae present in heavily colonized host cells, distant from the growing hyphal tips (Bailey, 1982; O'Connell et al., 1985). It may be possible to modify plants such that the phase of biotrophic fungal growth is curtailed leading to an earlier activation of defense responses and hence more successful lesion limitation by the plant. The switch from biotrophy to necrotrophy represents a fundamental nutritional change and hence introduction of genes encoding the production of entities

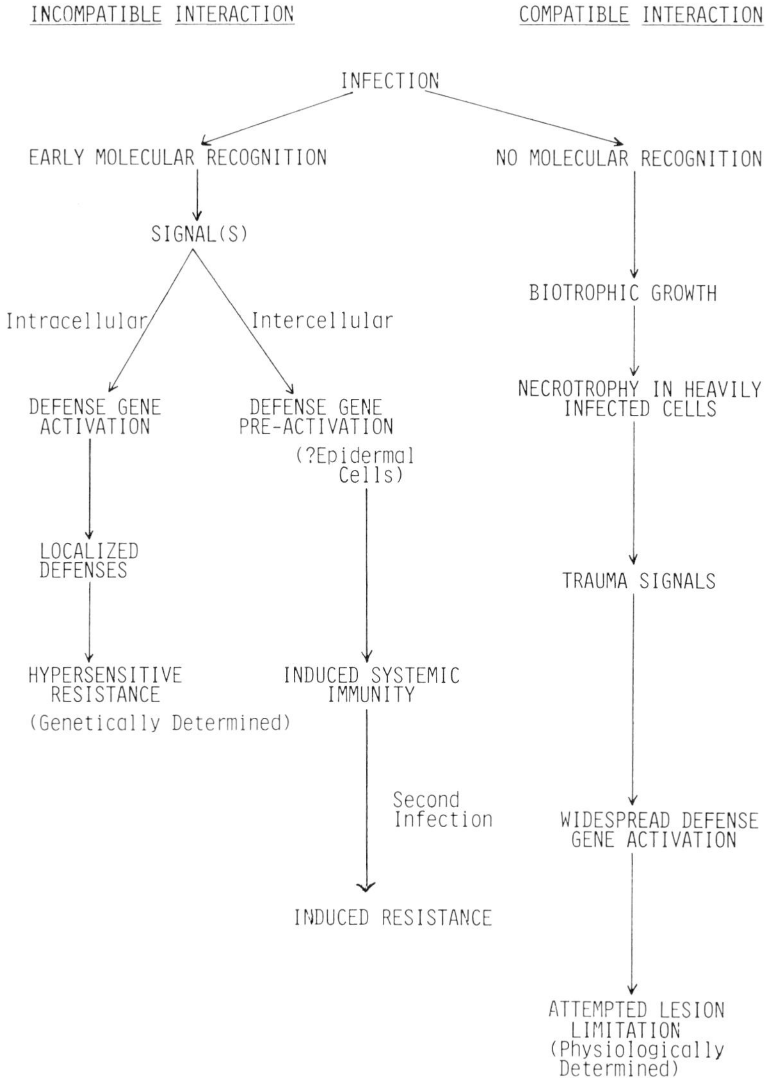

Figure 1. Activation of defense responses in race:cultivar specific interactions between *Colletotrichum lindemuthianum* and *Phaseolus vulgaris*.

such as siderophores (Neilands, this volume) to disrupt uptake and assimilation of nutrients by the fungus may have the result that the pathogen is less able to sustain biotrophic growth. This would cause earlier release of trauma signals arising from the onset of necrotrophy and hence an earlier activation of defense responses by the plant.

It is also possible to envision strategies to enhance the attempted lesion limitation response of plants either by modification of the trauma signal system which possibly involves release of endogenous elicitors (Hahn, Darvill and Albersheim, 1981; Hargreaves and Bailey, 1978; Jin and West, 1984), or by enhancement of the response to the signal in distant uninfected cells. Thus it may be possible to modify plants such that genes for different defense responses are placed in new regulatory circuits, for example so that genes encoding HRGPs are placed under control of a CHS promoter leading to an earlier and more rapid expression of the HRGP defense response following reception of the trauma signal(s). Likewise, chitinase normally accumulates in vacuoles (Boller, 1985) and it may be possible to modify a plant such that a chitinase gene is placed under the control of a CHS promoter and also is expressed in conjunction with a signal sequence derived from a cell wall protein such as HRGP, leading to very rapid synthesis of an extra-cellular chitinase in response to trauma signals. Finally it should be noted that whilst the more specialized pathogens become virulent by avoiding host surveillance mechanisms, some less specialized pathogens become virulent by virtue of their ability to metabolize and detoxify the phytoalexins of the host plant (Tegtmeier and VanEtten, 1982; VanEtten and Matthews, 1984). In these cases it may be possible to insert into the host genome genes encoding the synthesis of different phytoalexins, thereby confounding those non-specialized pathogens which have the ability to metabolize and detoxify the natural phytoalexins of that plant. In some instances it would require only a relatively small number of genes or perhaps even a single gene to be transferred under appropriate regulatory control to the recipient plant to effect this kind of improvement (Jin and West, 1982; Rolfs and Kindl, 1984).

While it is clear that the response of plants to infection is complex and multifaceted, the recent molecular analysis of these events, although far from complete, suggests a number of attractive and feasible strategies for enhancement of this trait. Although defense responses are clearly a multigenic series of traits, paradoxically this complexity offers a number of interesting and exciting possibilities for modifications to enhance disease resistance. Enhancement of resistance through modification of plant responses may be much less susceptible to negation by pathogen mutation than resistance conferred by transfer of disease resistance genes responsible for molecular recognition and hence may be more durable in the field.

The strategies outlined here may prove, in hindsight, to be naive, ineffective, impracticable or damaging to the plant by virtue of energy demand and resultant yield reduction. However, it should be noted that the concepts and tools of molecular biology currently being applied to plant genetic modification will also facilitate further detailed study of the molecular basis of agriculturally important traits. Hence fundamental studies of disease resistance may in the future give deeper insights leading to the development of more powerful and elegant strategies to reduce crop losses by disease.

Acknowledgements Research of the Plant Biology Laboratory, Salk Institute for Biological Studies was supported by USDA CRGO #84-CRCR-1-1464, ICI and the Samuel Roberts Noble Foundation. CLC is an NSF Research Fellow in Plant Biology. We thank Evelyn Wilson for preparation of the typescript. CJL thanks Drs. Noel Keen and Terrence Graham for stimulating discussions.

References

Bailey, J. A. 1982. *Physiological and biochemical events associated with the expression of resistance to disease.* Pp. 39–65 in R. K. S. Wood, ed., *Active Defense Mechanisms in Plants.* Plenum Press, New York.

Bell, A. A. 1981. *Biochemical mechanisms of disease resistance.* Ann. Rev. Plant Physiol. 32:31–81.

Bell, J. N., R. A. Dixon, J. A. Bailey, P. M. Rowell and C. J. Lamb. 1984. *Differential induction of chalcone synthase mRNA activity at the onset of phytoalexin accumulation in compatible and incompatible plant-pathogen interactions.* Proc. Nat. Acad. Sci. USA 81:3384–3388.

Boller, T. 1985. *Induction of hydrolases as a defense reaction against pathogens.* Pp. 247–262 in J. L. Key and T. Kosuge, eds., *Cellular and Molecular Biology of Plant Stress.* Alan R. Liss, New York.

Boller, T., A. Gehri, F. Mauch and U. Vogeli. 1983. *Chitinase in bean leaves: induction by ethylene, purification, properties, and possible function.* Planta 157:22–31.

Bolwell, G. P., J. N. Bell, C. L. Cramer, W. Schuch, C. J. Lamb and R. A. Dixon. 1985. *L-Phenylalanine ammonia-lyase from Phaseolus vulgaris. Characterization and differential induction of multiple forms from elicitor-treated cell suspension cultures.* Eur. J. Biochem. (in press).

Cartwright, D., P. Langcake, R. J. Pryce, D. P. Leworthy and J. P. Ride. 1977. *Chemical activation of host defense mechanisms as a basis for crop protection.* Nature 267:511–513.

Chen, J. A. and J. E. Varner. 1985. *Isolation and characterization of cDNA clones for carrot extensin and a proline-rich 33 kilodalton protein.* Proc. Nat. Acad. Sci. USA (in press).

Cooper, J. B., J. A. Chen and J. E. Varner. 1984. *The glycoprotein component of plant cell walls.* Pp. 75–88 in W. M. Dugger and S. Bartnicki-Garcia, eds., *Structure, Function, and Biosynthesis of Plant Cell Walls.*

Cramer, C. L., J. N. Bell, T. B. Ryder, J. A. Bailey, W. Schuch, G. P. Bolwell, M. P. Robbins, R. A. Dixon and C. J. Lamb. 1985a. *Co-ordinated synthesis of phytoalexin biosynthetic enzymes in biologically-stressed cells of bean (Phaseolus vulgaris L.).* EMBO J. 4:285–289.

Cramer, C. L., T. B. Ryder, J. N. Bell and C. J. Lamb. 1985b. *Rapid switching of plant gene expression induced by fungal elicitor.* Science 227:1240–1243.

Deverall, B. J. 1982. *Phytoalexins.* Pp. 1–20 in J. A. Bailey and J. A. Mansfield, eds., Blackie, Glasgow.

Dixon, R. A. 1980. *Plant tissue culture methods in the study of phytalexin induction.* Pp. 185–196 in D. S. Ingram and J. P. Helgeson, eds., *Tissue Culture Methods for Plant Pathologists.* Blackwell, Oxford.

Dixon, R. A. and D. S. Bendall. 1978. *Changes in the levels of enzymes of phenylpropanoid and flavonoid synthesis during phaseollin production in cell suspension cultures of Phaseolus vulgaris.* Physiol. Plant Pathol. 13:295–306.

Dixon, R. A. and K. W. Fuller. 1976. *Effects of synthetic auxin levels on phaseollin production and phenylalanine ammonia-lyase (PAL) activity in tissue cultures of Phaseolus vulgaris L.* Physiol. Plant Pathol. 9:299–312.

Dixon, R. A. and K. W. Fuller. 1978. *Effects of growth substances on non-induced and Botrytis cinerea culture filtrate-induced phaseollin production in Phaseolus vulgaris cell suspension cultures.* Physiol. Plant Pathol. 12:279–288.

Dixon, R. A., C. Gerrish, C. J. Lamb and M. P. Robbins. 1983. *Elicitor-mediated induction of chalcone isomerase in Phaseolus vulgaris cell suspension cultures.* Planta 159:561–569.

Esquerré-Tugayé, M. T., D. Mazan, B. Pelissier, D. Roby, D. Rumeau and A. Toppan. 1985. *Induction by elicitors and ethylene of proteins associated to the defense of plants.* Pp. 459–473 in J. L. Key and T. Kosuge, eds., *Cellular and Molecular Biology of Plant Stress.* Alan R. Liss, New York.

Hagmann, M. and H. Grisebach. 1984. *Enzymatic rearrangement of flavanone to isoflavone.* FEBS Lett. 175:199–202.

Hahlbrock, K. and H. Grisebach. 1979. *Enzymic controls in the biosynthesis of lignins and flavonoids.* Ann. Rev. Plant Physiol. 30:105–130.

Hahn, M. G., A. G. Darvill and P. Albersheim. 1981. *Host-pathogen interactions* XIX: *the endogenous elicitor, a fragment of a plant cell wall polysaccharide that elicits phytoalexin accumulation in soybeans.* Plant Physiol; 68:1161–1169.

Hargreaves, J. A. and J. A. Bailey. 1978. *Phytoalexin production by hypocotyls of Phaseolus vulgaris in response to constitutive metabolites released by damaged bean cells.* Plant Physiol. 13:89–100.

Jin, D. F. and C. A. West. 1984. *Characteristics of galacturonic acid oligomers as elicitors of casbene synthetase activity in castor bean seedlings.* Plant Physiol. 74:989–992.

Kúc, J. and F. L. Caruso. 1977. *Activated coordinated chemical defense diseases in plants.* Pp. 78–79 in P. A. Hedin, ed., *Host Plant Resistance to Pests.* American Chemical Society, Washginton.

Lamb, C. J., T. K. Merritt and V. S. Butt. 1979. *Synthesis and removal of phenylalanine ammonia-lyase activity in illuminated discs of potato tuber parenchyme.* Biochim. Biophys. Acta 582:196–212.

Lawton, M. A., R. A. Dixon and C. J. Lamb. 1980. *Elicitor modulation of the turnover of L-phenylalanine ammonia-lyase in French bean cell suspension cultures.* Biochim. Biophys. Acta 633:162–175.

Lawton, M. A., R. A. Dixon, K. Hahlbrock and C. J. Lamb. 1983. *Rapid induction of the synthesis of phenylalanine ammonia-lyase and of chalcone synthase in elicitor-treated plant cells.* Eur. J. Biochem. 129:593–601.

Leach, J. E., M. A. Cantrell and L. Sequeira. 1982. *Hydroxyproline-rich bacterial agglutinin from potato: extraction, purification and characterization.* Plant Physiol. 70:1353–1358.

Meiners, J. P. 1981. *Genetics of disease resistance in edible legumes.* Ann. Rev. Phytopathol. 19:189–209.

Moesta, P. and H. Grisebach. 1980. *Effects of biotic and abiotic elicitors on phytoalexin metabolism in soybean.* Nature 286:710–711.

O'Connell, R. J., J. A. Bailey and D. V. Richmond. 1985. *Cytology and physiology of infection of Phaseolus vulgaris by Colletotrichum lindemuthianum.* Physiol. Plant Pathol. (in press).

Pegg, G. F. and D. H. Young. 1982. *Purification and characterization of chitinase enzymes from healthy and Verticillium albo-atrum-infected tomato plants and from V. albo-atrum.* Physiol. Plant Pathol. 21:389–409.

Robbins, M. P. and R. A. Dixon. 1984. *Induction of chalcone isomerase in elicitor-treated bean cells: comparison of rates of synthesis and appearances of immunodetectable enzyme.* Eur. J. Biochem. 145:195–202.

Rolfs, C. H. and H. Kindl. 1984. *Stilbene synthease and chalcone synthease. Two different constitutive enzymes in cultured cells of Picea exceloa.* Plant Physiol. 75:489–492.

Rowell, P. M. and J. A. Bailey, 1983. *The influence of cotyledons, roots, and leaves on the susceptibility of hypocotyls of bean (Phaseolus vulgaris) to compatible races of Colletotrichum lindemuthianum.* Physiol. Plant Pathol. 23:245–256.

Ryder, T. B., C. L. Cramer, J. N. Bell, M. P. Robbins, R. A. Dixon and C. J. Lamb. 1984. *Elicitor rapidly induces chalcone synthase mRNA in Phaseolus vulgaris cells at the onset of the phytoalexin defense response*. Proc. Nat. Acad. Sci. USA 81:5724–5728.

Sarni, F., C. Grand and A. M. Boudet. 1984. *Purification and properties of cinnamoyl-CoA reductase and cinnamyl alcohol dehydrogenase from poplar stems (Populus X euramericana)*. Eur. J. Biochem. 139:259–265.

Sequeira, L. 1983. *Mechanisms of induced resistance in plants*. Ann. Rev. Microbiol. 37:51–79.

Sequeira, L. 1984. *Cross protection and induced resistance: their potential for plant disease control*. Trends in Biotechnology 2:25–29.

Stetler, G. L. and J. Thorner. 1984. *Molecular cloning of the hormone-responsive genes from the yeast Saccharomyces cerevisiae*. Proc. Nat. Acad. Sci. USA 81:1144–1148.

Tegtmeier, K. J. and H. D. VanEtten. 1982. *The role of pisatin tolerance and degradation in the virulence of Nectria haematococca on peas: A genetic analysis*. Phytopathology 72:608–612.

VanEtten, H. D. and P. S. Matthews. 1984. *Naturally occurring variation in the inducibility of pisatin demethylating activity in Nectria haematococca mating population vl*. Physiol. Plant Pathol. 25:149–160.

Whitmore, F. W. 1978. *Lignin-protein complex catalyzed by peroxidase*. Plant Sci. Lett. 13:241–245.

Zähringer, U., J. Ebel and H. Grisebach. 1979. *Induction of phytoalexin synthesis in soybean. Dimethylallylprophosphate: trihydroxypterocarpan dimethylallyl transferase from elicitor-induced cotyledons*. FEBS Lett. 101:90–92.

Biocontrol of insects – *Bacillus thuringiensis*

BRUCE C. CARLTON and JOSÉ M. GONZÁLEZ, JR.
Ecogen, Inc. Lawrenceville, New Jersey, NJ, U.S.A.

Abstract Biocontrol of insects is increasingly attractive as an alternative to chemical insecticides due to safety and specificity considerations and because of increased problems of resistance development to conventional chemicals. A better understanding of the genetics and physiology of biocontrol agents will greatly facilitate their future development through the application of genetic engineering methodologies. *Bacillus thuringiensis* (BT) is the most widely recognized and best understood of the microbial insect control products currently available. Many different strains, representing at least 30 sub-species, are now recognized on the basis of their surface flagellar antigens and insecticidal activities. These activities reside in phase-visible crystalline inclusions produced in sporulating cells, together with a highly refractile and heat resistant spore. The active toxin is generated in the insect gut after ingestion of the crystals, by solubilization and proteolytic activation of a high molecular weight protoxin protein. The genetic inheritance and diversity of genes specifying these proteins in different strains is of paramount importance in efforts aimed at strain improvement for increased efficacy on selected target insects. These efforts have, in part, focussed on the role of extrachromosomal plasmids in toxin crystal production. Most strains of BT harbor complex arrays of up to 13 plasmids, ranging in size from ~ 2 megadaltons to over 150 megadaltons. Curing (loss) of various plasmids, either spontaneously or by growth of the bacteria at elevated temperatures, has permitted the correlation of insecticidal toxin production with specific plasmids in several different strains of BT. In addition, the recent discovery of a natural conjugational system in BT has allowed the construction of hybrid strains containing a toxin-specifying plasmid derived from a donor of one type in a recipient strain background of another type, or even in a different *Bacillus* species. These studies have provided definitive evidence that the toxin-specifying plasmids do, in fact, carry the structural genes for relevant toxin proteins. Molecular cloning of several toxin coding genes by recombinant DNA methods has further facilitated the localizations of toxin genes to specific plasmids, using cloned gene segments as molecular probes in DNA hydridization studies. The results obtained from nucleotide sequencing of these genes provide a basis for comparison of different toxin proteins as related to their insecticidal activities. This information promises to contribute to an acceleration of efforts directed at the improvement of these insecticidal activities through applications of gene splicing and other molecular manipulations.

Introduction

Biological control of insects is not a recent innovation, although it is clear there has been a major resurgence of interest in this approach to plant protection. A variety of naturally-occurring microbial pathogens of insects have been described [see Kurstak and Tijssen (1982) for review]. We believe the recent resurgence of interest in this area to be due to several factors. First is the increasing challenges to the chemical pesticides based upon environmental concerns, evidence for human carcinogenicity and toxicity of these materials. The extreme hazards, both immediate and long-term, to man and other life forms posed by most chemical insecticides are no longer acceptable to advanced societies. Incidents such as the recent disaster in Bhopal, India serve as frightening examples of the inherent dangers of these chemicals to the quality of human life, and will undoubtedly accelerate legislative and executive actions to ban or severely restrict their use. A second factor that has led to increased interest in alternatives to chemical insecticides is that their use tends to be self-defeating. Insects have a remarkable ability to develop genetic resistance to even complex organic chemicals and the prolonged use of such insecticides, especially on crops grown in monoculture, like corn, potatoes, and cotton, can very rapidly produce resistant populations of insects. In some applications, such as for treatment of soil insects, natural soil microorganisms can adapt to the presence of these agents by developing enzymatic mechanisms for their rapid degradation. Chemical control of corn rootworm, the major insect pest of corn in the U.S., is plagued by this problem. A third factor involved in the renewed interest in biological insecticides is the discovery and refinement of methodologies for genetic engineering (i.e., gene splicing or recombinant DNA technology). This remarkable breakthrough in molecular biology makes it possible not only to isolate the genetic determinants for an endless array of cellular products, but more importantly to restructure them in ways that natural genetic processes cannot. Thus it is possible to isolate a gene from one organism and to insert it into the genetic background of a totally unrelated organism. For biological insect control agents, the way is now open to replace naturally-occurring first generation products with second generation forms genetically manipulated to induce improvements such as greater activity, wider host range, and extended field persistence.

Biological alternatives to chemical insecticides include a variety of microbial agents such as bacteria, viruses, and fungi. These agents essentially function as pathogens of the target insects, causing infections which utilize the insect as a propagation medium and leading to death of the host insect in the process. Persistence and proliferation of the infecting microbe as a maturing epizootic varies, depending upon the particular organism and the conditions under which the infection occurred. Viruses tend to be the most epizootic followed by fungal pathogens, with bacteria generally the least proliferative.

Of all the various microbial agents that have been evaluated, the most successful by far has been the bacterium *Bacillus thuringiensis* Berliner, more commonly referred to simply as BT. This soil-borne sporulating bacterium has been sold commercially for insect control since the late 1950's, and currently has a worldwide market of

\$50–55 Million. Its activity is directed primarily against the caterpillar stages of insects in the order Lepidoptera, including such economically-important insect pests as cabbage loopers, imported cabbageworms, armyworms, spruce and tobacco budworms, and the larval stage of the gypsy moth. The insecticidal activity of BT resides in a high molecular weight protein that is produced in sporulating cells in the form of a generally bipyramidal crystal, or parasporal body. The crystal is released from the mother cell, along with the spore, at the completion of sporulation. Commercial producers of BT, among them Sandoz, Inc., Abbott Laboratories, and Biochem Products, employ large fermentation tanks to grow the bacterium and typically recover the spores and crystals by centrifugation. The mixture, together with particulate cellular debris, is then formulated as wettable powders, oil emulsions, or as a water-miscible flowable compound. The various formulations are applied with conventional spray equipment, either aerially or with ground equipment depending on the application.

Unlike a typical organic chemical insecticide such as an organophosphate, a carbamate, or a synthetic pyrethroid, most of which act as contact poisons, BT must be ingested by a susceptible insect to exert its activity. The parasporal crystal, which is highly insoluble under usual physiological conditions, becomes solubilized when it encounters the alkaline ($pH > 10$) mid-gut contents of a susceptible insect. The solubilized protein, whose molecular weight is typically 135 kilodaltons (kd)–140 kd, is then activated to generate a toxin which acts by an as-yet not completely understood mechanism to bring about degradation of the mid-gut epithelium. This effect causes the insect to cease feeding almost immediately due to gut paralysis, which is usually followed by a more general paralysis and death within 2–6 days. Although the exact relationship between the inactive protoxin and the active toxin has not yet been established, it is thought that the activation most likely involves a proteolytic cleavage of the protoxin protein, generating a toxic product approximately half the size of the protoxin.

One of the intriguing features of BT is that it is not a single bacterial strain but rather an array of strains, each with its own particular properties. One such property which has been used as a major criterion for strain classification is that of the immunological determinants (antigens) of surface appendages (flagella) which these cells use for locomotion. The delineation of these flagellar antigens, as pioneered by de Barjac and associates (de Barjac and Bonnefoi, 1962; de Barjac, 1981) has served as the primary basis for classification of strains of BT into subspecies of varieties. A second immunological property, that of the protein contained in the parasporal crystal, has also been used as a classification aid (Krywienczyk and Angus, 1967). Perhaps the most intriguing characteristic of the array of different BT strains, particularly from the applications viewpoint, is that of the insecticidal activity residing in the parasporal crystal itself. Extensive bioassay analyses have been conducted on a variety of different insects, including economically important crop pests as well as beneficial insects and those without economic significance. *Bacillus thuringiensis* strains typically show activity on an array of insects, primarily on those in the order Lepidoptera. To date, nearly 150 different species of insects within this order have shown varying levels of susceptibility to the toxin contained within the parasporal crystal. Of particular interest is the fact that the relative activities of toxins produced by different strains of BT vary enormously when assayed against an array of target insects. This variability is most conveniently expressed as a ratio of activity directed against pairs of insects. As shown in Table 1, for example, the activity ratios against pairs of insects is highly variable, and this example includes only 2 of the now-recognized 27 flagellar serotypes of BT. Probably the most outstanding example of this great diversity in insecticidal activity is exemplified by the strain BT var. *israelensis* (BTI), of serotype H-14 (Goldberg and Margalit, 1977). This strain, and its derivatives, have no activity against any lepidopteran insect, tested at standard dosages, but have high activity on several species of dipteran insects including mosquitos and blackflies, some of which are important vectors of human diseases. A number of other strains from several serotypes have been shown to have at least modest activity on both lepidopteran and dipteran insects. There is a recent report (Krieg et al., 1983) of a strain producing a parasporal crystal active against

Table 1. Relative activities of *Bacillus thuringiensis* strains on selected lepidopteran insects

Strain type		Activity ratios			
Flagellar serotype	Crystal serotype	Tn/Hv	Bm/Hv	Hc/Bm	Ec/Pi
H1	thu	4–8	0.7–1.8	8–28	0.5–0.7
H3a3b	k-1	1.2–2.8	1–6	1.3–2.8	0–0.5
H1	k-1	1.5–2.7	5–29	0.04–1	–**
H3a3b	k-73	0.6	–*	–*	0.3

Abbreviations: Tn = *Trichoplusia ni*; Hv = *Heliothis virescens*; Bm = *Bombyx mori*; HC = *Hyphantria cunea*; EC = *Ephestia cautella*; Pi = *Plodia interpunctella*.

* K73 type crystals essentially inactive vs. *B. mori*.

** Varies; some strains inactive against one or the other insect, or both.

Data from Dulmage, H.T. et al. (1981).

coleopteran insects (beetles), suggesting that the spectrum of potential insecticidal activity may be broader yet. It would obviously be of great academic as well as practical interest to understand the basis for this variability in insecticidal activity among different BT strains. Such knowledge would facilitate the systematic genetic manipulation of the relevant strains to construct variants having enhanced activities on various target insects. Certainly the extensive natural genetic diversity already observed supports the notion that new strains can be constructed having novel insecticidal activities.

Microbials such as BT offer a number of distinctive advantages over chemical insecticides. Chief among them, because of the high specificity of the microorganisms, is the high degree of safety to humans, other mammals, birds, fish, etc. *Bacillus thuringiensis* (BT) can be sprayed on vegetable crops such as lettuce, cabbage, and other cole crops right up to the day of harvest, and there are no re-entry tolerance limitations for field workers. The high specificity also has the advantage of not interfering with natural predators and parasites of the target insects, so that the natural ecological balance between pests and beneficial species is not disturbed. This effect minimizes the number of required treatments to control the pest insect. A second advantage of a microbial like BT is that the development of resistance on the part of target insects appears, at least at present, to be much less rapid and less pronounced than for chemical insecticides. While the basis for this apparent difference is not understood, it seems likely that the mechanisms of action of the BT toxin may be sufficiently complex that multiple genetic mutations in the insect may be necessary for resistance to develop. Alternatively, the pathological mechanism may be linked to essential cellular functions so that mutations in these functions, required for resistance, may be lethal to the insect. In any event, if the current situation regarding resistance holds true in the future, this could be a major contributing factor to a greater acceptance of BT and other microbials in the future.

A third advantage of BT is that, as a bacterial fermentation product, it is potentially less expensive to produce than organic chemical insecticides. A single fermentation tank could be used to produce a variety of different microbial products (as is currently the case with BT and BTI, two quite distinct strains). Chemical insecticides, even those in the same general group, typically require a completely different and very costly plant.

Despite its stated advantages, over the more than 20 years of its commercial availability, BT has not achieved widespread acceptance. Current sales of the product are running about $10–15 million in the

U.S., and perhaps \$30–35 million worldwide. This is an insignificant fraction of the \$4 billion worldwide insecticide market. There are several limitations of BT that have contributed to this low level of acceptability, some real and some more perceived than real. First is the narrow spectrum of insecticidal activity, which is directed to certain lepidopteran insects and a few dipterans. Because BT is ineffective on such important economic insects as aphids, mites, white flies, various beetles, etc., its use is restricted, at the present, to those crop applications for which lepidopterans are the principal pests.

A second limitation influencing the acceptance of BT has been its relatively slow killing action. Unlike contact insecticides that kill quickly, BT acts as a stomach poison. It must be ingested by the insect larva, and migrate into the mid-gut, where solubilization and activation of the protoxin must occur. Although this process probably occurs within a few minutes following ingestion of the toxin crystals (Lüthy and Ebersold, 1981), followed fairly rapidly by a cessation of feeding, death of the insect does not occur for 3–5 days. This delayed response can lead to problems in perceived efficacy of the product, even though crop damage may be halted with BT as rapidly as it is with a synthetic chemical acting as a contact insecticide.

A third limitation of BT has been its low degree of field persistence and efficacy relative to its inherent potency as determined by laboratory bioassay. Due to its mode of action, its effectiveness clearly depends upon obtaining a thorough and uniform coverage on the plant foliage, and a level of persistence that will ensure that feeding insects will continue to ingest it. Numerous studies as well as considerable folklore have led to the notion that BT is rapidly inactivated in the field by exposure to the ultraviolet rays of sunlight. While it has certainly been well-documented that the spores undergo rapid UV-inactivation (Cantwell and Franklin, 1966), the inactivation of the toxin protein itself is not well-documented; in fact, there are reports to the contrary (Burges, Hillyer and Chanter, 1975; Leong, Cano and Kubénski, 1980; Sneh, Schuster and Broza, 1981). Despite these uncertainties, there is still a negative image of low field persistence projected for BT.

In the same vein, studies of the genetics of insecticidal toxin production, where the genes are located and how their expression is regulated have been initiated only recently, despite the fact that BT has been sold commercially for nearly 25 years. The recognition that BT is a diverse continuum of strains having a broad array of insecticidal activities is only now coming to be recognized. That the limitations inherent in the single strain used for commercial production (at least in the U.S.) may be circumvented by the identification and genetic manipulation of genes from other strains has been a recent revelation.

The assignment, by us and others, of toxin genes to extrachromosomal plasmids has significantly clarified how BT may evolve in the wild, and also has potential ramifications for strain improvement directed at insect control. In the remainder of this presentation we will review the approaches that have been used to identify and manipulate toxin-specifying plasmids in a variety of strains of BT.

Genetic analysis of BT toxin production

Plasmid arrays of Bacillus thuringiensis strains. The majority of the genetic work on BT strains to date has involved locating the δ-endotoxin gene on a plasmid and/or molecular cloning and characterization of this gene. These approaches to the genetic study of δ-endotoxin production were brought about by several reports in the literature implying that the toxin genes are extrachromosomal. For example, Hannay (1967) reported the loss of crystal production ("dissociation") in the Mattes strain of BT (BT var. *thuringiensis*, H-1); colonies varying in sporulation and in crystal formation were isolated after growing cells in nutrient broth and plating them on nutrient agar. Hannay distinguished between "unstable" strains (the Mattes strain) and "stable" strains of BT (not named) which did not "dissociate". The disappearance of toxin production was apparently permanent; Norris (1970) described the evidence for reversion of Cry^- BT strains to Cry^+ as unconvincing. Although these reports were not detailed genetic studies, they suggested a possible extrachromosomal location for crystal protein genes in some BT strains. The generation of Cry^- (non-virulent) strains at high or relatively high frequencies, without mutagenic treatment, and the absence of reversion, are characteristic of traits carried by plasmid or temperature phage genomes (Campbell, 1969). These reports probably stimulated the search in recent years for plasmids in BT with a role in crystal production. Work was also stimulated by the industrial/agricultural importance of the δ-endotoxin, which if plasmid-associated would be especially suited for modification using genetic engineering methods and the generation of improved strains (Levin, 1979).

The first report of the presence of plasmids in BT was by Zakharyan et al. (1976). Since then, the accumulating evidence has indicated that BT plasmids are ubiquitous, numerous, and span a broad range of sizes, as seen in reports where several strains of BT were examined for plasmid content (Miteva, 1978; González, Dulmage and Carlton, 1981; Lereclus et al., 1982; Kronstad, Schnepf and Whiteley, 1983; Carlton and González, 1985). Plasmid sizes range from 1.4 to

~ 150 megadaltons (Mdaltons), the number of plasmids per strain can be as high as 11 or 12, and some unusual extrachromosomal DNA elements are present mostly in the OC (open circular) or linear form (González et al., 1981). The plasmid array of each BT serotype or variety is unique and can usually be distinguished from that of another variety. Plasmids have also been found in strains of *B. cereus* (Bernhard, Schrempf and Goebel, 1978).

Figure 1 shows the plasmid arrays of three BT strains and two *B. cereus* strains. The results are typical of the plasmid number and range in sizes usually found (Carlton and González, 1985). The plasmids range from 1 to 12 per strain and from 2 to ~ 150 Mdaltons in size. The resemblance of the small plasmids of HD73-1 and HD263-1

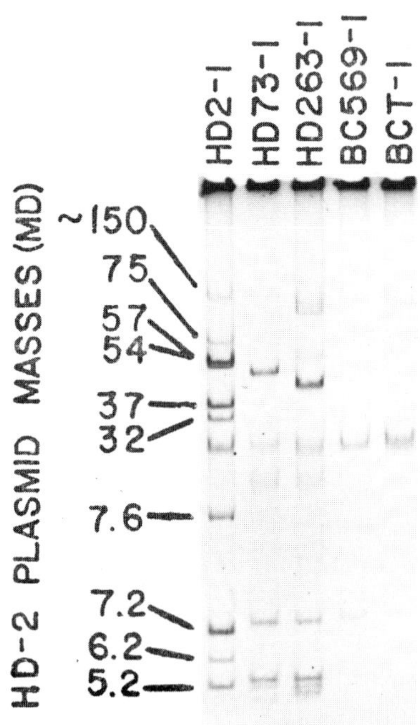

Figure 1. Plasmid patterns of three representative wild-type BT strains and two wild-type *B. cereus* strains. The plasmids were visualized by electrophoresis on a 0.5% vertical agarose gel by a simple method for plasmid content analysis described by Eckhardt (1978) and modified by González et al. (1981), which does not require isolation of plasmid DNA. Essentially, the electrophoresis of SDS through a layer of BT protoplasts in the well of a vertical gel lyses the cells gently, so that the chromosomal DNA remains largely intact at the gel origin, while the plasmid DNA molecules form bands and are resolved normally. The molecular weights (in Mdaltons) of the plasmids of HD2-1 are listed on the left margin as size references; the 7.6-Mdalton plasmid is present only as the OC form. HD263-1 carries a 1.4-Mdalton plasmid which ran off the bottom of the gel.

(both members of variety *kurstaki*) is typical of the intravarietal similarities observed between BT plasmid arrays (Kronstad et al., 1983; Carlton and González, 1985).

BT plasmids have been implicated in δ-endotoxin synthesis by means of three experimental approaches: plasmid "curing", plasmid transfer, and cloning of the toxin genes, as described below.

Correlation between loss of specific plasmids and loss of toxin production. Plasmid "curing", the isolation of variants lacking one or more of the plasmids present in the parental strain, has proven useful in localizing BT toxin genes on specific plasmids. Variants of BT that fail to produce toxin crystals (Cry^-) are isolated and their plasmid contents analyzed to determine whether loss of crystal synthesis is accompanied by loss of plasmids. Conversely, a set of variants cured of one or more plasmids is examined to assess whether any variants are Cry^-. If a given plasmid is always present in Cry^+ variants and absent in Cry^- variants, it can be concluded that the plasmid is probably required for crystal production (either because it carries the gene for δ-endotoxin, or, less likely, because it controls the expression of a toxin gene located elsewhere).

The first report of plasmid-toxin association by plasmid curing was that of Stahly et al. (1978). They reported a multiple CCC plasmid array in the commercially important strain HD-1 (var. *kurstaki*, H-3ab). Electron microscopy revealed six size classes, from 1.3 to 47.1 Mdaltons in size. Spo^+ Cry^- mutants of HD-1 were isolated from heat-shocked spores (frequency, 1 per 500–1000 colonies). All mutants had apparently lost all six plasmids of the parental strain. This comparison of $Spo^+ Cry^+$ and $Spo^+ Cry^-$ variants of the same strain implied involvement of the lost plasmids in crystal formation. It was not possible, however, to determine if all or only one of the plasmids was required, since all the plasmids were lost simultaneously.

González et al. (1981) implicated single plasmids in δ-endotoxin synthesis by applying a modified Eckhardt (1978) lysate-electrophoresis method to the analysis of five Cry^+ strains of BT: HD-2 (H-1, *thuringiensis*), HD-4 (H-3a, *alesti*), HD-8 (H-5ab, *galleriae*), HD-1 (H-3ab, *kurstaki*), and HD-73 (H-3ab, *kurstaki*). All five strains carried multiple CCC plasmid arrays (the arrays of HD-2 and HD-73 are shown in Fig. 1). Specific plasmids were implicated in δ-endotoxin production in each strain by the isolation of spontaneous $Spo^+ Cry^-$ mutants, which no longer made a detectable phase-refractile crystal on nutrient agar at 30°C. Comparison of the plasmid arrays of Cry^+ and Cry^- variants of the same strain showed that only one plasmid had changed (that is, either had been lost or underwent deletion or insertion of DNA) during loss of δ-endotoxin activity. The implicated plasmids were of a different size in each strain: 75 Mdaltons in HD-2, 50 Mdaltons in HD-73, ~ 105 Mdaltons in HD-4, ~ 130 Mdaltons in HD-8, and 29 Mdaltons in HD-1. The evidence implicating a single plasmid in HD-2 was especially convincing, being based on analysis of a rather large number (20) of independently derived Cry$^-$ mutants. In many cases, other plasmids present in these strains were also lost, but without any effect on crystal production. The presence or absence of δ-endotoxin activity in Cry$^+$ and Cry$^-$ variants was confirmed by direct bioassay against larvae of the appropriate target insects. Recent studies by T. Yamamoto, J. M. González, Jr. and B. C. Carlton (unpublished) and Carlton and González (1985), indicate that in HD-1 it is not the 29-Mdalton plasmid, but two of the larger plasmids, that control crystal synthesis.

The important mosquito-toxic H-14 variety, BTI, has been investigated in several laboratories, most of which implicated a large plasmid in toxin production. For example, Ward and Ellar (1983) used the method of Eckhardt (1978) and detected eight plasmids, ranging from 3.2 to ~ 110 Mdaltons in size, in BTI strain IPS 78. Growth of IPS 78 on nutrient agar at 42°C was used to generate 107 cured derivatives, most of which were Spo$^+$ Cry$^-$. Toxicities of Spo$^+$ Cry$^+$ and Spo$^+$ Cry$^-$ variants were confirmed by testing sporulated cultures against *Aedes aegypti* larvae. All Spo$^+$ Cry$^-$ variants had lost a large plasmid of 72 ± 2 Mdaltons that was present in the wild type and all Spo$^+$ Cry$^+$ variants. Loss of any of the other plasmids had no effect on crystal production. Spo$^+$ Cry$^-$ variants of six other BTI strains were isolated; all had lost the 72 ± 2 Mdalton plasmid. Our own results (González and Carlton, 1984) are consistent with the preceding and indicate that BTI toxin production requires the presence of a large transmissible plasmid, 75 Mdaltons in size. A Spo$^+$ Cry$^+$ BTI strain, HD-567, was examined for plasmid DNA by standard techniques (Gonzáles et al., 1981) and found to contain a multiple plasmid array. Eight plasmids were detected (sizes 3.3, 4.2, 4.9, 10.6, 68, 75, ~ 105, and ~ 135 Mdaltons), as well as a linear plasmid. Fifteen independently isolated, spontaneous Spo$^+$ Cry$^-$ mutants all had lost the 75-Mdalton plasmid, implicating it in δ-endotoxin production. The presence or absence of toxicity in Cry$^+$ and Cry$^-$ variants was confirmed by bioassay of sporulated cultures against *Aedes aegypti* larvae. Loss of the other BTI plasmids did not affect δ-endotoxin synthesis; a visible, toxic crystal was always produced.

Correlation between plasmids and δ-endotoxin synthesis from conjugal plasmid transfer experiments. The plasmid transfer system of

BT strains (González and Carlton, 1982) holds great promise for the future genetic analysis of BT, *B. cereus*, and possibly other *Bacillus* species. The plasmid transfer phenomenon was reported to be conjugation-like (DNA resistant and requiring cell-to-cell contact) and highly efficient. Plasmid transfer required only that the donor and recipient strains be grown together for 5 to 10 generations in nutrient broth, after which up to 80% of recipient cells had acquired one or more donor plasmids in the absence of any selection for plasmid markers. Both large and small donor plasmids transferred equally well. The 50-Mdalton plasmid of strain HD-73, implicated in toxin production by curing studies (González et al., 1981), was seen to transfer into a Cry$^-$ HD-2 strain, which became Cry$^+$ once more.

The BT plasmid transfer system was used by González, Brown and Carlton (1982) to prove that δ-endotoxin structural genes in three BT strains were located on large transmissible plasmids. The HD-73 50-Mdalton plasmid, the HD-263 44-Mdalton plasmid (both strains are var. *kurstaki*, H-3ab), and the HD-2 (var. *thuringiensis*, H-1) 75-Mdalton plasmid transferred into Cry$^-$ Hd-73, HD-2, and *B. cereus* recipient strains and converted them into crystal producers. Recipient strains usually acquired toxin plasmids at a high frequency (~ 10–20% of the cell population). The crystals were shown to be immunologically like these of the plasmid donor strain, even when the recipient strain had originally produced a diferent type of crystal. For example, an HD-2 recipient that had acquired the toxin plasmid from either HD-73 or HD-263 produced a *kurstaki*-type (not a *thuringiensis*-type) crystal. These results indicated that the implicated plasmids carry the actual δ-endotoxin structural genes.

González and Carlton (1984) extended these experiments to BTI. The 75-Mdalton plasmid implicated in BTI toxin production by curing studies was found to transfer efficiently into a plasmid-free, Cry$^-$ BTI recipient strain, converting it to Cry$^+$. A Spo$^+$ Cry$^+$ BTI strain was generated that carried the 75-Mdalton plasmid alone, indicating that this plasmid is both necessary and sufficient for toxin production. Other large BTI plasmids could transfer also, but did not convert the BTI recipient to Cry$^+$.

Table 2 summarizes the results of our investigation of plasmid-associated δ-endotoxin production in BT strains using both plasmid curing and plasmid transfer, and shows that BT strains carry variable numbers of plasmids. The toxin plasmids implicated to date are all relatively large (ranging from 44 to 130 Mdaltons) and many are transmissible into at least one BT recipient strain.

Table 2. Correlations between specific plasmids and delta-endotoxin production in various strains of *B. thuringiensis*

Prototype strain no.[a]	H-serotype (flagellar antigens) and subspecies	No. of plasmids	Size of plasmid (MDaltons) implicated in toxin production[b]	Type of implicating evidence[c]
HD-2	1, *thuringiensis*	10	75 [54]	A T
HD-225	1, *thuringiensis*	11	~ 110 (P1, P2)	D
HD-73	3ab, *kurstaki*	6	50	A, T
HD-74	3ab, *kurstaki*	7	55	A, T
HD-1	3ab, *kurstaki*	12	~ 115 (P1, P2) [44 (P1)]	A, D A, T
HD-263	3ab, *kurstaki*	11	~ 110 (P1, P2) 60 (P1) 44 (P1)	A A, I T
HD-4	3a, *alesti*	10	~ 105	A
HD-8	5ab, *galleriae*	4	~ 130	D
HD-9	6, *entomocidus*	4	~ 105	A
HD-536	8ac, *ostriniae*	4	68	A, T
HD-13	9, *tolworthi*	7	~ 110 44	A A
HD-146	10, *darmstadiensis*	5	70	A
HD-498	10, *darmstadiensis*	5	65	A
HD-542	12, *thompsoni*	4	~ 100	A
HD-567	14, *israelensis*	9	75	A, T
HD-500	14, *israelensis*	9	75	A
BT Col	21, *colmeri*	6	~ 100	D

[a] Refer to numbers in the strain collection of H. T. Dulmage, who provided the original cultures, except for BT Col, the *B. thuringiensis* subsp. *colmeri* isolate, provided by A. J. DeLucca II.

[b] Plasmids in brackets code for crystal synthesis conditionally only (dependent on growth medium, or temperature, or both). P1 and P2 distinguish between plasmids coding for P1 toxin only and plasmids encoding both P1 and P2 toxins, in strains that produce both types of crystal (González and Carlton, in preparation; Yamamoto et al., in preparation).

[c] A, Absence of the plasmid in Cry^- variant(s); D, deleted form of the plasmid detected in Cry^- variant(s); I, form of plasmid with insertion detected in Cry^- variant(s); T, transfer of the plasmid into a Cry^- *B. thuringiensis* or *B. cereus* recipient coverted it to Cry^+.

Cloning of toxin genes from BT. The third approach used to locate δ-endotoxin genes has involved the cloning of the genes from either total DNA or a BT strain or from isolated plasmid DNA. The first published reports of molecular cloning of a toxin gene from BT were by Schnepf and Whiteley (1981) and Whiteley et al. (1982). They used as the source of DNA a strain of HD-1 derived from a commercial preparation, Dipel. Fractionation of total plasmid DNA on sucrose gradients produced a "large plasmid fraction" (a mixture of 30- and 47-Mdalton plasmids) and a "small plasmid fraction" containing four plasmids less than 10 Mdaltons in mass. These fractions were digested with a restriction endonuclease and the resulting fragments were cloned into the plasmid pBR322, and transformed into *Escherichia coli*. By means of radioimmunoassay, using antibodies against the major crystal protein, a cross-reacting colony was recovered (ES12) that contained a hybrid plasmid (pES1) with an insert of about 12 kb (kilobases) from the large plasmid fraction. Southern hybridization assays showed that the insert contained sequences that hybridized with both the 47- and 30-Mdalton plasmids of the BT donor DNA. The *E. coli* cloned produced a protein of about the same size as the authentic BT protoxin and this protein cross-reacted with the antibody against the BT toxin crystals. Ingested extracts of the ES12 clone were toxic to larvae of *Manduca sexta*.

In their 1982 report. Whiteley et al. described the cloning of a second var. *kurstaki* toxin gene, from strain HD-73, which produces a toxin crystal somewhat different in its antigenicity and spectrum of insecticidal activity (Krywienczyk et al., 1978). "Large plasmid fraction" DNA was digested, ligated into pBR322, and cloned into *E. coli*. The transformant colonies were probed with an internal fragment of the *kurstaki* gene previously cloned from HD-1-Dipel. One clone (JWK1) had an insert of about 11 Mdaltons and hybridized in Southern blots to a 45-Mdalton plasmid (later changed to 50 Mdaltons by Kronstad et al., 1983) of the HD-73 donor strain. This clone produced a polypeptide of about 130 kdaltons that cross-reacted with antibodies to the HD-1-Dipel crystal protein. Extracts of the recombinant strain were toxic to tobacco hornworm larvae.

Genes for δ-endotoxin have also been cloned from strains of BT var. *thuringiensis* (serotype H-1) and var. *israelensis* (H-14). Klier et al. (1982) described the isolation of an *E. coli* clone containing the toxin gene out of BT strain "*berliner* 1715" (= var. *thuringiensis*, H-1) from an *E. coli* clone bank constructed by shearing of total BT DNA and blunt-end ligation into pBR322 (Klier, Kunst and Rapoport, 1979). Southern hybridization experiments with the cloned toxin gene indicated that this strain carried toxin genes in two locations: on a 42-Mdalton plasmid and also on the chromosome (or perhaps on a very large, undetected plasmid). *E. coli* cells carrying the cloned toxin gene were toxic to *Pieris brassicae*.

The mosquitocidal δ-endotoxin gene from BTI has been cloned into

E. coli by Ward, Ellar, and Todd (1984), and into *B. megaterium* by Sekar and Carlton (1985). In both instances, the purified 75-Mdalton plasmid, previously implicated in toxin synthesis by curing experiments (Ward and Ellar, 1983; González and Carlton, 1984), was used as the source of DNA for cloning. Recombinant plasmids carrying the cloned toxin gene, whether in *E. coli* or in *B. megaterium*, made the host cells toxic to *A. aegypti* larvae.

The availability of a cloned gene for the δ-endotoxin from any BT strain suggests a method for identifying the location of related δ-endotoxin genes in any other strain. By radioactive labeling of the gene sequence, or a fragment thereof, one should be able to determine whether a related gene is present on one or more specific plasmids by probing the DNA complements of other strains by the blotting procedure of Southern (1975). Kronstad et al. (1983) tested the plasmid arrays of 22 strains of BT (representing 14 varieties) with an internal fragment of the cloned toxin gene from HD-1-Dipel (Schnepf and Whiteley, 1981). The probe hybridized with one or more plasmids in 17 of the strains. In ten strains the probe annealed to a single plasmid, ranging from 33 to ~ 150 Mdaltons in size, while in six others it hybridized to two or three plasmids, suggesting the presence of more than one homologous toxin gene. In five of these strains, including *kurstaki* HD-1 and a strain similar to it, they detected hybridization to plasmids in the 40–60 Mdalton range and also to plasmids in the 100–150 Mdalton range. In two other strains resembling HD-1 in plasmid content, there was no hybridization in the 40–60 Mdalton range, possibly due to prior loss of the plasmid(s) responsible (as seen in the plasmid arrays in their paper). Four strains showed no hybridization to the probe, although they all possessed several plasmids. Finally, in one strain (var. *wuhanensis*), the probe did not anneal with any plasmids, although it hybridized strongly with a region believed to contain fragments of chromosomal DNA. In strains where only one plasmid annealed to the probe, such as *galleriae* HD-8, *thuringiensis* HD-2, *kurstaki* HD-73, *alesti*, and others, most results agreed with those listed in Table 2. The results with HD-1 and related strains agreed with those of Carlton and González (1985), and suggest that HD-1 may contain two or more toxin plasmids.

Genetic complexity of toxin synthesis in different BT strains

Recent findings indicate that while synthesis of δ-endotoxin in some BT strains is simple (e.g., one toxin plasmid controls production of a single type of δ-endotoxin protein), in other BT strains the situation is more complex: the strains contain two or more toxin plasmids, and more than one type of δ-endotoxin is synthesized. The complexities of δ-endotoxin production in three BT strains of var. *kurstaki* H-3ab, namely HD-73, HD-1, and HD-263, are a good example of the variations within a single variety.

In strain HD-73 a single entomocidal protein, of the P1 type, is produced (Yamamoto, 1983). A single large, transmissible plasmid of 50 Mdaltons has been shown to determine δ-endotoxin and crystal synthesis by studies involving plasmid curing (González et al., 1981), plasmid transfer (González et al., 1982) and cloning of the toxin gene (Whiteley et al., 1982). Thus HD-73 harbors a single toxin plasmid which codes for a single δ-endotoxin protein (or possible one or more very similar toxin proteins), and is the simplest of the three strains in its pattern of toxin expression.

δ-Endotoxin synthesis in the commercially-important H-3ab strain, HD-1, is more complex than in HD-73. Yamamoto and McLaughlin (1981) described the production by HD-1 of a second toxin protein, P2, which differs in size and toxicity spectrum from the major toxin protein, P1. In our initial study of HD-1 (González et al., 1981), a 29-Mdalton plasmid was implicated in toxin production, based on the isolation of a single Cry^- mutant, in which this plasmid had acquired an insertion of 4 to 5 Mdaltons of DNA. More extensive studies with other partially cured Cry^+ and Cry^- variants of HD-1 has convinced us that the insertion into the 29-Mdalton plasmid was fortuitous, and that HD-1 toxin production is controlled by two other plasmids, of 44 and ~ 115 Mdaltons (T. Yamamoto, J. M. González, Jr., and B. C. Carlton, unpublished observations; Carlton and González, 1985). The ~ 115-Mdalton plasmid codes for the P2 protein and one type of P1, and is nontransmissible, so that it has been implicated only by plasmid curing. The 44-Mdalton plasmid codes for a slightly different P1, and is both easily cured and transmissible to Cry^- BT recipients. The two plasmids also differ in that HD-1 variants carrying the ~ 115-Mdalton plasmid form visible crystals at 30°C on nutrient agar, while variants that have lost the ~ 115-Mdalton plasmid but retain the 44-Mdalton plasmid form a visible crystal at 25°C but not at 30°C. Crystal formation controlled by the 44-Mdalton plasmid appears to be affected by both the growth temperature and the nutrient medium. This temperature-sensitive crystal production was first observed by Minnich and Aronson (1984); however, they believed the 29-Mdalton plasmid to be responsible. Because of differences in methods of plasmid analysis between the two laboratories, it is possible that the same plasmid is being studied.

The third and most complex strain, HD-263, also produces

both P2 and P1 proteins. Plasmid curing and transfer studies of HD-263 (T. Yamamoto, J. M. González, Jr., and B. C. Carlton, unpublished observations) showed that a large (~ 110 Mdalton) nontransmissible plasmid carries both P2 and P1 genes, and a 44-Mdalton transmissible plasmid carries at least one P1 gene. However, the HD-263 44-Mdalton plasmid codes for a crystal even at 30°C (González et al., 1982), so that it is probably similar but not identical to the 44-Mdalton plasmid of HD-1. Curing experiments (J. M. González, Jr. and B. C. Carlton, unpublished observations) have implicated another HD-263 plasmid, of 60 Mdaltons, in the production of a P1 crystal. So in HD-263, three toxin plasmids code for perhaps four δ-endotoxin polypeptides.

The H-1 (var. *thuringiensis*) strain HD-2 may also carry two toxin plasmids. In addition to the 75-Mdalton toxin plasmid, implicated by curing and transfer studies (González et al., 1981; González et al., 1982), a second HD-2 plasmid, of 54 Mdaltons, also codes for a bipyramidal crystal (and therefore for some type of δ-endotoxin). Transfer studies have shown that the 54-Mdalton plasmid transfers readily from HD-2 into BT and *B. cereus* recipients (J. M. González, Jr., and B. C. Carlton, unpublished observations). The transconjugants are, however, only conditionally Cry^{+}, in that few or no crystals appear on nutrient agar, but appear reproducibly when the transcipients are grown on Difco starch agar. The conditions required for crystal synthesis by the 54-Mdalton HD-2 plasmid were similar but not identical to those required by the HD-1 44-Mdalton plasmid (whether in the original hosts, HD-2 and HD-1, or in transconjugant strains).

Discussion and conclusions

The characterization of *B. thuringiensis*, both as a biological entity of considerable intrinsic interest and as a biocontrol agent, has advanced considerably during the past few years. Genetic approaches aimed at unravelling the mode of inheritance of insecticidal toxin production have produced several observations that are worthy of consideration. First, each BT strain has its own characteristic array of extrachromosomal plasmids that can be conveniently displayed by agarose gel electrophoresis for purposes of monitoring strain purity and stability. The major importance of this observation is that genetic analyses of toxin production have shown that the genes specifying the toxin proteins are generally encoded on moderate to large (50–200 kb) plasmids. The few exceptions to plasmid-coded toxin genes are, we believe, most likely due to events in which a toxin-coding plasmid has undergone a recombinational insertion into the bacterial chromosome.

The significance of the rather large sizes of the toxin-coding plasmids is not obvious. Certainly all of the implicated plasmids are considerably larger than the 4 kb estimated to code for the typical protoxin genes. It may be that there are regulatory aspects associated with the low copy numbers estimated for these large plasmids relative to the smaller ones, although there is no evidence one way or the other on this as yet. Despite their large sizes, many toxin-coding plasmids are capable of conjugal transmission from one cell to another, including into other serotypes as well as into strains of *B. cereus*. This ability of toxin plasmids to transcend flagellar serotype boundaries is of significance, we feel, in explaining the occurrence of strains producing a toxin crystal of one type in an atypical serotypic background. It is also of considerable interest as a tool for constructing novel combinations of toxin plasmids in strain improvement programs.

Another observation of note is that several strains possess two or more toxin-coding plasmids, leading to the production of heterogeneous bipyramidal toxin crystals or even two or more distinct crystals. This multiplicity of toxin genes in a single strain is undoubtedly one basis for the wide diversity of insecticidal activities observed between different strains. A "blending" of qualitatively different toxin proteins in different amounts can lead to quantitatively-variable insecticidal activities on various target insects.

The observations cited above will be important for the future genetic engineering of BT to develop strains having improved or novel activities on various insect pests. The occurrence of toxin genes on conjugally-transmissible plasmids offers an approach to constructing novel combinations of toxin genes derived from widely-diverse strains. This approach can also be used, in conjunction with the selective curing of various plasmids, to substitute a less-desirable toxin gene with a better one. If such manipulations are carried out in a background strain that has previously been optimized for fermentation production of the toxin, potential problems with using toxin genes of diverse origin should be minimized. Furthermore, the manipulation of naturally-occurring toxin plasmids by natural conjugational transfer in BT itself should minimize problems in registering such derived strains as new products. Such strains, we contend, should be viewed in the same light as natural isolates having a multiplicity of toxin genes.

Finally, and as additional genetics, physiological, and biochemical characterizations of various BT strains and their toxins become available, it will undoubtedly be possible to refine their activities through more sophisticated genetic manipulations such as site-direct mutagenesis of cloned genes or segments thereof. The recent publication of the complete sequences of two toxin genes (Schnepf, Wong and Whiteley, 1985; Adang et al., 1985) and a partial sequence of another (Shibano et al., 1985) provide a basis for comparison of gene and toxin protein structures as related to their insecticidal activities which will most certainly accelerate in the near future. The application of gene splicing methodologies to the genetic dissection and restructuring of various segments of the BT insecticidal toxin genes promises to produce a new generation of BT strains having insect control properties that extend far beyond the naturally occurring isolates currently known.

Acknowledgement The authors gratefully acknowledge the advice and counsel of Dr. Howard T. Dulmage.

References

Adang, M. J., M. J. Staver, T. A. Rocheleau, J. Leighton, R. F. Barker and D. V. Thompson. 1985. *Characterized full-length and truncated plasmid clones of the crystal protein of Bacillus thuringiensis subspecies kurstaki HD-73 and their toxicity to Manduca sexta.* Gene 36:289–300.

Bernhard, K., H. Schrempf and W. Goebel. 1978. *Bacteriocin and antibiotic resistance plasmids in Bacillus cereus and Bacillus subtilis.* J. Bacteriol. 133:897–903.

Burges, H. D., S. Hillyer and D. O. Chanter. 1975. *Effect of ultraviolet and gamma rays on the activity of δ-endotoxin protein crystals of Bacillus thuringiensis.* J. Invertebr. Pathol. 25:5–9.

Campbell, A. M. 1969. *Episomes.* Harper and Row, New York.

Cantwell, G. E. and B. A. Franklin. 1966. *Inactivation by irradiation of spores of Bacillus thuringiensis var. thuringiensis.* J. Invertebr. Pathol. 8:256–258.

Carlton, B. C. and J. M. González, Jr. 1985. *Plasmids and delta-endotoxin production in different subspecies of Bacillus thuringiensis.* Pp. 246–252 in J. A. Hoch and P. Setlow, eds., *Molecular Biology of Microbial Differentiation 1985.* Am. Soc. Microbiol., Washington, DC.

de Barjac, H. and A. Bonnefoi. 1962. *Essai de classification bio-chimique et serologique de 24 souches de Bacillus du type B. thuringiensis.* Entomophaga VII:5–31.

de Barjac, H. 1981. *Identification of H-serotypes of Bacillus thuringiensis.* Pp. 35–43 in H. D. Burges, ed., *Microbial Control of Pests and Plant Diseases 1970–1980.* Academic Press, New York.

Dulmage, H. T. 1981. *Insecticidal activity of isolates of Bacillus thuringiensis and their potential for pest control.* Pp. 193–222 in H. D. Burgess, ed., *Microbial Control of Pests and Plant Diseases.* Academic Press, London.

Eckhardt, T. 1978. *A rapid method for the identification of plasmid desoxyribonucleic acid in bacteria.* Plasmid 1:584–588.

Goldberg, L. J. and J. Margalit. 1977. *A bacterial spore demonstrating rapid larvicidal activity against Anopheles sergentii, Uranotaenia unguiculata, Culex univitattus, Aedes aegypti, and Culex pipiens.* Mosquito News 37:355–358.

González, J. M., Jr. and B. C. Carlton. 1982. *Plasmid transfer in Bacillus thuringiensis.* Pp. 85–95 in U. N. Streips, S. H. Goodgal, W. R. Guild and G. A. Wilson, eds., *Genetic Exchange: A Celebration and a New Generation.* Dekker, New York.

González, J. M. Jr. and B. C. Carlton. 1984. *A large transmissible plasmid is required for crystal toxin production in Bacillus thuringiensis var. israelensis.* Plasmid 11:28–38.
González, J. M., Jr., H. T. Dulmage and B. C. Carlton. 1981. *Correlation between specific plasmids and δ-endotoxin production in Bacillus thuringiensis.* Plasmid 5:351–365.
González, J. M., Jr., B. J. Brown and B. C. Carlton. 1982. *Transfer of Bacillus thuringiensis plasmids coding for δ-endotoxin among strains of B. thuringiensis and B. cereus.* Proc. Nat. Acad. Sci. USA 79:6951–6955.
Hannay, C. L. 1967. *The dissociation of Bacillus thuringiensis.* Can. J. Microbiol. 13:1566–1568.
Klier, A., F. Kunst and G. Rapoport. 1979. *Structure of cloned ribosomal DNA cistrons from Bacillus thuringiensis.* Nuc. Acids Res. 7:997–1010.
Klier, A., F. Fargette, J. Ribier and G. Rapoport. 1982. *Cloning and expression of the crystal protein genes from Bacillus thuringiensis strain berliner 1715.* EMBO J. 1:791–799.
Krieg, A., A. M. Huger, G. A. Langenbruch and W. Schnetter. 1983. *Bacillus thuringiensis var. tenebrionis: ein neuer, gegenüber Larven von Coleopteren wirksamer Pathotyp.* Z. ang. Ent. 96:500–508.
Kronstad, J. W., H. E. Schnepf and H. R. Whiteley. 1983. *Diversity of locations for Bacillus thuringiensis crystal protein genes.* J. Bacteriol. 154:419–428.
Krywienczyk, J. and T. A. Angus. 1967. *A serological comparison of several crystalliferous insect pathogens.* J. Invertebr. Pathol. 9:126–128.
Krywienczyk, J., H. T. Dulmage and P. G. Fast. 1978. *Occurrence of two serologically distinct groups within Bacillus thuringiensis serotype 3ab var. kurstaki.* J. Invertebr. Pathol. 31:372–375.
Kurstak, E. and P. Tijssen. 1982. *Microbial and viral pesticides: modes of action, safety, and future prospects.* Pp. 3–32 in E. Kurstak, ed., *Microbial and Viral Pesticides.* Dekker, New York.
Leong, K. L. H., R. J. Cano and A. M. Kubinski. 1980. *Factors affecting Bacillus thuringiensis total field persistance.* Environ. Entomol. 9:593–599.
Lereclus, D., M.-M. Lecadet, J. Ribier and R. Dedonder. 1982. *Molecular relationships among plasmids of Bacillus thuringiensis: conserved sequences through 11 crystalliferous strains.* Mol. Gen. Genet. 186:391–398.
Levin, B. R. 1979. *Problems and promise in genetic engineering in its potential applications to insect management.* Pp. 171–175 in M. A. Hoy and J. J. McKelvey, Jr., eds., *Genetics in Relation to Insect Management.* The Rockefeller Foundation, New York.
Lüthy, P. and H. R. Ebersold. 1981. *Bacillus thuringiensis delta-endotoxin: histopathology and molecular mode of action.* Pp. 235–267 in E. W. Davidson, ed., *Pathogenesis of Invertebrate Microbial Diseases.* Rowman and Allanheld, Totowa, New Jersey.
Minnich, S. A. and A. I. Aronson. 1984. *Regulation of protoxin synthesis in Bacillus thuringiensis.* J. Bacteriol. 158:447–454.
Miteva, V. I. 1978. *Isolation of plasmid DNA from various strains of Bacillus thuringiensis and Bacillus cereus.* Dokl. Bolg. Akad. Nauk. 31:913–916.
Norris, J. R. 1970. *Sproeformers as insecticides.* J. Appl. Bacteriol. 33:192–206.
Schnepf, H. E. and H. R. Whiteley. 1981. *Cloning and expression of the Bacillus thuringiensis crystal protein gene in Escherichia coli.* Proc. Nat. Acad. Sci. USA 78:2893–2897.
Schnepf, H. E., H. C. Wong and H. R. Whiteley. 1985. *The amino acid sequence of a crystal protein from Bacillus thuringiensis deduced from the DNA base sequence.* J. Biol. Chem. 260:6264–6267.
Sekar, V. and B. C. Carlton. 1985. *Molecular cloning of the delta-endotoxin gene of Bacillus thuringiensis var. israelensis.* Gene 33:151–158.
Shibano. Y., A. Yamagata, N. Nakamura, T. Iizuka, H. Sugisaki and M. Takanami. 1985. *Nucleotide sequence coding for the insecticidal fragment of the Bacillus thuringiensis crystal protein.* Gene 34:243–251.
Sneh, B., S. Schuster and M. Broza. 1981. *Insecticidal activity of Bacillus thuringiensis strains against the Egyptian cotton leaf worm Spodoptera littoralis (LEP.: NOCTUIDAE).* Entomophaga 26:179–190.

Southern, E. M. 1975. *Detection of specific sequences among DNA fragments separated by gel electrophoresis*. J. Mol. Biol. 98:503–517.

Stahly, D. P., D. W. Dingman, L. A. Bulla, Jr. and A. I. Aronson. 1978. *Possible origin and function of the parasporal crystals in Bacillus thuringiensis.* Biochem Biophys. Res. Commun. 84:581–588.

Ward, E. S. and D. J. Ellar. 1983. *Assignment of the δ-endotoxin gene of Bacillus thuringiensis var. israelensis to a specific plasmid by curing analysis*. FEBS Lett. 158:45–49.

Ward, E. S., D. J. Ellar and J. A. Todd. 1984. *Cloning and expression in Escherichia coli of the insecticidal δ-endotoxin gene of Bacillus thuringiensis var. israelensis*. FEBS Lett. 175: 377–382.

Whiteley, H. R., J. W. Kronstad, H. E. Schnepf and J. P. DesRosier. 1982. *Cloning the crystal protein gene of B. thuringiensis in E. coli*. Pp. 131–144 in A. T. Ganesan, S. Chang, and J. A. Hoch, eds., *Molecular Cloning and Gene Regulation in Bacilli*. Academic Press, New York.

Yamamoto, T. 1983. *Identification of entomocidal toxins of Bacillus thuringiensis by high performance liquid chromotography*. J. Gen. Microbiol. 129:2295–2603.

Yamamoto, T. and R. E. McLaughlin. 1981. *Isolation of a protein from the parasporal crystal of Bacillus thuringiensis var. kurstaki toxic to the mosquito larva, Aedes taeniorhynchus.* Biochem. Biophys. Res. Commun. 103:414–421.

Zakharyan, R. A., A. S. Agabalyan, L. A. Chil-Akopyan, N. S. Gasparyan, K. A. Bankunts, P. E. Tatevosyan and E. K. Afrikyan. 1976. *On the possible role of extrachromosomal DNA in the formation of the entomocidal endotoxin of Bacillus thuringiensis*. Dokl. Akad. Nauk. Arm. SSR 63:42–47.

Spectrum of cellular immune responses in host resistance

DEAN BEFUS*
Host Resistance Programme, Department of Pathology, McMaster University, Hamilton, Ontario L8N 3Z5, Canada

Abstract The complexity of antigen-specific and non-specific interactions involved in host resistance and pathology is unravelling at an exciting rate. Following infection, macrophage-dependent antigenic stimulation activates a diversity of T and B lymphocytes. This initiates a series of specific antibody and cellular responses which deal directly with the infectious insult and/or call in a range of other, non-specific effector cells such as macrophages, neutrophils, eosinophils, mast cells/basophils, and natural killer cells. These cells are protective through direct anti-parasite activities or by establishing a microenvironment compatible with cohabitation between host and parasite. They also amplify and modulate effector functions and facilitate repair mechanisms by activating local epithelial cells, goblet cells, smooth muscle cells, and even the nervous system. At the forefront of immunology are our efforts to dissect the chemical mediators of host protection such as tumor necrosis factor, lymphotoxin, oxidative pathways, toxic enzyme systems and phospholipid metabolites. Such chemical mediators and others including interferon and interleukins 1, 2 and 3, can be isolated, purified and studied on cloned cells to define their activities. The challenge that remains is to dissect the in vivo activities which facilitate host survival and productivity in the face of constant exposure to infectious agents.

Introduction

As in so many other fields, there is an excitement about the potential of the innovative tools of biotechnology to stimulate discovery and create a new level of fundamental understanding of host defence mechanisms and infectious disease. Our manipulation of the genome of various pathogens will continue to uncover the basis of pathogenicity (Mekalanos et al., 1983) and provide new approaches to vaccination, such as the use of deletion mutants (Kaper et al., 1984) or synthetic peptide sequences (Bittle et al., 1982). These and many other recent developments withinfectious agents will be reviewed in other presentations of this symposium. My objective is to construct a conceptual overview of the complex cellular and molecular events in mammalian immune responses, and to discuss their integration into an effective network which provides host resistance to infectious diseases.

In the dissection of the underlying mechanisms of host resistance, the new biotechnical tools provide the important opportunity to analyze the characteristics and functions of homogeneous populations of cells and specific molecules. Rather than developing concepts upon

*Present address: Department of Microbiology and Infectious Disease, University of Calgary, Health Sciences Centre, Calgary, Alberta T2N 4N1, Canada.

the tenuous, and often rapidly crumbling base of data provided using complex mixtures of cells or molecular soups, it is now possible to study relatively large quantities of highly purified molecules or cells and thus assess their function in host defence. For example, clones of antigen-specific T lymphocytes have been quintessential in the recent avalanche of information about the nature of the recognition receptor(s) for antigen-major histocompatibility complex (MHC) on T lymphocytes (Meuer et al., 1984). Similarly, the recent expression in *Escherichia coli* of cDNA for interleukin-1 (IL-1); Lomedico et al., 1984) is a major step forward in understanding the synthesis, secretion, activation and function of this macrophage-derived factor which interacts with a diversity of cell types throughout the body (Dinarello, 1984).

This brief review will discuss the cellular and molecular basis for the recognition of specific antigens from infectious agents or other sources, and consider the subsequent activation of cellular pathways of immune responses. After the recognition of the pathogen, numerous pathways, some of which involve direct messages from specific antigen reactive cells, and some which do not, lead to the activation of other cells which do not possess endogenous antigen-specific receptors. Together, these antigen-specific and non-specific cells amplify and regulate the essential pathways of host defence and tissue repair. However, the ultimate effector mechanisms of defence are largely unknown. Some possible mechanisms will be discussed, the conceptual framework will be laid and obvious gaps in our knowledge identified. Because of the magnitude of the topic and space limitations, this review cannot be comprehensive. However specific examples will be used to illustrate how new biotechnical developments will aid the development of a fundamental understanding of host resistance, and facilitate future application of immunotherapeutic and immunoprophylactic methodologies.

Specificity: antigen recognition by lymphocytes

Two basic categories of lymphocytes are recognized, namely B and T lymphocytes. B lymphocytes express immunoglobulin molecules on their surface with exquisite capabilities of specific antigen recognition. By contrast, until recently the nature of the antigen recognition structures on T lymphocytes was unknown. These thymus-dependent T lymphocytes are now known to express a unique, complex surface receptor(s) which recognizes both specific antigen as well as self MHC products. The new revolution in biotechnology has rapidly uncovered the molecular basis of B cell and T cell-specific receptors and further understanding is evolving at an exponential rate.

B lymphocytes. In mammals, lymphocytes of the B lineage originate predominantly in the bone marrow. They express on their surface antigen-specific receptors, namely immunoglobulins. These glycoprotein receptors are composed of disulphide-linked heavy and light chains, each with a constant and variable region. The 3-dimensional structure of the variable regions of the heavy and light chains forms the specific antigen binding site of the antibody molecule. Encoded within the genome and enriched by genetic transpositions and somatic mutations, there is a diverse repertoire ($\geqslant$ 1 million for the mouse) of unique antigen recognition structures (Honjo, 1983).

Once a B lymphocyte has been activated by combination with its complementary antigen in the context of T lymphocytes and antigen presenting macrophages, it will proliferate and either differentiate into a plasma cell synthesizing and secreting specific antibody, or become a sensitized B memory lymphocyte. The latter respond rapidly upon re-exposure to the antigen. Five distinct classes of antibody molecules can be produced, each one distinguishable by the qualities of the constant region of its heavy chain. Through genetic transposition a particular antibody binding site may appear on antibody of any of these five classes (Honjo, 1983). Because the constant regions of each class are distinct and bear the domains of the molecules which direct functions (Fc end) other than antigen recognition ($F(ab)^2$ end), the class of antibody is important in the sequence of events that ensues following antigen-antibody combination.

Antibodies may: (1) directly neutralize pathogens or their toxins, (2) interact with the complement cascade to generate enzymatic activities which are lytic or activate various cell types, or (3) activate cells by Fc receptor mediated mechanisms. Cells activated by complement fragments or Fc receptors provide a spectrum of responses ranging from phagocytosis, lysosomal enzyme release, production of superoxide anion, singlet oxygen, hydroxyl radical and hydrogen peroxide, and prostaglandin and leukotriene production to the release of numerous other molecular species, e.g. IL-1, which modulate immune and inflammatory responses (see below).

The five classes of antibodies are IgM, IgG, IgD, IgA and IgE, and their characteristics and functions have been widely reviewed (e.g. Paul, 1984; Nisonoff, 1984). IgM production occurs rapidly in a primary immune response, but is rapidly replaced in quantity by IgG. IgM normally exists as a large molecular ratio (Mr), approximately 900,000, pentameric structure which is a potent neutralizing and inhibitory antibody, particularly good at complement fixation. The four subclasses of IgG make up the major serum immunoglobulins and have a range of functions associated with complement activation, cellular Fc receptors and antigen binding. IgD is not abundant in the circulation but is important as a marker, and perhaps an agent, of B lymphocyte differentiation.

IgA has two subtypes, but is best known because of its abundance and unique molecular characteristics in mucosal secretions. In these secretions it is found primarily as a dimeric or oligomeric structure possessing an additional glycoprotein of epithelial origin, namely secretory component (SC). Secretory component provides IgA with enhanced resistance to proteolysis, as well as facilitating its transport into secretions. Secretory IgA provides a protective barrier against mucosal colonization and invasion. Moreover, in the tissues and circulation, IgA activates various cells through IgA Fc receptors and is important in removing antigens from the circulation by a hepatic SC-mediated mechanism (Bienenstock and Befus, 1983).

IgE is the least abundant immunoglobulin in the circulation. However, mast cells and basophils possess high affinity receptors for IgE and can be sensitized by individual antibody molecules for prolonged periods. When specific antibodies on the surface of a mast cell or basophil are cross-linked by antigen, immediate hypersensitivity allergic reactions ensue. IgE is also involved in mediating antibody-dependent cellular cytotoxic reactions by macrophages and eosinophils (Capron et al., 1982).

T lymphocytes. T lymphocytes are thought to be educated with regard to their antigen recognition properties in the thymus. They learn to see antigens in the context of MHC molecules. In the last decade, research in this field has sought to understand whether a single or two separate receptors are involved, and its (their) precise nature. Moreover, T lymphocytes are comprised of functionally distinct subpopulations, many of which can be distinguished by their surface phenotype (Ballieux and Heijnen, 1983). Some populations (killer T cells) see antigen in the context of Class I MHC molecules, whereas others see antigen concurrently with Class II MHC molecules (T helper and T suppressor cells; Robertson, 1984).

Significant progress in the definition of the T cell receptor(s) has been made in the last 3 to 5 years by the use of two approaches. Firstly, monoclonal antibodies to surface components of cloned, antigen-specific T cells were selected for reactivity solely with the particular cloned T cell (Meuer et al., 1984). Given that each antigen-specific cell clone will have many T cell markers in common with other T cell clones, but will also have a unique antigen receptor, this selection strategy was aimed at the identification of monoclonal antibodies to the unique T cell receptors. With these anti-receptor monoclonal antibodies, the T cell receptor was isolated and chemically characterized. Such approaches in human and mouse systems have established that the T cell receptor complex is composed in part of a 80,000 to 90,000 M_r heterodimer (alpha and beta chains). In humans, the T3 marker is closely associated with this heterodimer and the analogue (L3T4) is present in the mouse.

A second approach, namely the selection from genomic libraries of T lymphocyte-specific sequences whose products are likely to be expressed on the cell surface, has complemented the studies in which the receptors were isolated using monoclonal antibodies. Alpha and beta chain sequences have been isolated and analyzed (Patten et al., 1984; Fabbi et al., 1984; Yoshikai et al., 1984) and each appeared to bear variable and constant sequences analogous to immunoglobulin from B lymphocytes. An additional chain, gamma, has also been isolated. Its role is unclear, but interestingly, mRNA for gamma chain is relatively abundant in the prenatal thymus but decreases thereafter. By contrast, levels of the beta chain mRNA are relatively constant in prenatal and adult thymocytes, whereas levels of alpha chain mRNA increase from prenatal to adult times (Raulet et al., 1985).

Taken together with experiments of cellular immunology (Kappler et al., 1981; Hedrick et al., 1982), these analyses of the T cell receptor favour the model that a single receptor complex is involved in dual recognition of antigen and MHC molecules (Patten et al., 1984).

The subpopulations of T lymphocytes that are identified by their functional potential and/or surface phenotype include: helpers, suppressors, inducers, contrasuppressors, cytotoxic cells, delayed hypersensitivity cells, etc. The range of surface markers used to categorize and study T cell subpopulations in the mouse include Thy, Lyt, Qa, and Ia (especially IJ) (Scollay and Shortman, 1983). In the human, analogous markers conclude the Leu or OKT series (Ballieux and Heijnen, 1983). The various pathways which have been proposed for the interactions of these T cell subpopulations amongst themselves and with B lymphocytes, macrophages and many other cells active in inflammatory responses and host defences have been reviewed elsewhere (Green, Flood and Gershon, 1983; Nabholz and MacDonald, 1983; Singer and Hodes, 1983). However, some T cell activities such as influences on goblet cell differentiation (Miller and Nawa, 1979) and perhaps hypothalamic neuronal firing (Besedovsky et al., 1983) have received less attention, but may be equally relevant in host resistance and disease.

Cells without endogenous antigen recognition structures

In addition to T and B lymphocytes with their specific antigen recognition structures, a large spectrum of other cells without such endogenous antigen-specific receptors are recruited and activated as important elements in host defence. Cells commonly acknowledged to be important in inflammation include circulating monocytes and tissue macrophages, neutrophils, eosinophils, mast cells and basophils, and natural killer cells and perhaps natural cytotoxic cells. Goblet, epithelial, endothelial and smooth muscle cells, as well as neurons must also be involved in sites of infection and inflammation.

The mechanisms which recruit and activate these different cells are varied. Many cell types bear complement receptors and can be activated and/or stimulated by specific fragments of the complement cascade (see Table 1, Fearon and Wong, 1983). Other chemotactic factors can be derived from lymphocytes, mast cells, eosinophils, macrophages and other cells, and include arachidonic acid metabolites, particularly leukotriene B4 (LTB4), and various oligopeptides (Snyderman and Pike, 1984). Because many of these cells possess Fc receptors for different classes of antibody, e.g. IgG, IgA, IgE, a common mode of cell activation is for complexes of antigens and antibodies to interact at the cell surface Fc receptor, thereby activating the cell. These receptor-ligand interactions have a degree of specificity which is dependent upon the recognition sites of the individual antibody molecules on the cell surface.

Recently, an analogous aquisition of specificity, but of T cell origin, has been defined for mast cells (Askenase and van Loveren, 1983). Some antigen-activated T cells produce an antigen-specific factor, fundamentally unlike IgE in many of its characteristics, but like IgE because it is capable of sensitizing mast cells to make them specifically responsive to the stimulating antigen. Precise characteristics and functions of this T cell-derived factor await molecular dissection using tools of genetic biotechnology. Interesting results to date strongly suggest that the factor activates mast cells in a manner unlike IgE-mediated activation (van Loveren, Kraiuter-Kops and Askenase, 1984). The powerful tools provided by molecular biology will also help to determine whether this type of T cell-derived, antigen-specific factor is restricted to mast cell sensitization, or can also sensitize other cells, such as macrophages, eosinophils or neutrophils to recognize antigen.

The possible functions that these different cells perform in the expression of host defences are too many to mention completely in this forum (see Befus and Bienenstock, 1982, 1984) and undoubtedly many important functions have yet to be uncovered. Activities such as phagocytosis, antibody-dependent cellular cytotoxicity, lysosomal enzyme release, altered vascular flow and permeability, and smooth muscle modulation (e.g. bronchus, intestine) are obvious (e.g. Larsen and Henson, 1983). However, mucous secretion (Shelhamer, Maron and Kaliner, 1980), control of myelopoiesis (Griffin et al., 1984), immunoregulation (Beer and Rocklin, 1984; Goetzl, Payan and Goldman, 1984) and tissue repair (Rao et al., 1983) are other examples of the broad categories of activities that cells of immune and inflammatory responses orchestrate in host defence. Overall, the array of cells called upon following infectious insult act to prime the immune system, activate effector pathways of host resistance through generation of a microenvironment incompatible with pathogen survival (or at least pathogenicity), and facilitate tissue repair.

Molecular bases of host defence

The precise molecular bases of host defences are unclear, but some chemical mediators for cell-to-cell communication and of the integration of antigen-specific and non-specific activities have been identified. Moreover, numerous candidate molecules which effect parasite death or inactivation are also being studied.

Mediators which modulate host responses. A number of molecules which modulate immune responses are known including metabolites of arachidonic acid such as LTB_4 and PGE_2, interferons (alpha, beta, gamma; Trinchieri and Perussia, 1985) and neuropeptides, e.g. substance P, somatostatin (Payan, Levine and Goetzl, 1984) and vasoactive intestinal polypeptide (Ottaway, 1984). However, given space limitations only interleukins (IL) 1, 2, and 3 will be discussed herein. Exciting advances have been made with these molecules recently and the next few years promise to be most enlightening.

Interleukin-1 is synthesized primarily by monocytes and macrophages, although keratinocytes, gingival and corneal epithelial cells, mesangial cells and astrocytes also produce a similar molecule (Dinarello, 1984). Native human IL-1 occurs in four forms, two of about 33,000 M_r (pI 5, pI 7) and two of 15,000 M_r (pI 5, pI 7) (Wood et al., 1985). The gene for murine IL-1 has recently been cloned in *E. coli* (Lomedico et al., 1984). It appears that IL-1 is synthesized as a 270 amino acid polypeptide (approximately 33,000 M_r) and can be converted to a 156 amino acid (carboxy terminus, 15,000 M_r) form.

The IL-1 family of molecules has been identified as having numerous activities and at various times has been given acronyms like lymphocyte activating factor, endogenous pyrogen, leukocyte endogenous mediator, mononuclear cell factor, synovial factor, proteolysis inducing factor, epidermal cell derived thymocyte activating factor, hepatocyte stimulating factor, acute phase mediator, and fibroblast activating/inhibiting factor. Interleukin-1 obviously inteiacts directly or indirectly with lymphocytes, the hypothalamus, bone marrow, neutraophils, fibroblasts, hepatocytes, muscle and undoubtedly many other cells important in immune and inflammatory responses.

In contrast to IL-1, IL-2 appears to be less exhaustive in its repetoire of collaborators (Cantrell and Smith, 1984). Interleukin-2 is produced by T lymphocytes as a 15,000 M_r glycoprotein and stimulates T and B lymphocyte proliferation. It is important in clonal expansion of lymphocytes, in differentiation events such as T killer cell development, and in augmentation of natural killer activity (NK cells) and mediator production by lymphocytes, e.g. interferon and B cell growth factors. The gene for IL-2 has been cloned in *E. coli* (Kashima et al., 1985). Unlike IL-1, the receptor for IL-2 is known in mouse, rat and human systems and careful studies of receptor-ligand interaction have been conducted (Cantrell and Smith, 1984). Further definition of cell activation by IL-2 promises to facilitate careful clinical investigations with this interleukin.

Interleukin-3 is derived from activated T lymphocytes and is recognized to have the ability to stimulate growth and some differentiation events in a spectrum of myelopoietic stem cells (Ihle et al., 1982). For example, it is a mast cell growth factor and a stimulator of macrophage/granulocyte colonies, erythroid burst-forming units and colony-forming units (spleen) (CFU-S; Rennick et al., 1985). Moreover, it maintains natural cytotoxic (NC) but not NK cells in culture (Djeu et al., 1983). The murine gene for IL-3 has been introduced into monkey cells and a 15,000 M_r polypeptide produced (Yokota et al., 1984). However, native murine IL-3 is about 28,000 M_r and presumably the discrepancy between the native molecule and molecules from transfected cells relates to post-translation modifications. In vivo studies of the clinical relevance of IL-3 and myelopoiesis, host defences and pathology are eagerly awaited.

Possible effector molecules. The ultimate effector molecules of parasite death or expulsion from the host also include many candidates. Much effort has focused upon peroxidases, hydrogen peroxide and active oxygen species such as hydroxyl radical, singlet oxygen and superoxide (Locksley and Klebanoff, 1983), as well as enzymes such as arginase (Schneider and Dy, 1985). In this chapter, tumor necrosis factor (TNF) and lymphotoxin (LT) will be reviewed.

Both TNF and LT possess the interesting and highly relevant capability of killing tumor cells but not normal cells (Pennica et al., 1984; Gray et al., 1984), although the mechanisms underlying this selectivity are unknown. Tumor necrosis factor is derived from activated macrophages, whereas LT is from activated lymphocytes. The human genes for both have recently been cloned in *E. coli* and the products are distinct, having only about 30% homology (Pennica et al., 1984; Gray et al., 1984). Now that these cloned products are available, detailed studies of reaction in vitro and preliminary clinical investigations can be undertaken. With such tools, new understandings of defence mechanisms and pathogenesis will arise, and innovative strategies for immunotherapy and immunoprophylaxis will be spawned.

Conclusions

The last 25 years have witnessed the birth and exponential expansion of modern immunology. However, its foundations were laid by earlier scientists and clinicians trying to dissect the mysteries of infectious diseases. Exciting developments have occurred within the last 5 years in the field of immunology and host defence as cells have been isolated, cultured and cloned, and specific recognition structures and mediators identified. Molecular genetics and recombinant DNA technology have identified some of the genes involved and provided large quantities of relatively purified molecules for studies in vitro and in vivo. Origins, controls of biosynthesis, and sites and modes of action can now be systematically dissected in isolated systems. In addition, preliminary animal and human trials can be initiated with at least some of these mediators to investigate when and how they may be relevant in vivo.

Given the complexity of cellular and molecular interactions in the immune system and in defence mechanisms, it will not present an easy task to construct comprehensive conceptual models of the activities of these mediators in vivo. Molecular biologists and immunologists will continue to develop new tools. However, to facilitate the most efficient utilization of these tools to understand in vivo relevance and develop therapeutic or prophylactic modalities, strong teams of basic biologists, cellular immunologists and molecular biologists/immunologists must be assembled. Only with such teams can the range of skills necessary be brought to bear upon the challenges of the in vivo action of mediators of host defences.

Acknowledgements J. Butera and G. Hardy provided skilled editorial assistance. Financial support was provided by Medical Research Council of Canada, The Rockefeller Foundation, and The Canadian Foundation for Ileitis and Colitis.

References

Askenase, P. W. and H. Van Loveren. 1983. *Delayed-type hypersensitivity: activation of mast cells by antigen-specific T-cell factors initiates the cascade of cellular interactions.* Immunol. Today 4:259–264.

Ballieux, R. E. and C. J. Heijnen. 1983. *Immunoregulatory T cell subpopulations in man: dissection by monoclonal antibodies and Fc-receptors.* Immunol. Rev. 74:5–28.

Beer, D. J. and R. E. Rocklin. 1984. *Histamine-induced suppressor-cell activity.* J. Allergy Clin. Immunol. 73:439–452.

Befus, A. D. and J. Bienenstock. 1984. *Induction and expression of mucosal immune responses and inflammation to parasitic infections.* Pp. 71–108 in J. J. Marchalonis, ed., *Contemporary Topics in Immunobiology: Immunobiology of Parasites and Parasitic Infections.* Plenum Press, New York.

Besedovskyl, H., A. Del Ray, E. Sorkin, M. da Prada, R. Burri and C. Honegger. 1983. *The immune response evokes changes in brain noradrenergic neurons.* Science 221:564–566.

Bienenstock, J. and A. D. Befus. 1983. *Some thoughts on the biologic role of IgA.* Gastroenterology 84:178–185.

Bittle, J. L., R. A. Houghten, H. Alexander, T. M. Shinnick, J. G. Sutcliffe, R. A. Lerner, D. J. Rowlands and F. Brown. 1982. *Protection against foot-and-mouth disease by immunization with a chemically synthesized peptide predicted from the viral nucleotide sequence.* Nature 298:30–33.

Cantrell, D. A. and K. A. Smith. 1984. *The interleukin-2 T-cell system: a new cell growth model.* Science 224:1312–1316.

Capron, A., J. P. Dessaint, M. Capron, M. Joseph and G. Torpier. 1982. *Effector mechanisms of immunity to schistosomes and their regulation.* Immunol. Rev. 61:41–66.

Dinarello, C. A. 1984. *Interleukin-1 and the pathogenesis of the acute-phase response.* N. Engl. J. Med. 311:1413–1418.

Djeu, J. Y., E. Lanza, S. Pastore and A. J. Hapel. 1983. *Selective growth of natural cytotoxic but not natural killer effector cells in interleukin-3.* Nature 306:788–791.

Fabbi, M., O. Acuto, J. E. Smart and E. L. Reinherz. 1984. *Homology of Ti α-subunit of a T-cell antigen-MHC receptor with immunoglobulin.* Nature 312:269–271.

Fearon, D. T. and W. W. Wong. 1983. *Complement ligand-receptor interactions that mediate biological responses.* Ann. Rev. Immunol. 1:243–271.

Goetzl, E. J., D. G. Payan and D. W. Goldman. 1984. *Immunopathogenetic roles of leukotrienes in human diseases.* J. Clin. Immunol. 4:79–84.

Gray, P. W., B. B. Aggarwal, C. V. Benton, T. S. Bringham, W. J. Henzel, J. A. Jarrett, D. W. Leung, B. Moffat, P. Ng, L. P. Svedersky, M. A. Palladino and G. E. Nedwin. 1984. *Cloning and expression of cDNA for human lymphotoxin, a lymphokine with tumour necrosis activity.* Nature 312:721–724.

Green, D. R., P. M. Flood and R. K. Gershon. 1983. *Immunoregulatory T-cell pathways.* Ann. Rev. Immunol. 1:439–463.

Griffin, J. D., S. C. Meuer, S. F. Scholssman and E. L. Reinherz. 1984. *T cell regulation of myelopoiesis: analysis at a clonal level.* J. Immunol. 133:1863–1868.

Hedrick, S. M., L. A. Matis, T. T. Hecht, L. E. Samelson, D. L. Longo, E. Heber-Katz and R. B. Schwartz. 1982. *The fine specificity of antigen and Ia determinant recognition by T cell hybridoma clones specific for pigeon cytochrome C.* Cell 30:141–152.

Honjo, T. 1983. *Immunoglobulin genes.* Ann. Rev. Immunol. 1:499–528.

Ihle, J. H., L. Rebar, J. Keller, J. C. Lee and A. Hapel. 1982. *Interleukin 3: possible roles in the regulation of lymphocyte differentiation and growth.* Immunol. Rev. 63:5–32.

Kaper, J. B., H. Lockman, M. M. Baldini and M. M. Levine. 1984. *Recombinant nontoxinogenic*

Vibrio cholerae strains as attenuated cholera vaccine candidates. Nature 308:655–658.

Kappler, J. W., B. Skidmore, J. White and P. Marrack. 1981. *Antigen-inducible, H-2 restricted, interleukin-2-producing T cell hybridomas. Lack of independent antigen and H-2 recognition.* J. Expl. Med. 153:1198–1214.

Kashima, N., C. Nishi-Takaoka, T. Fujita, S. Taki, G. Yamada, J. Hamuro and T. Taniguchi. 1985. *Unique structure of murine interleukin-2 as deduced from cloned cDNAs.* Nature 313:402–404.

Larsen, G. L. and P. M. Henson. 1983. *Mediators of inflammation.* Ann. Rev. Immunol. 1:335–359.

Locksley, R. M. and S. J. Kelbanoff. 1983. *Oxygen-dependent microbicidal systems of phagocytes and host defense against intracellular protozoa.* J. Cell. Biochem. 22:173–185.

Lomedico, P. T., U. Gubler, C. P. Hellmann, M. Dukovich, J. G. Giri, Y. C. E. Pan, K. Collier, R. Semionow, A. O. Chua and S. B. Mizel. 1984. *Cloning and expression of murine interleukin-1 cDNA in Escherichia coli.* Nature 312:458–462.

Mekalanos, J. J., D. J. Swartz, G. D. N. Pearson, N. Harford, F. Groyne and M. de Wilde. 1983. *Cholera toxin genes: nucleotide sequence, deletion analysis, and vaccine development.* Nature 306:551–557.

Meuer, S. C., O. Acuto, T. Hercend, S. F. Schlossman and E. L. Reinherz. 1984. *The human T-cell receptor.* Ann. Rev. Immunol. 2:23–50.

Miller, H. R. P. and Y. Nawa. 1979. *Immune regulation of intestinal goblet cell differentiation. Specific induction of nonspecific protection against helminths.* Nouv. Rev. Fr. Hematol. 21:31–45.

Nabholz, M. and H. R. MacDonald. 1983. *Cytolytic T lymphocytes.* Ann. Rev. Immunol. 1:273–306.

Nisonoff, A. 1984. *Introduction to Molecular Immunology*, Second Edition. Sinauer Associates, Inc., Sunderland, Massachusetts.

Ottaway, C. A. 1984. *In vitro alteration of receptors for vasoactive intestinal peptide changes the in vivo localization of mouse T cells.* J. Expl. Med. 160:1054–1069.

Patten, P., T. Yokota, J. Rothbard, Y-H. Chien, K. I. Arai and M. M. David. 1984. *Structure, expression, and divergence of T-cell receptor-chain variable regions.* Nature 312:40–46.

Paul, W. E. 1984. P. 809 in *Fundamental immunology.* Raven Press, New York.

Payan, D. G., J. D. Levine and E. J. Goetzl. 1984. *Modulation of immunity and hypersensitivity by sensory neuropeptides.* J. Immunol. 132:1601–1604.

Pennica, D., G. E. Nedwin, J. S. Hayflick, P. H. Seeburg, R. Derynck, M. A. Palladino, W. J. Kohr, B. B. Aggarwal and D. V. Goeddel. 1984. *Human tumour necrosis factor: precursor structure, expression, and homology to lymphotoxin.* Nature 312:724–729.

Rao, P. V. S., M. M. Friedman, F. M. Atkins and D. D. Metcalfe. 1983. *Phagocytosis of mast cell granules by cultured fibroblasts.* J. Immunol. 130:341–349.

Raulet, D. H., R. D. Garman, H. Saito and S. Tonegawa. 1984. *Developmental regulation of T-cell receptor gene expression.* Nature 314:103–107.

Rennick, D. M., F. D. Lee, T. Yokota, K. I. Arai, H. Cantor and G. J. Nabel. 1985. *A cloned MCGF cDNA encodes a multilineage hematopoietic growth factor: multiple activities of interleukin 3.* J. Immunol. 134:910–914.

Robertson, M. 1984. *The capture of the snark.* Nature 312:16–17.

Schneider, E. and M. Dy. 1985. *The role of arginase in the immune response.* Immunol. Today 6:136–140.

Scollay, R. and K. Shortman. 1983. *Thymocyte subpopulations: and experimental review, including flow cytometric cross-correlations between the major murine thymocyte markers.* Thymus 5:245–295.

Shelhamer, J. H., Z. Maron and M. Kaliner. 1980. *Immunologic and neuropharmacologic stimulation of mucous glycoprotein release from human airways in vitro.* J. Clin. Invest. 66:1400–1408.

Singer, A. and R. J. Hodes. 1983. *Mechanisms of T cell-B cell interaction*. Ann Rev. Immunol. 1:211–241.

Snyderman, R. and M. C. Pike. 1984. *Chemoattractant receptors on phagocytic cells*. Ann Rev. Immunol. 2:257–281.

Trinchieri, G. and B. Perussia. 1985. *Immune interferon: a pleiotropic lymphokine with multiple effects*. Immunol. Today 6:131–136.

Van Loveren, H., S. Kraiuter-Kops and P. W. Askenase. 1984. *Different mechanisms of release of vasoactive amines by mast cells occur in T cell-dependent compared to IgE-dependent cutaneous hypersensitivity responses*. Eur. J. Immunol. 14:40–47.

Wood, D. D., E. K. Bayen, M. B. Goldring, M. Gowen, D. Hamerman, J. L. Humes, E. J. Ihrie, P. E. Lipsky and M. J. Staruch. 1984. *The four biochemically distinct species of human interleukin-1 all exhibit similar biologic activities*. J. Immunol. 134:895–903.

Yokota, T., F. Lee, D. Rennick, C. Hall, N. Arai, T. Mosmann, G. Nabel, H. Cantor and K. I. Arai. 1984. *Isolation and characterization of mouse cDNA clone that expresses mast-cell growth-factor activity in monkey cells*. Proc. Nat. Acad. Sci. USA 81:1070–1074.

Yoshikai, Y., D. Anatoniou, S. P. Clark, Y. Yanagi, R. Sangster, P. Van den Elsen, C. Terhorst and T. W. Mak. 1984. *Sequence and expression of transcripts of the human T-cell receptor B-chain genes*. Nature 312:512–524.

Genetic control of host resistance to diseases

JOAN K. LUNNEY
Helminthic Diseases Laboratory, Animal Parasitology Institute, Agricultural Research Service, U.S. Department of Agriculture, Building 1040, BARC-East, Beltsville, MD 20705, U.S.A.

Abstract Resistance to disease can be attributed to multiple host genes. One important set of genes that has been demonstrated to be associated with resistance/susceptibility to such diverse diseases as arthritis of the spine, herpes virus induced Marek's diseases, and diabetes, is the major histocompatibility complex (MHC). The MHC codes for cell surface proteins that regulate a wide range of host cellular and humoral immune responses. For this paper, the epidemiologic and breeding data showing MHC association for several diseases is presented, and the potential mechanisms involved in causing this genetic regulation is discussed. In addition, the importance of MHC genes in predicting responses to small proteins and, for the future, to genetically engineered peptide vaccines is examined. Then, as a background for analyzing the molecular biology of MHC genes, the genetic structure of the MHC and the current status of monoclonal antibody reagents that react with specific MHC cell surface antigens is reviewed. DNA cloning of individual MHC genes and the potential for restriction fragment length polymorphism mapping of MHC genes is analyzed. Finally, future prospects for altering immune responses associated with diseases by producing transgenic animals are discussed.

Introduction

Defense against disease is a major function of the immune system. Dr. Befus described the complex cellular and molecular interactions that are involved in immune responses to foreign agents. In this paper I will highlight some important genetic factors that direct the immune system in distinguishing self tissue from foreign agents and their products. The major histocompatibility complex genes code for the dominant cell surface antigens involved in regulating these immune interactions and, because of their importance, will be the focus of this presentation.

The major histocompatibility complex antigens

The major histocompatibility complex (MHC) antigens are familiar to many of us because they are "transplantation antigens", the antigens that should be matched between donor and recipient to get successful bone marrow or kidney transplantation with minimun use of immunosuppressive drugs (vanRood et al., 1985; Thomas, 1985). Transplantation, however, is a phenomenon of modern medical science and not the reason the body expresses these antigens. Research in the last 10 years has demonstrated that these antigens are crucial to the body's response to foreign materials such as infectious agents or protein vaccines (Paul, 1984).

The MHC antigens can be divided into 3 classes of gene products (Klein, 1984). The class I MHC antigens are cell surface proteins that are expressed on most cells of the body and that are especially important in regulating responses to viral infections. The class II MHC genes also code for cell surface glycoproteins which are expressed only on certain immunologically active cells such as B cells and macrophages. The class II MHC antigens are involved in regulating both humoral and cellular immune responses to most foreign products and also in establishing the relative magnitude of the response to any one foreign antigen. The class III MHC genes code for selected serum proteins including the complement components which are known to be involved in lytic processes.

There are several important properties of MHC antigens that should be emphasized. First, most individuals express products of several genes for each class of MHC antigens. The genetic structures of the MHC of humans, termed human lymphocyte antigens or HLA, swine, termed SLA, and chicken, referred to as the B complex, are illustrated in Fig. 1 (Bodmer and Bodmer, 1984; Lunney, Davis, and Pescovitz, 1985; Longenecker and Mosmann, 1981). For human and swine, at least two loci each for the class I and class II antigens have been described, for the chicken only one locus for each class has been described thus far.

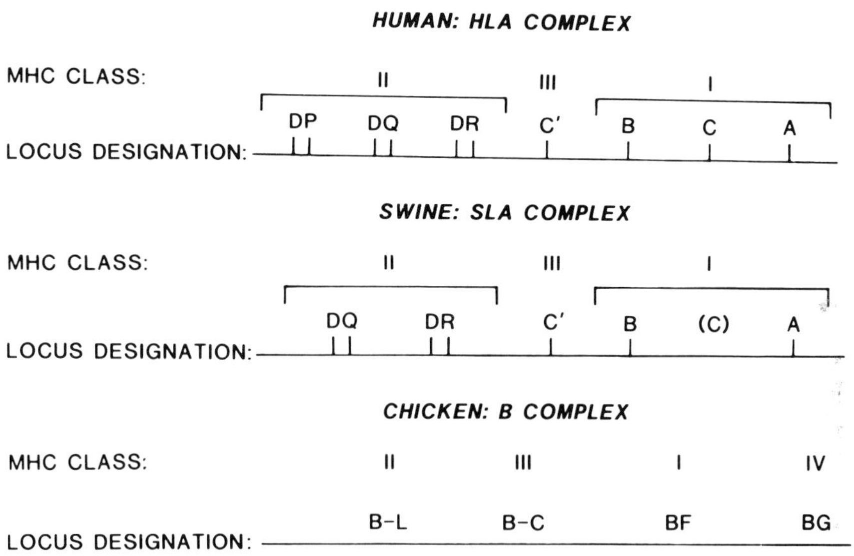

Figure 1. Genetic structure of the major histocompatibility complex (MHC) of humans, swine and chicken. The letters refer to the locus designation; the Roman numerals above the letters refer to the class of MHC antigen coded by that gene.

Second, most outbred individuals are heterozygous, i.e., they express 2 different allelic products for each locus. Third, the MHC antigens are very polymorphic, i.e., for any one locus there may be as many as 30 different alleles, the products of which differ by as much as 20% in amino acid sequence (Hood, Steinmetz and Malissen, 1983; Mengle-Gaw and McDevitt, 1985); in contrast sickle cell hemoglobin varies by only one amino acid ($\sim 0.5\%$) from its allelic product. In summary, the polymorphism of the MHC antigens endows the individual with significant genetic diversity, an important property for the immune system as it responds to a broad range of viruses of foreign antigens.

MHC antigens and disease associations

Laboratories around the world have tried to identify genes involved in modulating diseases responses. Studies in humans have resulted in the identification of a number of genes that are each involved in specific diseases, as well as a set of genes, the major histocompatibility complex, or HLA, genes that are implicated in a large number of unrelated diseases (Zmijewski, 1984). These associations were found by epidemiological studies comparing patient populations to age and sex matched controls. As shown in Table 1, the incidence of certain diseases is increased significantly in individuals that express specific HLA antigenic alleles. The most striking example for class I associated diseases is ankylosing spondylitis, arthritis of the spine, that is 87 times more prevalent in HLA-B27^{+} individuals than in HLA-B27^{-} individuals. Other diseases from thyroiditis to specific types of cancer show similar, though less striking, increases in incidence among individuals bearing specific HLA class I antigen alleles (Table 1).

Disease associations are far more numerous for class II HLA antigens. The examples cited in the bottom half of Table 1 range from ragweed antigen allergies that are associated with HLA-DR2 to juvenile arthritis that is increased in HLA-DR5^{+} and HLA-DRw8^{+} individuals. The latter illustrates the changes that are occuring in this field, i.e., as medical science becomes more sophisticated physicians realize that what used to be a single disease should properly be three or four separate diseases. So now, juvenile arthritis pauciarticular (in a few joints) has been demonstrated to be more prevalent in HLA-DR5^{+} individuals. However, since typing for HLA-DQ and HLA-DP antigens is only now being established (Albert, Baur and Mayr, 1985), associations with these antigens have not yet been analyzed.

HLA class II antigen association is even more clearly demonstrated with diabetes where the incidence of insulin-dependent diabetes mellitus (IDDM) is increased in HLA-DR3^{+} and HLA-DR4^{+} individuals.

Table 1. Prominent HLA-disease associations[a]

Disease	HLA	Relative risk[b]
Class I MHC-associated diseases		
Hemochromatosis (idiopathic)	A3	8.2
	B14	4.7
Endometrial cancer	Aw23	13.0
Vitiligo		
Moroccan Jews	B13	9.3
Yeminites	Bw35	13.9
Ankylosing spondylitis	B27	87.4
Reiter's disease	B27	37.0
Subacute thyroiditis (de Quervain)	Bw35	13.7
Class II MHC-associated diseases		
Allergy (Ragweed Ra5 response)	DR2	19.0
Ocular histoplasmosis	DR2	11.7
Chronic active hepatitis	DR3	17.0
Dermatitis Herpetiformis	DR3	15.4
Goodpasture's disease	DR3	15.0
Insulin-dependent diabetes mellitus	DR3	3.3
Primary biliary cirrhosis	DR3	7.6
Sjorgren's disease	DR3	9.7
Adult rheumatoid arthritis	DR4	4.2
Insulin-dependent diabetes mellitus	DR4	6.4
Pemphigous	DR4	14.4
Juvenile rheumatoid arthritis (pauciarticular)	DR5	5.2
Testicular cancer	Dw7	8.3
Juvenile rheumatoid arthritis (all types)	DR28	3.6

[a] Data adapted from Zmijewski (1984).

[b] Relative risk indicates how many times more frequently a disease develops in individuals carrying the HLA antigen as compared to the frequency of that disease in individuals lacking the antigen (Svejgaard, Platz and Ryder, 1983).

For this disease gene dosage becomes significant since, as illustrated in Table 2, individuals who are HLA-DR3,3 homozygous are 98 times more likely to get the disease whereas HLA-DR3,X individuals are only 4.6 times more at risk (Svejgaard et al., 1980). Similarly, HLA-DR4,4 individuals are 77 times more at risk for IDDM compared to HLA-DR4,X individuals who are 15 times more at risk and HLA-DR3,4 individuals who are 174 times more likely to get the disease. Since a double dose of either one of the susceptibility alleles significantly increases probability of having the disease, IDDM exhibits neither true recessive nor true dominant inheritance but intermediate type genetic regulation associated with specific MHC antigens. In addition, a decreased incidence of IDDM associated with certain MHC antigens has been reported, HLA-DR2$^+$ individuals are 1/20 to 1/4 as likely to

Table 2. HLA-DR associations with insulin dependent diabetes millitus

HLA antigen	Relative risk
DR3,X	4.6
DR3,3	97.9
DR3,4	173.6
DR4,X	15.4
DR4,4	76.9
DR2,X	0.2

Data adapted from Svejgaard, Platz and Ryder (1980).

become diabetic. This indicates that there is an HLA associated disease resistance gene. It is these resistance genes that will be very important, and that must be identified, for future research on animal diseases.

Mechanism of MHC disease associations

Several different mechanisms seem to be involved in MHC gene associations with diseases. They include antigen mimicry, MHC associated susceptibility genes, and MHC associated immune response genes.

Antigen mimicry implies that an infectious agent tries to mimic the structure of MHC antigens of the host so that the individual will not recognize the invading organism as foreign and, therefore, not initiate an immune response against the organism. This mechanism seems to be active with HLA-B27 and *Klebsiella* organisms in the initiation of ankylosing spondylitis. Studies have shown that antibodies to certain *Klebsiella* organisms will bind to HLA-B27 antigens from individuals with this disease (Geczy et al., 1983). This might mean that when HLA-B27$^+$ individuals are infected with *Klebsiella* they do not respond to it as foreign, thus allowing the bacteria to replicate and trigger the arthritic response.

MHC linked susceptibility genes have clearly been identified for the class III MHC serum proteins because complement and 21-hydroxylase deficiencies map to the class III genes. Recent cloning experiments have demonstrated that individuals with these deficiencies have mutations in the class III genes (White, New and Dupont, 1985).

The third, and more general mechanisms of MHC gene association with diseases is HLA linked immune response genes. For this, the association of particular disease antigens, e.g., viral proteins or bacterial antigens, with the HLA antigens cause inappropriate, or total absence of, stimulation of immune responses. To demonstrate this idea more clearly, an analysis of the T cell receptor and its recognition struction is necessary.

The T cell receptor and antigen recognition

Research in the last two years has defined the structure of the receptor on T cells which is responsible for the initiation of the complex cellular interactions which lead to successful immune responses (Davis, 1985; Schwartz, 1985). The T cell receptor, as diagrammed in Fig. 2, is composed of 2 polypeptide chains that, as a complex with MHC antigens, interact to recognize foreign materials such as viruses or vaccines. Thus, when a virus infects a target cell it uses the host protein synthesis machinery to make viral proteins some of which are expressed on the target cell surface along with the class I MHC antigens. Circulating T cells now notice that the virus infected cell expresses a new surface determinant consisting of foreign viral protein associated with class I self MHC proteins. The subset of T cells which express a T cell receptor that can effectively interact with this viral protein-class I antigen complex then initiates the complex set of cellular interactions that lead to a cytolytic response against virus infected cells. Similarly, after

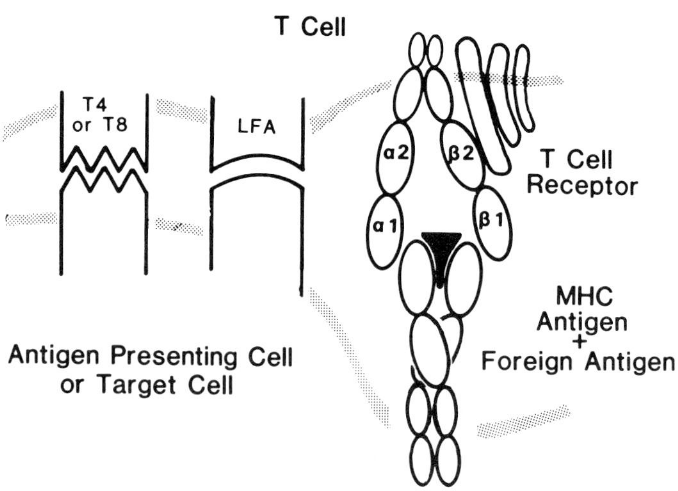

Figure 2. Model for the interaction of the T cell receptor molecule with the complex of major histocompatibility complex (MHC) antigen and foreign antigen. The T cell receptor structure is based on current gene cloning data. Accessory T cell molecules such as the helper T cell structure, T4, the cytotoxic T cell structure, T8, or the monomorphic lymphocyte antigen, LFA, are illustrated because of their importance in enhancing the interaction of the T cell with the antigen presenting cell or the target cell. The three polypeptides of the T3 molecule are shown as they associate with the T cell receptor. The MHC antigen structure indicated is that for class II antigens. Similar interactions are expected for virus modified class I MHC antigens.

vaccinations, macrophages pick up vaccine antigens and "process" them so that small peptides are now expressed, or presented, on the macrophage cell surface as a complex with class II MHC antigens. This complex then stimulates specific subsets of T cells which eventually cooperate to produce a protective immune response. Examples of these MHC restricted interactions have been reported in species from humans (Sasazuki et al., 1983) to mice (Schwartz, 1985). In fact, the increased resistance to herpes virus associated Marek's disease is clearly associated with inheritance of a specific MHC, or B, complex type (Longenecker et al., 1976; Hala et al., 1981).

Having T cells recognize complexes of MHC antigen and foreign antigen to initiate immune responses is clearly advantageous to the individual because protein antigens, that might by themselves not be immunogenic, now become immunogenic by being presented on the cell surface in association with MHC antigens. Understanding of the T cell recognition system also helps us comprehend how MHC linked immune response genes cause their effects. Some foreign antigens may not form effective immune stimulation complexes with a given set of MHC antigens and therefore not stimulate a response. In this way, they evade the immune system. The implications of this potential lack of response for vaccine design will be discussed below.

Importance of MHC in vaccine responses

Genetic control of the effectiveness of foreign antigens in stimulating immune responses has been known for many years (reviewed in Paul, 1984). Our studies using MHC inbred miniature swine of four different SLA haplotypes illustrate this point. Animals were immunized with two low molecular weight proteins, lysozyme and myoglobin, or with the synthetic polymer (T,G)-A–L, and their antibody responses were analyzed. The results outlined in Table 3 demonstrate that swine of all SLA types were able to make antibodies to myoglobin whereas only animals of SLA^a or SLA^c haplotypes responded to (T,G)-A–L and only animals of SLA^a and SLA^d haplotypes responded to lysozyme (Lunney, VanderPutten and Pescovitz, 1984). These responses could be mapped genetically using SLA^g swine who express class I MHC antigens of the *c* haplotype and class II antigens of *d* haplotype. Because the SLA^g animals respond to lysozyme and not to (T,G)-A–L, the same pattern as the SLA^d swine, this maps the ability to mount an antibody response to these foreign proteins to SLA-DR^d DQ^d, i.e., to the class II genes of the SLA complex. The exact SLA-D region locus, SLA-DR or SLA-DQ, that determines each of these responses is currently under investigation using monoclonal antibody (mAb) blocking of proliferative

Table 3. SLA linked immune response gene control of antibody production

SLA haplotype	SLA allelic product (Locus) DR	DQ	B	A	Antibody production[a] to Myoglobin	Lysozyme	(T,G)-A–L
a	a	a	a	a	+++	+	+++
c	c	c	c	c	+++	+++	–
d	d	d	d	d	+++	–	+++
g	d	d	c	c	+++	–	+++

[a]Antibody production was assayed by ELISA technique (Lunney, VanderPutten and Pescovitz, 1984).

responses. In other species, it has been demonstrated that different class II MHC genes control different antigen specific responses (Paul, 1984).

An important aspect of these findings is in their implication for designing modern vaccines. As the molecule used for vaccination becomes smaller it expresses fewer antigenic epitopes, this increases the probability that the immune responses to the vaccine will be influenced by genetic factors. Therefore, potential vaccines must be screened for their ability to stimulate immune responses in every individual of a species. Methods of circumventing this MHC linked immune response gene control involve coupling small antigens or peptides to large carrier proteins or synthesizing these molecules as part of larger transcripts such as the beta-galactosidase fusion proteins. With these large protein constructs most individuals will make an immune response. Researchers, however, must demonstrate that there is an appropriate response to the original antigenic epitope so that the full protective immune response is stimulated.

Structure of MHC genes

Because of the importance of MHC antigens in controlling a wide range of immune responses and thereby regulating disease susceptibility/resistance, there has been a great interest in analyzing and cloning these genes. The genes have been cloned extensively in humans and mice and to a lesser degree in rats and rabbits (Kaufman et al., 1984; Hood, Steinmetz and Malissen, 1983; Mengle-Gaw, L. and McDevitt, H.O., 1985; Howard, 1983; Tykocinski et al., 1984). Very little cloning has been reported so far for agriculturally important species although several laboratories are now actively making cDNA and genomic libraries to probe for class I and II MHC genes in avian and bovine species.

The major exception for cloning MHC genes for agricultural species is the swine. In 1982, Singer, Rudikoff and their colleagues reported the first cloning of a class I SLA gene (Singer et al., 1982). Since that time they have cloned 6 separate class I genes and characterized the regulation of their expression in L cells (Satz and Singer, 1984). Their research and recent work from Chardon and his colleagues (1985) in France have clearly demonstrated that swine have fewer class I genes than mice or humans; swine have 8–10 MHC class I genes while mice have 25–35 (Hood, Steinmetz and Malissen, 1983) and humans have around 20 (Orr, 1983). This difference in number of genes is interesting not only for evolutionary considerations but also because recent data have demonstrated that one mechanism of producing diversity in MHC is gene conversion (Schulze et al., 1983). This results when a small piece of DNA (10–30 bp) is transferred from a specific region of one gene to the homologous region of a second gene. Gene conversion now appears to be the mechanism by which both class I and class II MHC mutants have arisen (Miyada et al., 1985; Mengle-Gaw and McDevitt, 1985), and may be an important means of generating new genetic alleles. This mechanism may be an evolutionary necessity. Viruses have adapted by changing their surface antigen. Thus, individuals might be able to respond adequately to infections with these new viruses because their MHC antigens have been altered by gene conversion, using this MHC gene pool to generate genetic diversity.

Cloning of SLA class II genes is just now underway (D. H. Sachs, personal communication). The availability of cloned SLA class I genes and DNA transfection technology has resulted in mouse cell lines that express individual class I genes of pigs (Singer et al., 1982) and, more recently, with embryo manipulation, the production of a transgenic mouse, i.e., a mouse that resulted from an embryo that had been microinjected with a class I SLA gene (Frels et al., 1985). The importance of these findings will be discussed below.

Serologic identification of individual MHC alleles

Because the MHC antigens are codominantly expressed cell surface antigens, identification of specific MHC allelic gene products can be performed on peripheral blood lymphocytes (PBL) (Zmijewski, 1984). In the human, HLA typing has traditionally been performed using sera from multiparous women as the source of antibodies. These sera are complex because the mother will make antibodies to all of the father's HLA antigens. Therefore, HLA typing involves testing with a large panel of such sera, used at specific dilutions where each serum is operationally monospecific. This procedure has not been significantly altered by the advent of mAb technology. Although there are now many mAb that recognize specific HLA allelic products, there are also many HLA specificities for which no mAb are available (Albert, Baur and Mayr, 1985). Because most mAb are made in other species (mice or rats), the mAb most frequently recognize common determinants expressed by several HLA antigens rather than single allele specific determinants.

For agriculturally important species, similar procedures for MHC typing and antibody production have been used with the problems noted above, as well as with the additional problem of having fewer researchers that are developing reagents. In swine for example, for SLA typing, the major laboratory that has developed reagents is in France (Vaiman, Chardon, and Renard, 1979), but because of foot and mouth disease, the reagents are not easily available to U.S. researchers. For the SLA inbred miniature swine some operationally specific antibodies are available but most antibodies react with all of the class I or all of the class II antigens of a given haplotype (Pennington et al., 1981). Most monoclonal reagents recognize monomorphic determinants of given allelic products (Davis, Perryman and McGuire, 1984; Lunney, Davis and Pescovitz, 1985). Panels of reagents have also been developed for the bovine (Spooner, 1982) and the chicken (Briles and Briles, 1982).

Identification of MHC alleles in genomic DNA

Because of the difficulties with using serologic reagents to do MHC typing, many laboratories have turned to restriction fragment length polymorphisms (RFLP) for analyzing MHC alleles. Basically the procedure involves digesting cellular genomic DNA with a variety of restriction endonucleases, separating the resulting fragments by electrophoresis, transferring them by the Southern blot technique and then hybridizing them with MHC specific probes. Chardon and his colleagues (1985) have used this procedure to do the first reported RFLP typing of outbred swine. Their results demonstrate the usefulness, as well as the complexity, of this technique. Probing with an MHC class I cDNA, 7 to 10 restriction fragments were seen, many of which varied for each different SLA haplotype. In addition some animals that theoretically had the same SLA class I antigens, had similar, but not identical, RFLP patterns. This probably resulted because all class I type genes are hybridized in this procedure including unexpressed genes. In summary, animals could be identified as having a particular haplotype or set of class I SLA genes. The exact RFLP that represents each class I SLA gene, or allele, was not identified in this procedure because the genes for these alleles have not yet been cloned and sequenced. In the future, it will be important to identify specific sequences that could be used as allele specific oligonucleotide probes.

Implications of RFLP mapping to disease associations

RFLP mapping will be an important tool for future mapping of genes associated with disease resistance in agriculturally important animals. In small inbred populations mAb reagents for every MHC allele will probably be available to identify each specific allelic product associated with disease resistance. However, the more important studies will involve large outbred populations for which it would probably be impractical to develop serologic typing reagents. Therefore, RFLP analyses with multiple restriction enzymes will have to be employed to identify specific allelic genes that are associated with resistance to a given disease. For example, one could screen Spanish pigs that have survived an African swine fever virus infection to identify genes that correlate with resistance to reinfection. Once identified the MHC gene could be cloned and sequenced.

Potential use of transgenic animals

Genes associated with disease resistance would first be tested in vitro, if possible using DNA transfected cell lines, to verify that the cloned gene product did induce the resistance phenotype. New transgenic animals could then be produced by microinjecting the cloned DNA into embryos, as has been done in agricultural species with the growth hormone gene (Hammer et al., 1985) or with SLA genes in transgenic mice (Frels et al., 1985). Thus, new progeny, developed by the now standard methods of embryo transfer and with the new methods of embryo manipulation, would be more resistant to the disease in question and yet retain the full complement of genes that had been bred into each of its parent's germ plasm. The implications of this technique for increasing disease resistance, as well as many other characteristics, are obvious. There are, of course, limitations to this technique; introduction of new genes will likely bring the positive effect of added resistance factors to the new transgenic offspring but may bring as yet unknown negative side effects. In addition, the efficiency of this technique must be improved (Hammer et al., 1985). Hopefully, the positive effects that could result from these studies will stimulate research in this exciting area.

Summary

In conclusion, MHC genes are significant factors in the regulation of disease resistance/susceptibility. MHC genes have been studied in a variety of agriculturally important species and the problems associated with their identification at the cellular and DNA levels have been discussed. The potential of these genes for designing vaccination procedures and for future modification of the germ plasm has been emphasized.

References

Albert, E. D., M. P. Baur and W. R. Mayr. 1985. *Histocompatibility testing, 1984*. Springer-Verlag, New York.

Bodmer, J. and W. Bodmer. 1984. *Histocompatibility 1984*. Immunol. Today 5:251–254.

Briles, W. E. and R. W. Briles. 1982. *Identification of haplotypes of the chicken major histocompatibility complex (B)*. Immunogenetics 15:449–459.

Chardon, P., M. Vaiman, M. Kirszenbaum, C. Geffrotin, C. Renard and D. Cohen. 1985. *Restriction fragment length polymorphism of the major histocompatibility complex of the pig*. Immunogenetics 21:161–171.

Davis, M. 1985. *Molecular genetics of T-cell receptor beta chain*. Ann. Rev. Immunol. 3:537–560.

Davis, W. C., L. E. Perryman and T. C. McGuire. 1984. *The identification and analysis of major functional populations of differentiated cells*. Pp. 121–150 in N. J. Stern and H. R. Gamble, eds., *Hybridoma Technology in Agricultural and Veterinary Research*. Rowman and Allanheld, Totowa, New Jersey.

Frels, W. I., J. A. Bluestone, R. J. Hodes, M. R. Capecchi and D. S. Singer. 1985. *Expression of a microinjected porcine class I MHC gene in transgenic mice*. Science 228:577–580.

Geczy, A. F., K. Alexander, H. V. Bashir, J. P. Edmonds, L. Upfold and J. Sullivan. 1983. *HLA-B27, Klebsiella and ankylosing spondylitis: biological and chemical studies*. Immunol. Rev. 70:23–50.

Hala, K. R. B. and G. Wick. 1981. *Chicken major histocompatibility complex and disease*. Scand. J. Immunol. 14:607–617.

Hammer, R. E., V. G. Pursel, C. E. Rexroad, R. J. Wall, D. G. Bolt, K. M. Ebert, R. D. Palmiter and R. L. Brinster. 1985. *Production of transgenic rabbits, sheep, and pigs by microinjection*. Nature 315:680–683.

Hood, L., M. Steinmetz and B. Malissen. 1983. *Genes of the major histocompatibility complex of the mouse*. Ann. Rev. Immunol. 1:529–568.

Howard, J. C. 1983. *The major histocompatibility complex of the rat: a partial review*. Metabolism 32:41–55.

Kaufman, J. F., C. Auffray, A. J. Korman, D. A. Shackelford and J. Strominger. 1984. The *class II molecules of the human and murine major histocompatibility complex*. Cell 36:1–13.

Klein, J. 1984. *Immunology: The Science of Self-Nonself Discrimination*. John Wiley & Sons, Somerset, New Jersey.

Longenecker, B. M. and T. R. Mosmann. 1981. *Structure and properties of the major histocompatibility complex of the chicken. Speculations on the advantages and evaluation of polymorphism*. Immunogenetics 13:1–23.

Longenecker, B., M. F. Pazderka, J. S. Gavora, J. L. Spencer and R. F. Ruth. 1976. *Lymphoma induced by Herpes virus: resistance associated with a major histocompatibility gene*. Immunogenetics 3:401–407.

Lunney, J. K., D. VanderPutten and M. D. Pescovitz. 1984. *MHC linked immune response gene control of humoral and cellular responses to (T, G)-A–L and lysozyme in miniature swine*. Fed. Proc. 43:1821.

Lunney, J. K., W. C. Davis and M. D. Pescovitz. 1985. *Identification of monoclonal antibodies that recognize the SLA-DwQ antigens, a second set of swine Ia antigens*. Fed. Proc. 44:554.

Mengle-Gaw, L. and H. O. McDevitt. 1985. *Genetics and expression of mouse Ia antigens*. Ann. Rev. Immunol. 31:367–396.

Miyada, C. G., C. Klofett, A. A. Reyes, E. McLaughlin-Taylor and R. B. Wallace. 1985. *Evidence that polymorphism in the murine major histocompatibility complex may be generated by the assortment of subgene sequences*. Proc. Nat. Acad. Sci. USA 82:2190–2894.

Orr, H. T. 1983. *Use of Southern blotting to analyze the size and restrictions fragment polymorphism of HLA class I DNA in the human population*. Transplant. Proc. 15:1900–1906.

Paul, W. E. 1984. *Fundamental Immunology*. Raven Press, New York.

Pennington, L. R., J. K. Lunney and D. H. Sachs. 1981. *Transplantation in miniature swine. VIII: recombination within the MHC of miniature swine*. Transplantation 31:66–71.

Sasazuki, T., Y. Nishimura, M. Muto and N. Ohta. 1983. *HLA genes controlling immune response and disease susceptibility*. Immunol. Rev. 70:51–68.

Satz, M. L. and D. S. Singer. 1984. *Effect of mouse interferon on the expression of a porcine major histocompatibility gene introduced into mouse L cells*. J. Immunol. 132:496–501.

Schulze, D. H., L. R. Pease, S. S. Geier, A. A. Reyes, L. A. Sarmiento, R. P. Wallace and S. G. Nathenson. 1983. *Comparison of the cloned $H\text{-}2K^{bml}$ variant gene with the $H\text{-}2K^{b}$ gene shows a cluster of seven nucleotide differences*. Proc. Nat. Acad. Sci. USA 80:2007–2010.

Schwartz, R. H. 1985. *T. lymphocyte recognition of antigen in association with gene products of the major histocompatibility complex*. Ann. Rev. Immunol. 3:237–261.

Singer, D. S., R. D. Camerini-Otero, M. L. Satz, B. Osborne, D. Sachs and S. Rudikoff. 1982. *Characterization of a porcine genomic clone encoding a major histocompatibility antigen: expression in mouse L cells*. Proc. Nat. Acad. Sci. USA 79:1403–1406.

Spooner, R. L. 1982. *Proceedings of the Second International Bovine Lymphocyte Antigen (BoLA) Workshop*. Animal Blood Group and Biochemical Genetics 13:33–53.

Svejgaard, A., P. Platz and L. P. Ryder. 1980. *Insulin-dependent diabetes mellitus*. Pp. 638–641 in P. I. Terasaki, ed., *Histocompatibility Testing 1980*. UCLA Tissue Typing Laboratory, Los Angeles.

Svejgaard, A., P. Platz and L. P. Ryder. 1983. *HLA and disease, 1982-A survey*. Immunol. Rev. 70:193–215.

Thomas, E. D. 1985. *Current status of bone marrow transplantation*. Transpl. Proc. 17:428–431.

Tykocinski, M. L., P. N. Marche, E. E. Max and T. J. Kindt. 1984. *Rabbit class I MHC genes: cDNA clones define full-length transcripts of an expressed gene and a putative psuedogene*. J. Immunol. 133:2261–2269.

Vaiman, M., P. Chardon and C. Renard. 1979. *Genetic organization of the pig SLA complex. Studies on nine recombinants and biochemical and lysostrip analysis*. Immunogenetics 9:353–361.

Van Rood, J. J., G. F. J. Hendricks, J. D'Amaro, J. W. Gratama, M. Jager, A. Vanes and G. G. Persign. 1985. *HLA matching, patient survival, and immune response genes in renal transplantation*. Transpl. Proc. 17:681–685.

White, P. C., M. I. New and B. Dupont. 1985. *Genetics of HLA linked congential adrenal hyperplasia*. Pp. 41–44 in J. W. Streilein et al., eds., *Advances in Gene Technology: Molecular Biology of the Immune System*. ICSU Press.

Zmijewski, C. M. 1984. *HLA and disease*. CRC Crit. Rev. Clin. Lab. Sci. 20:285–370.

Genomic and antigenic variation of African swine fever virus

E. VIÑUELA, R. BLASCO, M. AGÜERO, B. GARCÍA BARRENO[†] and L. ENJUANES
Centro de Biología Molecular (CSIC-UAM), Universidad Autónoma de Madrid, Facultad de Ciencias, Canto Blanco, 28049 Madrid, Spain

Abstract African swine fever virus is morphologically like an iridovirus, but its DNA structure, the enzyme activities present in the virus particles and the host RNA polymerase II – independent RNA synthesis make African swine fever virus similar to poxvirus. The viral DNA ends show high-variability regions that could be related to the mechanism of virus evasion of the host immune system. A comparison of the binding properties of a panel of monoclonal antibodies to different African swine fever virus field isolates and to virus passaged in porcine macrophages have shown that the virus changes readily in most of the proteins recognized by the monoclonal antibodies.

Introduction

African swine fever (ASF) virus is a menace to the pig population in the world because there is no vaccine, the virus is transmitted by ticks and different virus forms produce either acute disease with a mortality close to 100% or inapparent infections which are difficult to detect (reviews, Wardley et al., 1983; Viñuela, 1985).

The virus is widely distributed through sub-Saharan Africa, Portugal, Spain, Sardinia, Brazil and Haiti. Figure 1 shows the dates of some ASF outbreaks.

A major problem for the ASF control is the absence of neutralizing antibodies in both the virus hosts (several genera of the *Suidae* family) and virus-resistant animal species inoculated with the virus (DeBoer, 1967). The reasons for this atypical immune response are unknown. Among other possibilities, antigenic variation of some undetermined critical antigens may account for the ability of ASF virus to evade the host immune system.

Properties of ASF virus

African Swine Fever (ASF) virus is a large icosahedral virus of about 200 nm with a lipoprotein envelope around a core of about 90 nm, a capsid and an external envelope acquired in the budding process through the host plasma membrane (Breese and DeBoer, 1966; Carrascosa et al., 1984). The virus particle contains about 34 structural

[†]Centro Nacional de Microbiología, Virología e Inmunología Sanitarias, Majadahonda, Madrid, Spain.

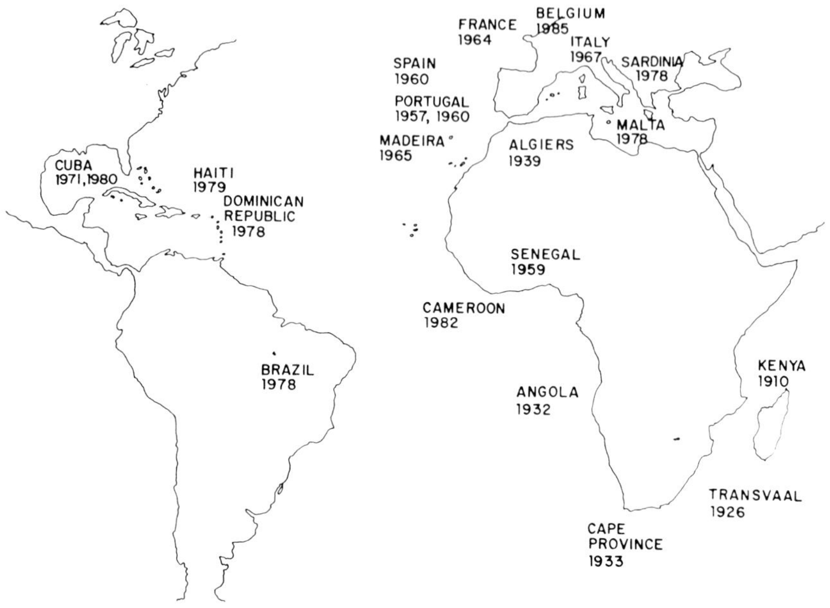

Figure 1. ASF outbreaks.

proteins (Carrascosa et al., 1985). They contain a complete transcription system that is able to cap, methylate and polyadenylate messenger RNAs that are indistinguishable from those isolated from infected cells shortly after infection (Kuznar, Salas and Viñuela, 1980, 1981; Salas, Kuznar and Viñuela, 1981, 1983; M. L. Salas, J. Rey, J. M. Almendral, A. Talavera and E. Viñuela, unpublished observations).

The genome is a double-stranded DNA with about 170 kilobase pairs (kbp) (Enjuanes, Carrascosa and Viñuela, 1976; Almendral et al., 1984), terminal inverted repeats of 2.4 kbp (Sogo et al., 1984) and cross-links (Ortín, Enjuanes and Viñuela, 1979; Almendral et al., 1984). The nucleotide sequence of the ASF DNA ends has shown that they consist of a 37 nucleotide-long hairpin loop composed almost entirely of A and T residues, that were incompletely paired. The loops at each DNA end are present in about equimolar forms that, when compared in opposite polarities, were inverted and complementary (flip-flop), as in the case of poxvirus DNA (A. González, A. Talavera, J. M. Almendral and E. Viñuela, unpublished observations). Following the hairpin loops there is a terminal inverted repetition of 2440 nucleotides which consists of unique sequences interspersed with 42 direct repeats in tandem of 34 nucleotides, five repeats of 24 and three of 33 nucleotides (González, personal communication).

The replication cycle of ASF virus begins with the penetration of the virion into the host cell by adsorptive endocytosis, at least in VERO cells (A. Alcamí, A. L. Carrascosa and E. Viñuela, unpublished observations). The early phase of virus replication, defined as that occurring prior to DNA synthesis, starts with the early transcription that does not require both host RNA polymerase II and protein synthesis (Salas, personal communication; J. Salas, M. L. Salas and E. Viñuela, unpublished observations). Early virus mRNAs hybridize to about 60% of the viral DNA. After the onset of DNA replication new RNA species arise which hybridize with DNA regions not transcribed in infected cells in the presence of protein or DNA synthesis inhibitors (Salas, personal communication). The late RNA synthesis is also independent from host RNA polymerase II. Late mRNAs code for most of the structural proteins (Salas, personal communication).

Virus progeny appear in VERO cells by 10 hours and maximum yields are produced by 36–48 hours. About half of ASF virus particles remains cell-associated. The released fraction acquires an additional envelope from the host plasma membrane (Breese and DeBoer, 1966).

Although ASF virus morphology resembles the *Iridoviridae*, in its DNA structure, virions-associated enzymes, and mode of replication, it behaves like a *Poxvirus*. However, since ASF virus does not have the morphology of a poxvirus it does not fit into that group (Fig. 2). African swine fever virus multiplies in both ticks and swine, and is thus an 'arthropod-borne' virus, but the only one to contain DNA.

<table>
<tr><th></th><th>VACCINIA VIRUS</th><th>ASF VIRUS</th><th>FROG VIRUS 3</th></tr>
<tr><td>DNA STRUCTURE</td><td colspan="2">Unique sequence with hairpin loops and terminal inverted repeats</td><td>Circularly permuted sequence with direct terminal repeats</td></tr>
<tr><td rowspan="2">VIRUS STRUCTURE</td><td>Complex symmetry</td><td colspan="2">Icosahedral symmetry</td></tr>
<tr><td>Structural glycoprotein</td><td colspan="2">No structural glycoprotein</td></tr>
<tr><td rowspan="3">RNA SYNTHESIS</td><td colspan="2">The virion contains the enzymes required for mRNA synthesis</td><td>The virion does not contain the enzymes required for mRNA synthesis</td></tr>
<tr><td colspan="2">RNA synthesis is host cell RNA polymerase II - independent</td><td>RNA synthesis is host cell RNA polymerase II - dependent</td></tr>
<tr><td colspan="2">Polyadenylated mRNAs</td><td>No polyadenylated mRNAs</td></tr>
</table>

Figure 2. Comparative properties of vaccinia virus, ASF virus and frog virus 3 (Viñuela, 1985).

Genetic variation

A comparison of the Sal I restriction maps of 22 ASF virus field isolates (8 African, 10 European and 4 American) revealed the existence in ASF virus DNA of a central highly conserved region that spans the majority of the genome (~ 125 kbp) and two variable regions close to the DNA ends (Fig. 3).

The virus isolates were classified into five groups according to the Sal I sites present in the central region (Table 1). All the European and American isolates had identical central regions and, therefore, lie within the same group. Most of the African isolates showed more differences and this might be due to the fact that in Africa the virus has been circulating for a long time and may have diverged extensively. The uniformity of the non-African isolates, relative to that of the African ones, might be due to the limited number of viruses that entered Europe during 1957–1960 and to the comparatively short time available for those virus to evolve.

The analysis of nine virus clones isolated in 1971 from the spleen of an infected pig in Badajoz, Spain, revealed that all of them shared a

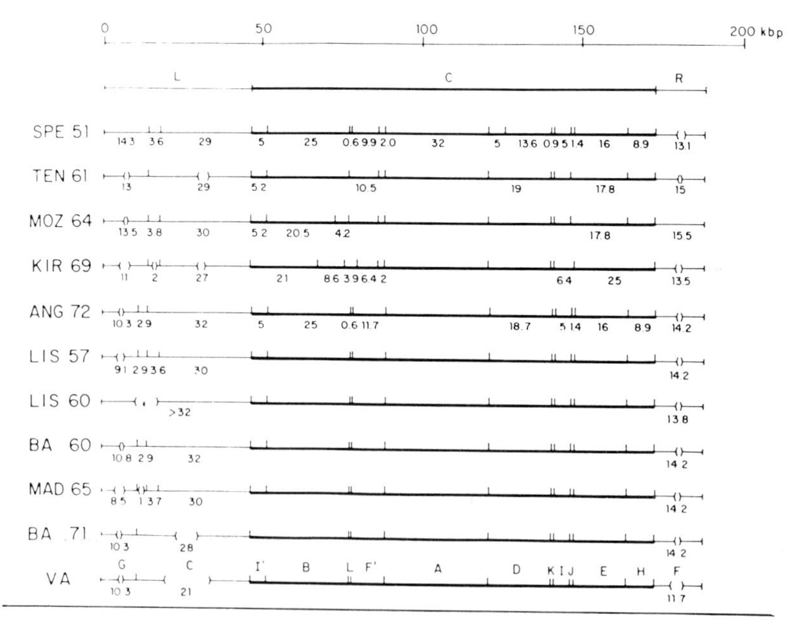

Figure 3. Sal I restriction maps of ASF virus field isolates. Each virus is named by a code that indicates the origin and the isolation year. SPE, Spencer; TEN, Tengani; MOZ, Mosambique; KIR, Kirawira; ANG, Angola; LIS, Lisboa; BA, Badajoz; MAD, Madeira; VA, BA71 adapted to VERO cells.

Table 1. ASF virus field isolates with the same Sal I sites in the DNA constant region[a]

Group	Origin	Isolates
I	Africa	SPE51, WH55
II	Africa	UGA59, KAT67, ANG72
	Europe	LIS57, LIS60, BA60, FRA64, MAD65, ITA67, PO70, BA71, B78, CC83
	America	BRA78, BRA79, DR79, HAI81
III	Africa	TEN61
IV	Africa	MOZ64
V	Africa	KIR69

[a]The nomenclature of the field virus isolates is indicated in the legend to Fig. 3. Other viruses not shown in Fig. 3 are FRA, France; ITA, Italy; PO, Pontevedra; B, Barcelona; BRA, Brazil; DR, Dominican Republic; HAI, Haiti; SAR, Sardinia; CC, Cáceres; HW, virus isolated from a warthog in East Africa.

common central region and some showed small size variations in both DNA ends. These differences are probably due to an homologous and unequal recombination between the direct repeats present in the terminal inverted repetition.

The most frequent and extensive changes detected among field virus isolates were located at 10–20 kbp from the left end of the genome. This region contains repeated sequences which could be associated with the mechanism of variation generation. In order to understand the significance of those changes, in particular their possible relation to the virus escape from the host's immune system, it will be necessary to determine the function of the proteins encoded in the high-variability regions.

Antigenic variation

The availability of a panel of monoclonal antibodies that recognize at least nine ASF virus structural proteins has allowed binding and competition radioimmunoassay experiments using the monoclonal antibodies to different virus isolates (Sanz et al., 1985; B. García Barreno, A. Sanz, M. L. Nogal, E. Viñuela and L. Enjuanes, unpublished data).

A comparison of seven clones isolated from the spleen of a pig infected in Badajoz (Spain) in 1971 with the reference virus showed the existence of 1 clone with large changes in proteins p150 and p27 and small changes in proteins p37 and p14, 2 clones with small changes in proteins p150 and p27, and 4 clones without changes with respect to the reference virus.

The virus isolates passaged in swine macrophages in vitro changed faster and more extensively than a VERO-adapted virus passaged in VERO cells. An ASF virus field isolate passaged in porcine macrophages 10 to 20 times changed antigenically more than a strain of VERO cell-adapted virus passaged in those cells. In the case of the virus passaged in macrophages, the changes were seen in uncloned samples. In contrast, in the virus passaged in VERO cells the changes were seen only after clone analysis.

We have detected changes in 6 out of the 9 proteins analyzed. In the African isolates the most variable proteins were p150, p27, p14 and p12. In the non-African isolates the most variable proteins were p150 and p14; protein p12 did not change. Protein p72, the main antigen for ASF diagnosis by an enzymoimmunoassay, was very stable.

The distribution of ASF virus field isolates in antigenic homology groups indicated the coexistence in the field of a complex variety of ASF virus serotypes. With the available data, no clear-cut separation of virus isolates from different continents could be established, since virus isolates from Africa, America and Europe were present in the same antigenic homology groups (Table 2).

Table 2. Distribution of ASF virus field isolates in antigenic homology groups[a]

1	2	3	4	5	6
FRA64	HIN54-HAD$^+$	HIN54-HAD$^-$	SPE51	TEN61	BA60
ITA67	UGA59-VIR	UGA-59-ATT$^+$	WH55		
PO70	LIS60	KIR69			
B78	BA71				
BRA78	SAR78				
DR79	BRA79				
HAI81	CC83				
SAR82					

[a]The nomenclature of the field virus isolates is indicated in the legends to Fig. 3 and Table 1. HIN54-HAD$^+$ and HAD$^-$, Hinde, positive and negative hemadsorbing virus; UGA-VIR and ATT, Uganda, virulent and attenuated virus.

In the field isolates we have detected antigenic changes in 6 out of the 10 ASF virus proteins for which there were available monoclonal antibodies, suggesting a broad distribution of the variability.

The mechanism by which the ASF virus escape the host's immune system is unknown. Antigenic variation is a mechanism used by many parasites to evade the immune defenses of the infected animals. An analysis of this possibility in the case of ASF virus is difficult until the virus components involved in the adsorption and penetration processes of the virus into the target cell are known. Because of the lack of neutralizing antibodies, the identification of these components will not be straightforward. A possibility is to identify the external virus components, some of which probably will be involved in the interaction with the susceptible cells. This approach has been used in our laboratory by using immunoelectron microscopy with monoclonal antibodies, controlled liberation of structural proteins by treatment of virions with detergents and iodination of the virus followed by analysis of radioactive proteins. These experiments have shown that proteins p35, p24, p17, p14 and p12 are exposed proteins in the virion (A. L. Carrascosa, J. F. Santarén, and E. Viñuela, unpublished observations).

Acknowledgements We are grateful to M. L. Salas, J. Salas, A. Talavera, A. Alcamí, A. L. Carrascosa, J. F. Santarén, M. del Val., P. González, J. Rey, A. González, V. Calvo and J. M. Almendral for allowing us to comment on their unpublished observations.

References

Almendral, J. M., R. Blasco, V. Ley, A. Beloso, A. Talavera and E. Viñuela. 1984. *Restriction site map of African swine fever virus DNA*. Virology 133:258–270.

Breese, S. R., Jr. and C. J. DeBoer. 1966. *Electron microscope observations of African swine fever virus in tissue culture cells*. Virology 28:420–428.

Carrascosa, J. L., J. M. Carazo, A. L. Carrascosa, N. García, A. Santisteban and E. Viñuela. 1984. *General morphology and capsid fine structure of African swine fever virus particles*. Virology 132:160–172.

Carrascosa, A. L., M. del Val, J. F. Santaren and E. Viñuela. 1984. *Purification and properties of African swine fever virus*. J. Virol. 54:337–344.

DeBoer, C. J. 1967. *Studies to determine neutralizing antibody in sera from animals recovered from African swine fever and laboratory animals inoculated with African virus with adjuvants*. Arch. Ges. Virus Forsch. 20:164–179.

Enjuanes, L., A. L. Carrascosa and E. Viñuela. 1976. *Isolation and properties of the DNA of African swine fever virus*. J. Gen. Virol. 32:479–492.

Kuznar, J., M. L. Salas and E. Viñuela. 1980. *DNA-dependent RNA polymerase in African swine fever virus*. Virology 101:169–175.

Kuznar, J., M. L. Salas and E. Viñuela. 1981. *Nucleoside triphosphate phosphohydrolase activities in African swine fever virus*. Arch. Virol. 96:307–320.

Ortin, J., L. Enjuanes and E. Viñuela. 1979. *Crosslinks in African swine fever virus DNA*. J. Virol. 31:579–583.

Salas, M. L., J. Kuznar and E. Viñuela. 1981. *Polyadenylation, methylation, and capping of the RNA synthesized in vitro by African swine fever virus*. Virology 113:484–491.

Salas, M. L., J. Kuznar and E. Viñuela. 1983. *Effect of rifamycin derivatives and coumermycin Al on in vitro RNA synthesis by African swine fever virus*. Arch. Virol. 77:77–80.

Sanz, A., B. García-Barreno, M. L. Nogal, E. Viñuela and L. Enjuanes. 1985. *Monoclonal antibodies specific for African swine fever virus DNA*. Virology 56:199–206.

Sogo, J. M., J. M. Almendral, A. Talavera and E. Viñuela. 1984. *Terminal and internal inverted repetitions in African swine fever virus DNA*. Virology 113:271–275.

Viñuela, E. 1985. *African swine fever virus*. Curr. Top. Microbiol. Immunol. 116:151–170.

Wardley, R. C., C. de M. Andrade, D. N. Black, F. L. Castro Portugal, L. Enjuanes, W. R. Hess, C. Mebus, A. Ordas, D. Rutili, J. Sanchez Vizcaino, J. D. Vigario, P. J. Wilkinson, J. F. Moura Nunes and G. Thomson. 1983. *African swine fever virus*. Arch. Virol. 76:73–90.

Genetic manipulation of *Escherichia coli* K99 pilus production

KAREL NEWMAN, JR.,[1] RICHARD ISAACSON,[2,3] and JEAN PETRE[4]
[1]*Salsbury Laboratories, Inc., Charles City, Iowa, U.S.A.*
[2]*University of Michigan, Ann Arbor, Michigan U.S.A. and*
[4]*Solvay and Cie, Brussels, Belgium*

Abstract A 7.15 kb BamHI restriction fragment containing the gene encoding K99 pilus expression was removed from the wild-type 87.8 kb pK99 plasmid and inserted into a BamHI restriction cleaved pBR322 recipient vehicle. The newly developed plasmid, designated pIX12, was first used to transform recipient *E. coli* strain RH202 for final transfer to strain 711. The recombinant strain so developed was found to produce several-fold more pilus protein than the original plasmid donor. Pilus protein is apparently expressed at the surface of the organism, and may be readily removed with shearing forces. The resulting effluent is concentrated by ultrafiltration and may be subsequently quantitated by ELISA employing either an avidin-biotin based dual epitope recognition assay or conventional polyclonal antibody based assay. Product is formulated on the basis of computer generated concentration values. The resulting compositions containing the engineered protein induced the development of maternal antibodies capable of protecting offspring against mortality and severe morbidity.

Introduction

It has long been recognized that disease causing enteropathogenic *Escherichia coli* require the participation of colonization factors to permit the development of pathologic manifestations. Early work of Smith and Linggood (1971, 1972) demonstrated the requirement of fimbrial appendages (pili) for colonization of the gut to occur, which was subsequently confirmed by Moon et al. (1977). The reported findings indicated that the K99 pilus did promote adherence of organisms to the upper intestinal tract of several different animal species including swine, sheep and cattle.

The K99 pilus is a helical structure composed of a regular arrangement of pilin subunits. K99 antigen is a basic protein with an approximated pI of 9.5. The molecular weight of the protein is 18.2 K based upon amino acid sequence analysis (De Graaf and Mooi, 1983) and migrates by denaturing sodium dodecyl sulfate polyacrilamide electrophoresis with a corresponding apparent molecular weight of 18–19 K. One disulfide bond is permitted within the molecular structure based upon compositional analysis.

The identity of the receptor for the K99 pilus is still not clear. Reports from one laboratory demonstrate that sialic acid groups can

[3]Present address: Pfizer Central Research, Groten, CT, U.S.A.

compete with cellular sites for K99 binding (Lindahl and Wadstrom, 1983). However, a second laboratory has reported the isolation and elucidation of a K99 binding site of equine erythrocytes which includes a galactose residue (Smit et al., 1984). Final clarification of either observation will be important since it will provide insight into the molecular basis of the colonization process.

The organization of the pK99 plasmid and the organization of the K99 operon has been previously reported (De Graaf and Mooi, 1983; Isaacson and Richter, 1983). The wild-type plasmid has been found to regularly exist as a 87.8 kb element in most field isolates studied (Isaacson and Richter, 1983). Copy number of the plasmid appears to be stringently limited to one per chromosomal equivalent. The coding region responsible for expression of the K99 gene product is estimated at no more than 6.6 kb (Isaacson, Betzold and Petre, unpublished observations). In addition to the expression of K99 protein, the plasmid also encodes the production of 6 additional polypeptides with molecular weights ranging from 19 K to 76 K (De Graaf and Mooi, 1983). Restriction maps were reported by Isaacson and Richter (1983) and De Graaf and Mooi (1983).

Immunization of dams with compositions containing the K99 pilus protein has been demonstrated to effectively protect suckling pigs against neonatal disease by challenge with an organism expressing the same pilus type (Jayappa et al., 1983; Morgan et al., 1978; Nagy et al., 1978; Simonson et al., 1983). One mechanism which may serve partially or entirely as being responsible for affording protection against colonization may be agglutination. Studies have demonstrated that monoclonal antibodies capable of agglutinating piliated cells can afford passive protection against colibacillosis (Sherman et al., 1983; laboratory observation). Data suggest the colostral immunoglobulin levels are significantly elevated following vaccination with compositions of defined pilus content (Simonson et al., 1983).

Production and assessment of pilus antigen

Bacterial strains. E. coli strain B41 was used as the plasmid donor and contained the 87.5 kb pK99 plasmid. *E. coli* K12 strains 711 and RH202 served as hosts in the described recombinant studies.

Conditions of laboratory culture. Cultures were grown in Trypticase-soy broth at 37°C with continuous aeration. When minicells were required, the plasmid was first transfered into strain DS 410 by transfection and cultures were grown at 37°C in Tris Salts in medium the presence of [^{35}S]-methionine plus supplemental glucose and amino acids.

Plasmid preparation. The pK99 plasmid was initially moved from strain B41 to strain 711 by conjugal transfer. Strain 1474 was a result of this transfer and contained a 78.8 kb plasmid conferring tetracycline resistance as well as the 87.8 kb plasmid. Strain 1474 was subjected to further selection on Bochner's plates for loss of tetracycline resistance (Bochner et al., 1980). Strain 1252 was found to possess only the 87.8 kb plasmid. Plasmid DNA was isolated by published procedures (Ish-Horowitz and Burke, 1981). Liberated DNA was first precipitated with ethanol and then subjected to isopycnic centrifugation in cesium chloride-ethidium bromide density gradients.

Restriction endonuclease digestion and gene insertion. Restriction endonuclease was used in strict accordance with the manufacturer's instructions. The BamHI fragments of pK99 were ligated into alkaline phosphatase treated, BamHI cleaved pBR322. The ligated plasmid DNA was used to transform strain RH202. Screening of transformants was initially performed by culture on medium containing ampicillin, and subsequently screened for pilus expression by simple slide agglutination with a monospecific polyclonal antiserum raised in rabbits against the K99 antigen. One K99 expressing recombinant plasmid was designated pIX12, and contained a 7.15 kb insert as well as pBR322. Plasmid pIX12 was transferred to strain 711.

Transposition mutagenisis. The K99 coding region was first transferred to plasmid pACYC184. Mutagenesis was achieved by employing the transposon Tn3 of plasmid pSC204. Cultures were subsequently incubated at 43°C in the presence of ampicillin to eliminate the temperature sensitive pSC204 plasmid and to select for transposition events. The sites of Tn3 insertion in 85 independently occurring clones were determined by restriction enzyme analysis. The effect of transpositional events on K99 expression was determined by slide agglutination.

Antigen quantitation scheme. Antigen content in harvest material was determined by ELISA relative to a purified standard pilus preparation. In the conventional analysis of pilus antigen content, serial two-fold dilutions of antigen were sandwiched between immobilized anti-K99 polyclonal IgG and alkaline phosphatase-labelled IgG. For the two site epitope recognition assay, a biotinylated capture monoclonal antibody (Mab) was simultaneously mixed with an enzyme-labelled second Mab in the presence of serially diluted antigen samples and immobilized avidin; the assay scheme is depicted in Fig. 1. Retained

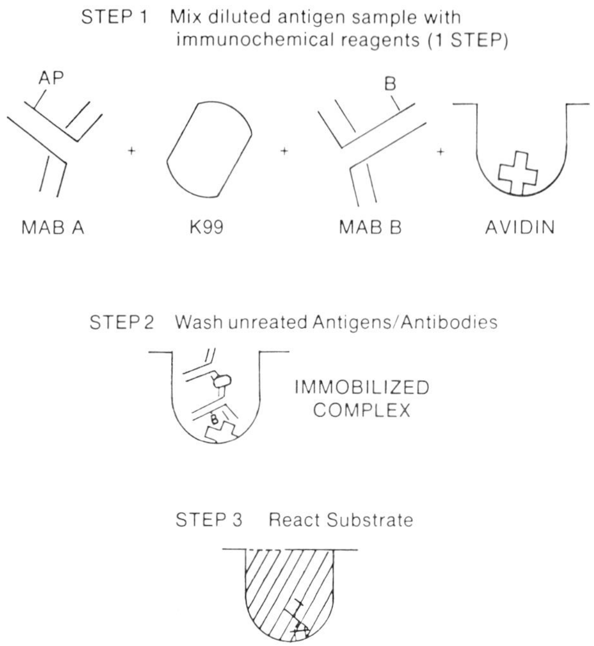

Figure 1. Monoclonal antibody-based two-site recognition assay procedure.

enzyme was measured with Sigma 104™ substrate in 1 M diethanolamine buffer, pH 9.6. Resulting absorbance values were subjected to logit-fit analysis, with the resulting linear portion of the curve subjected to computer-assisted comparison against the standard antigen curve.

Assessment of immune response to bacterin prepared with engineered protein. Vaccine efficacy and selected immunologic parameters were studied as previously described (Simonson et al., 1983).

Results and discussion

The scheme employed in developing the recombinant *E. coli* strain is depicted in Figs. 2 and 3. The steps have been previously outlined in production and assessment of pilus antigen.

The restriction endonuclease map of pIX12 is shown in Fig. 4. Applicable restriction cutting sites for the inserted K99 gene are shown along with the site of flanking BamHI insertion sites. The location of Tn3 observable insertion sites, is also given in Fig. 4.

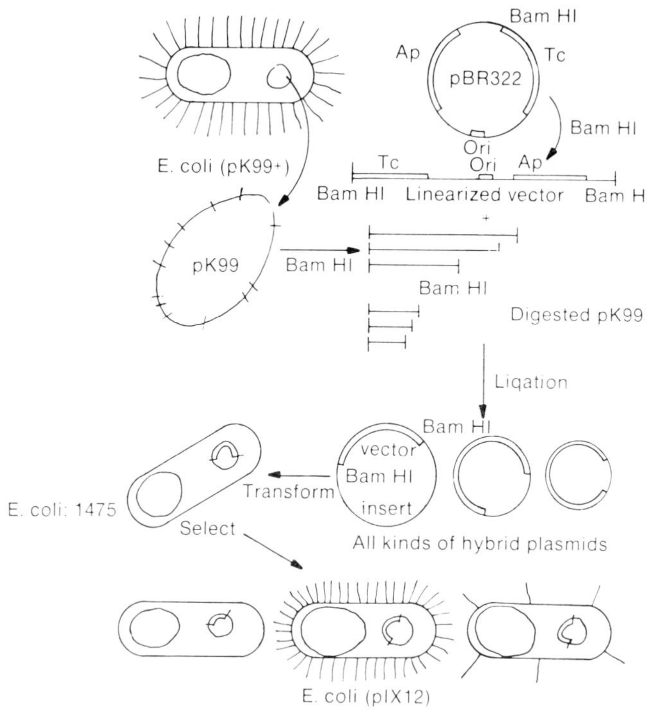

Figure 2. Diagrammatic representation of steps involved in cloning K99 gene.

The expression of K99 antigen in recombinants has been determined to be at a level of 2.5 to 5 fold better than the levels determined for strain B41 (Table 1). We suggest this to be due to the existence of

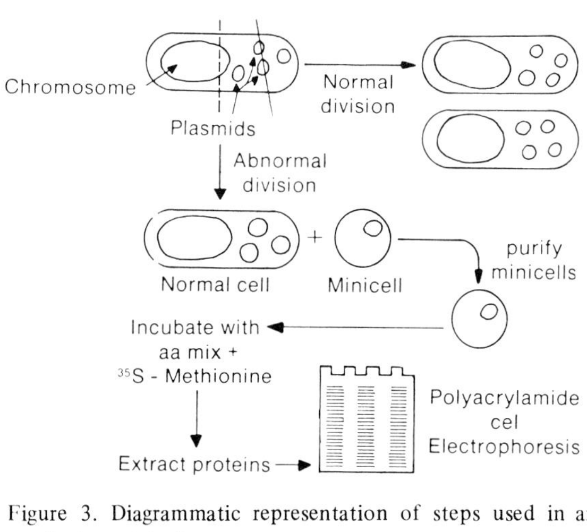

Figure 3. Diagrammatic representation of steps used in analysing pIX12 gene products by employing minicells.

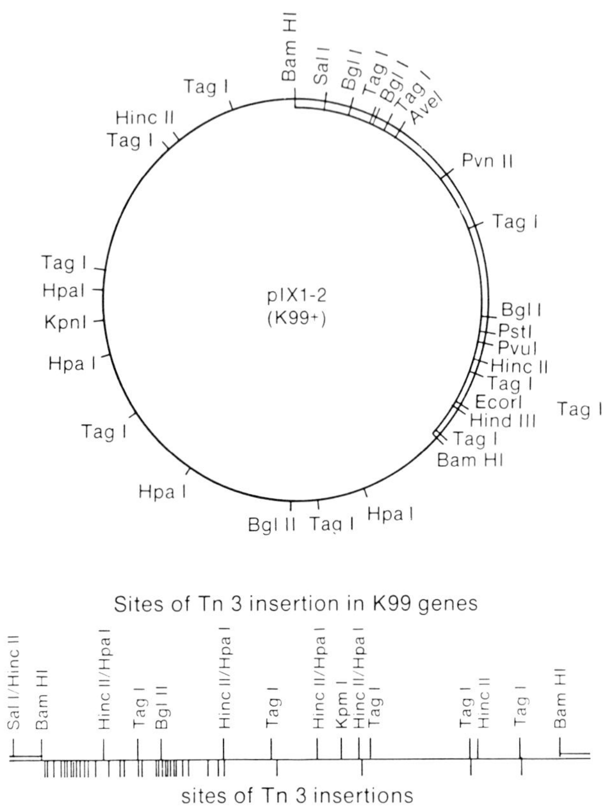

Fig. 4. Map of the pIX12 plasmid.

multiple gene copies of K99 in the transformant associated with the high copy number of pBR322. Loss of the associated plasmid was minimized by initial cultivation of the seed culture with ampicillin, before fermentor inoculation. It has however been observed that a certain proportion of cells will lose their plasmid during a typical production run which excludes ampicillin; the resulting yields were still economically satisfactory.

Pili were removed by a shearing process. The pilus laden supernatant was concentrated by ultrafiltration and the immunoreactive pilus content was measured by ELISA. One ELISA procedure described

Table 1. K99 pilus protein content of ultrafiltrate

Strain	Mean recovery	High	Low
B41	284 mcg/ml	875 mcg/ml	27 mcg/ml
711 w/pIX12	1183 mcg/ml	3185 mcg/ml	592 mcg/ml

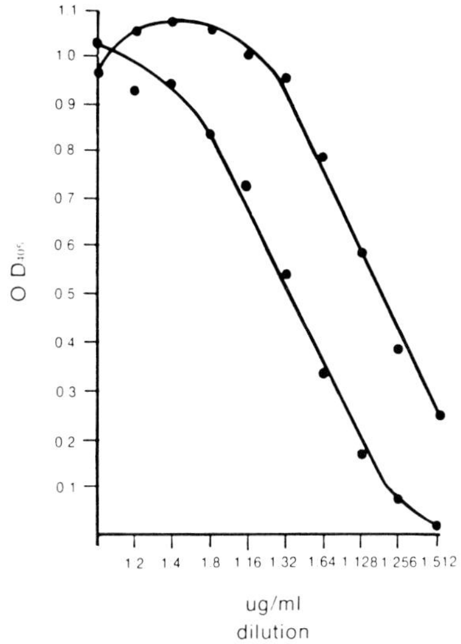

Figure 5. Antigen titration curve from the two-site ELISA.

herein employs the use of an avidin coated plate. A typical reaction curve is represented in Fig. 5. Note that in the case of antigen excess, the response curve decreases consistent with the notion that at saturation, proportionately less antigen is available for the development of antibody-antigen-antibody complexes. The described system can however be effectively applied toward the quantitation of K99 as well as a number of different antigens providing the proper monoclonal antibodies are available.

The vaccine prepared from the engineered organism was found to be effective in reducing the severity of diarrhea and limiting mortality in the offspring of vaccinated dams (Tables 2, 3 and 4). Mean morbidity was diminished to just over 50% in vaccinates compared to 100% in

Table 2. Morbidity following challenge of vaccinate and control animals

Challenge Strain	Day after Challenge	Percent Scouring[1]	
		Vaccinates	Controls
K99	1	51.2	100.0
	2	36.1	87.9
	3	6.1	46.7
	4	3.0	24.0
	5	3.0	0

[1]Calucalted from surviving animals in group

Table 3. Mortality following challenge of vaccinate and control animals

	Animal No.	No. Piglets	Challenge	Mortality	Group Survival
	166	9	K99	0	
Vaccinates	167	4	K99	0	
	168	12	K99	1	
	169	9	K99	0	97.1%
	159	9	K99	3	
Non-	160	11	K99	5	
Vaccinates	161	8	K99	8	
	162	10	K99	6	
	163	10	K99	5	43.8%

challenged control animals; it should be also noted that the severity of morbidity in scouring neonates from the vaccinated group was in fact lower than that of the corresponding control group. *E. coli* K99

Table 4. Antigen specific immunoglobulin content of fluids from vaccinate and control animals

Concentration of Antigen Specific Antibodies (mcg/ml) Antibody Class

Antigen	Sample	IgG		IgA		IgM	
		V[1]	C[2]	V	C	V	C
K88	Pr[3]	16.8	24.5	5.9	8.3	12.7	NA
	P+[4]	144.3	10.9	68.8	19.2	385.3	NA
	Col[5]	473.4	163.8	161.5	21.2	207.0	NA
K99	Pr	7.1	7.2	1.3	1.8	NA	NA
	P+	65.2	3.2	24.2	6.9	NA	NA
	Col	210.6	11.2	77.9	12.3	NA	NA
987P	Pr	9.2	8.5	7.6	8.9	NA	NA
	P+	657.6	64.6	130.3	33.4	NA	NA
	Col	2383.7	107.6	22.8	19.7	NA	NA

[1]Vaccinated
[2]Control
[3]Prebleed Sera
[4]Post-Vaccination Sera
[5]Colostrum at Farrow
NA - Not Available

(ETEC) could be readily isolated from scouring animals in either group but no effort was made to quantitatively differentiate shedding levels. A particularly dramatic effect was noted when mortality was examined. Nearly total protection against death loss was noted in animals allowed to suckle on vaccinated animals. When antibody levels in the serous and colostral fluids of vaccinated dams were examined, it was found that immunoglobulin levels against the K99 antigen had been elevated.

References

Bochner, B. R., H. C. Huang, G. L. Schleven and B. N. Ames. 1980. *Positive selection for loss of tetracycline resistance*. J. Bacteriol. 143:926–933.

De Graaf, F. K. and F. R. Mooi. 1983. *Organization of genes involved in the biosynthesis of the K88 and K99 fimbriae*. VIDO 4:191–201.

Isaacson, R. E. and P. Richter. *Molecular and genetic analysis of K99 plasmids*. VIDO 4:179–185.

Ish-Horowitz, D. and J. F. Burke. 1981. *Rapid and efficient cosmid cloning*. Nuc. Acids Res. 9:2989–2998.

Jayappa, H. P., J. G. Strayer, R. A. Goodnow, P. C. Fusco, S. W. Wood, P. Haneline, H-J. Cho and C. C. Brinton. 1983. *Experimental infection and field trial evaluations of a multiple-pilus phase-cloned bacterin for the simultaneous control of neonatal colibacillosis and mastitis in swine*. VIDO 4:518–535.

Lindahl, M. and T. Wadstrom. 1983. *Comparative properties of K99 and 3P⁻ (F41); two sialic acid specific hemagglutinins of enterotoxigenic E. coli (ETEC)*. VIDO 4:225–234.

Moon, R. L., B. Nagy, R. E. Isaacson and I. Orskov. 1977. *Occurrence of K99 antigen on Escherichia coli isolated from pigs and colonization of pigs' ileum by K99⁺ enterotoxigenic Escherichia coli from calves and pigs*. Inf. Immun. 15:614–620.

Morgan, R. L., R. E. Isaacson, H. W. Moon, C. C. Brinton and C-C. To. 1978. *Immunization of suckling pigs against enterotoxigenic E. coli-included diarrheal disease by vaccinating dams with purified 987P or K99 pili; protection correlates with pilus homology of vaccine and challenge*. Inf. Immun. 22:771–777.

Nagy, B., H. W. Moon, R. E. Isaacson, C-C. To and C. C. Brinton. 1978. *Immunization of suckling pigs against enterotoxigenic Escherichia coli infection by vaccinating dams with purified pili*. Inf. Immun. 21:269–274.

Sherman, D. M., S. D. Acres, P. L. Sadowski, J. A. Springer, B. Bray, T. J. G. Raybould and C. C. Muscoplat. 1983. *Protection against fatal enteric colibacillosis by orally administered Escherichia coli K99-specific monoclonal antibody*. Inf. Immun. 42:653–658.

Simonson, R. R., R. E. Isaacson, C. R. Jacobs and K. Z. Newman. 1983. *The class specific antibody in passive protection against Escherichia coli diarrhea in pigs using a subunit genetically engineered pili vaccine*. VIDO 4:548–557.

Smit, H., W. Gaastra, J. P. Kanereling, J. F. Vliegenthart and F. K. De Graaf. 1984. *Isolation and structural characterization of the equine erythrocyte receptor for enterotoxigenic Escherichia coli K99 fimbrial adhesin*. Inf. Immun. 46:578–584.

Smith, H. W. and M. A. Linggood. 1971. *Observations on the pathogenic properties of the K99, HLY, and ENT plasmids of Escherichia coli with particular reference to porcine diarrhea*. J. Med. Microbial. 4:467–485.

Smith, H. W. and M. A. Linggood. 1972. *Further observations on Escherichia coli enterotoxins and of a K antigen possessed by calf and lamb strains*. J. Med. Microbiol. 5:243–250.

Vaccinia virus vectors

BERNARD MOSS
Laboratory of Viral Diseases, National Institute of Allergy and Infectious Diseases, Bethesda, MD 20205, U.S.A.

Abstract Vaccinia virus has been used as a vector to express foreign genes. Infectivity is retained and animals inoculated with the recombinant virus mount humoral and cell-mediated immune responses. Protection against infection with herpes simplex virus, influenza virus, hepatitis B virus, rabies virus and vesicular stomatitis virus has been obtained. Live vaccinia virus recombinants may provide a new approach to the development of veterinary vaccines.

Introduction

Recombinant DNA technology has provided new approaches to the immunoprophylaxis of veterinary disease. Proteins synthesized in vitro by prokaryotic or eukaryotic expression vectors can potentially be used as subunit vaccines. Alternatively, genetic engineering may be employed to attenuate viruses that can then be employed as safe live vaccines. A third alternative, the use of live attenuated viruses as vectors for the expression of genes of unrelated infectious agents, also has been considered and is the subject of this review.

A variety of viruses have been developed as vectors (Rigby, 1983). Small DNA viruses, such as SV40, are simple to manipulate but cannot accommodate significant amounts of additional DNA without deletion of essential genes that are necessary for autonomous replication. The larger DNA viruses, including members of the adenovirus, herpesvirus, and poxvirus families either have nonessential DNA that could be deleted or can accommodate additional DNA without interfering with packaging. Vaccinia virus, a member of the poxvirus family, is particularly suited as a vector for vaccine purposes because of its wide host range, large capacity for foreign DNA, and previous use for immunoprophylaxis of smallpox. Thus far, genes from many viruses including hepatitis B (Mackett, Smith and Moss, 1982; Paoletti et al., 1984), influenza (Panicali et al., 1983; Smith, Murphy and Moss, 1983) herpes simplex type 1 (Paoletti et al., 1984; Cremer et al., 1985), vesicular stomatitis (Mackett et al., 1985), and rabies (Wiktor et al., 1984) as well as a protozoan parasite (Smith et al., 1984) have been expressed in vaccinia vectors.

Molecular biology of vaccinia virus

A knowledge of the basic features of poxvirus replication will facilitate an understanding of the strategy used for insertion and expression of foreign genes (Moss, 1985). Vaccinia is a large, complex virus with a double-stranded DNA genome of about 185,000 base pairs. The development of poxviruses within the cytoplasm of infected cells is a characteristic feature of the family and sets it apart from other DNA viruses. This property is correlated with the encoding of enzymes necessary for expression and replication of the virus genome. A DNA-dependent RNA polymerase, as well as enzymes for capping and methylation of the 5′ end of mRNA and polyadenylation of the 3′ end, are packaged within the core of the infectious virus particle. Thus, early mRNAs, which code for about 100 polypeptides, are made within minutes after infection. A DNA polymerase is one of the early gene products and replication commences after about 1.5 h. This sets the stage for expression of late genes which code for many of the structural proteins.

The vaccinia virus transcriptional system recognizes specific DNA sequences that are located immediately upstream of the RNA start sites of viral genes (Mackett et al., 1982; Weir and Moss, 1984; Bertholet, Drillien and Wittek, 1985). Furthermore different signals are recognized at early and late times after infection. Evidently, promoters of other virus families are not recognized at all (Puckett and Moss, 1983).

The early and late genes of vaccinia virus are distributed throughout the length of the genome. Remarkably, there are many genes that are not required for vaccinia to replicate efficiently in tissue culture cells. Thus, spontaneous mutants with large deletions have been isolated (Moss, Winters and Cooper, 1981; Nakano, Panicali and Paoletti, 1981). One very well characterized vaccinia gene, thymidine kinase (TK), is not essential at least in rapidly dividing tissue culture cells.

Strategy for using vaccinia virus as an expression vector

Since vaccinia virus has developed specific transcriptional signals, we considered that these would be necessary to obtain efficient expression of foreign genes (Mackett, Smith and Moss, 1984). Thus, our strategy has been to construct chimeric genes containing the coding sequence for a desired protein fused to transcriptional regulatory elements excised from an active vaccinia virus gene. In order to retain infectivity, the chimeric gene must be inserted into a non-essential region of the vaccinia virus genome. Because of the large genome size, it is not possible to carry out this last step by enzymatic methods in vitro. Instead, we depend on homologous recombination in cells infected with vaccinia virus and transfected with a plasmid containing the chimeric gene flanked by vaccinia DNA segments which surround the intended insertion site. This scheme is illustrated in Fig. 1. Since only about one out of a thousand progeny virus particles are recombinants, a simple

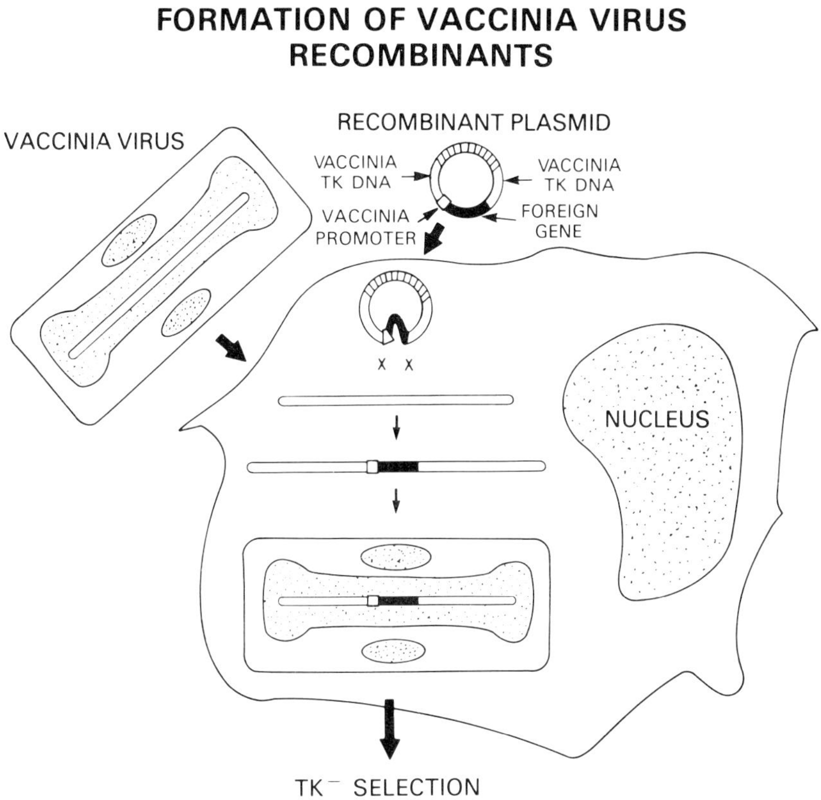

Figure 1. Formation of vaccinia virus recombinants. Plasmids containing the vaccinia virus thymidine kinase gene interrupted by the insertion of a vaccinia virus promoter have been constructed. A foreign gene has been inserted into a restriction endonuclease site just downstream of the promoter. In cells that have been infected with vaccinia virus and transfected with the plasmid, homologous recombination occurs. The thymidine kinase negative phenotype of the recombinant virus provides a simple method of selection.

way of distinguishing them from parental virus would be useful. In general, we have flanked the foreign gene with DNA sequences derived from the vaccinia TK gene which directs insertion into the TK locus. Consequently, all recombinants have a TK^- phenotype. The latter can be distinguished from wild-type vaccinia virus by carrying out the plaque assay in TK^- cells in the presence of 5-bromodeoxyuridine. Under these conditions, the synthesis of TK by wild-type virus causes phosphorylation of the thymidine analog and its lethal incorporation into DNA. In practice, about half of the virus plaques found under these conditions are recombinants and the remainder are spontaneous TK^- mutants. These can be readily distinguished, however, by DNA hybridization or binding of specific antibody.

To facilitate the general use of the vector system, we have constructed a series of plasmids that have all of the essential elements including convenient restriction endonuclease sites for insertion of foreign protein coding sequences. The time and level of expression of the foreign gene is determined by the choice of vaccinia promoter. For many of the recombinants, we have used a promoter that contains both early and late regulatory signals providing for continued expression during the growth cycle (Mackett et al., 1982).

Influenza virus

Influenza is a segmented RNA virus that causes respiratory infections in man and other animals. The virus has a lipid envelope containing two proteins: the hemagglutinin (HA) and neuraminidase. Immunity to influenza virus has been correlated with the presence of sub-type specific antibodies to these surface proteins. A cell-mediated immune response also may be important for recovery from infections.

The entire coding sequence for an influenza type A HA was inserted into an appropriate plasmid to form a chimeric gene which was then inserted into the thymidine kinase locus of vaccinia virus (Smith et al., 1983). Expression of the HA was initially detected by binding specific antiserum to the virus plaques. Subsequently, immunoprecipitation and polyacrylamide gel electrophoresis was used to demonstrate that a polypeptide of the correct size was made (Smith et al., 1983). Evidence for correct processing was obtained by incorporation of radioactively labeled glucosamine and by immunofluorescent staining of the cell surface. Thus, by all of these parameters the influenza HA made by recombinant vaccinia virus was indistinguishable from the authentic protein.

Antibodies to the HA were produced upon inoculation of a variety of experimental animals including rabbits, hamsters and mice with the recombinant virus. Furthermore, influenza A virus specific cytotoxic T cells were produced following in vitro stimulation of spleen cells from mice primed by intravenous inoculation with recombinant vaccinia virus (Bennink et al., 1984). Therefore, a recombinant vaccinia virus was able to stimulate both humoral and cell-mediated immune responses.

Experimental animals inoculated intradermally with the vaccinia virus recombinant were subsequently protected against lower respiratory infection with influenza virus. This was measured by determining the yield of influenza from the lungs following challenge (Table 1, Smith et al., 1983).

Herpesviruses

Herpesviruses are large DNA viruses that infect man and animals. The

Table 1. Protection of hamsters against influenza infection[a]

Inoculum[b]	Antibody titer[c]	Response to influenza challenge Mean log titer
VV	≤ 2	4.4 ± 0.6
$VV_{Inf\ HA}$	104	2.6 ± 0.1[d]
Influenza	49	2.6 ± 0.1[d]

[a] Data taken from Smith et al., 1983.
[b] 10 animals were inoculated intradermally with vaccinia virus (VV), vaccinia virus recombinant ($VV_{Inf\ HA}$) or intravenously with influenza virus.
[c] Hemagglutination inhibition titers are expressed as reciprocal dilutions.
[d] For calculation of mean titers, animals from which virus was not recovered were assigned maximum possible values of $10^{2.5}$.

most thoroughly studied member of this family is herpes simplex virus type 1 (HSV-1), which cause acute and latent infections in man. At least five glycoproteins are contained within the complex envelope structure. One of these, glycoprotein D (gD), has been shown to stimulate protective immunity to a lethal HSV-1 challenge. The gD gene has been inserted into vaccinia virus (Paoletti et al., 1984; Cremer et al., 1985). Recombinant virus was shown to synthesize a glycosylated polypeptide that was immunoprecipitated with HSV-1 antiserum and comigrated with the authentic product of HSV-1 infected cells (Cremer et al., 1985). Mice immunized with the recombinant vaccinia virus were protected against a lethal challenge with HSV-1 (Table 2) and against the development of latent HSV-1 infection of the trigeminal ganglion (Cremer et al., 1985). The recombinant also protected against lethal infection with HSV-2 evidently because of the similarity between gD proteins of the two serotypes of herpes simplex virus.

Vesicular Stomatitis Virus

The final example to be considered here is vesicular stomatitis virus (VSV). This negative stranded RNA virus infects cattle, horses, and pigs

Table 2. Protection of mice against herpes simplex types 1 and 2[a]

Inoculum[b]	Antibody titer[c]	Mortality[b] HSV-1	HSV-2
None	<4	58/72 (81%)	39/40 (98%)
VV	<4	9/14 (64%)	39/42 (90%)
$VV_{HSV\text{-}1\ gD}$	62	2/92 (2%)	2/59 (3%)

[a] Data taken from Cremer et al., 1985.
[b] 6 to 8 week old BALB/c mice were vaccinated with 1×10^8 pfu of vaccinia virus (VV) or vaccinia recombinant ($VV_{HSV\text{-}1\ gD}$)
[c] Geometric mean reciprocal neutralization titer.
[d] Mice were challenged intraperitoneally with 1×10^8 pfu of HSV-1 or 2×10^6 pfu of HSV-2.

Table 3. Protection of cattle against VSV[a]

Inoculum[b]	Antibody titer[c]	Vesicular lesion[d]
none	0	+
none	0	+
VV_{HBs}	0	+
VV_{HBs}	30	+
$VV_{VSV\ G}$	640	+
$VV_{VSV\ G}$	2560	–
$VV_{VSV\ G}$	480	+
$VV_{VSV\ G}$	1920	–
$VV_{VSV\ G}$	1280	–
$VV_{VSV\ G}$	2560	–

[a] Data taken from Mackett et al., 1985.
[b] Intradermal inoculation with 4×10^8 pfu of vaccinia virus recombinant expressing hepatitis B virus surface antigen (VV_{HBs}) or VSV G protein ($VV_{VSV\ G}$).
[c] Reciprocal neutralization titers on day 44.
[d] Challenge on day 44 with 10^2 pfu of VSV on dorsal surface of tongue.

and is of some economic importance in the United States, Mexico and parts of South America. Lesions are found on the tongue and oral mucosa as well as on the udders and coronary band of the hooves in cattle. Neither attenuated nor inactivated vaccines are currently in use within the United States. Since antibodies to the glycoprotein have been correlated with protective immunity, the corresponding gene was inserted into vaccinia virus (Mackett et al., 1985). As in the other examples, polypeptides from infected cells that were labeled with ^{35}S-methionine or ^{3}H-glucosamine were immunoprecipitated and shown to comigrate with the authentic product. In addition, the glycoprotein was incorporated into the plasma membrane and could be stained with fluorescent antibodies. Intradermal vaccination with the recombinant vaccinia virus protectively immunized mice against a lethal intravenous infection with the New Jersey serotype of VSV. Furthermore, cattle vaccinated with the recombinant virus produced neutralizing antibodies. Animals that developed antibody titers of 1280 or more were protected against an intradermalingual challenge of 10^2 pfu of VSV (Table 3) but not against higher challenge doses (Mackett et al., 1985). It is possible, however, that even a 10^2 pfu challenge is higher than that naturally encountered; field testing would be needed to further evaluate the vaccine.

Conclusions

Vaccinia virus recombinants can potentially be used against a variety of infectious diseases. Advantages of this type of vaccine includes (1) stimulation of humoral and cell-mediated immunity, (2) ability to insert multiple genes, (3) economy of manufacture, and (4) simplicity of administration.

References

Bennink, J. R., J. W. Yewdell, G. L. Smith, C. Moller and B. Moss. 1984. *Recombinant vaccinia virus primes and stimulates influenza virus HA-specific CTL*. Nature 311:578–579.

Bertholet, C., R. Drillien and R. Wittek. 1985. *One hundred base pairs of 5' flanking sequence of a vaccinia virus late gene are sufficient to temporally regulate late transcription.* Proc. Nat. Acad. Sci. USA 82:2096–2100.

Cremer, K. J., M. Mackett, C. Wohlenberg, A. L. Notkins and B. Moss. *Vaccinia virus recombinant expressing herpes simplex type 1 glycoprotein D prevents latent herpes in mice*. Science 228:737–739.

Mackett, M., G. L. Smith and B. Moss. 1982. *Vaccinia virus: a selectable eukaryotic cloning and expression vector*. Proc. Nat. Acad. Sci. USA 79:7415–7419.

Mackett, M., G. L. Smith and B. Moss. 1984. *A general method for the production and selection of infectious vaccinia virus recombinants expressing foreign genes*. J. Virol. 49:857–864.

Mackett, M., T. Y. Yilma, J. A. Rose and B. Moss. 1985. *Vaccinia virus recombinants express vesicular stomatitis virus genes and protectively immunize mice and cattle*. Science 227: 433–435.

Moss, B. 1985. *Replication of poxviruses*. in B. N. Fields, R. M. Chanock, and B. Roizman, eds., *Virology*. Raven Press, New York.

Moss, B., E. Winters and J. A. Cooper. 1981. *Deletion of a 9,000 base pair segment of the vaccinia virus genome that encodes non-essential polypeptides.* J. Virol. 40:387–395.

Nakano, E., D. Panicali and E. Paoletti. 1981. *Two major DNA variants present in serially propagated stocks of the WR strain of vaccinia virus*. J. Virol. 37:1000–1010.

Panicali, D., S. W. Davis, R. L. Weinberg and E. Paoletti. 1983. *Construction of live vaccines using genetically engineered poxviruses: biological activity of recombinant vaccinia virus expressing influenza virus hemagglutinin*. Proc. Nat. Acad. Sci. USA 80:5364–5368.

Paoletti, E., B. R. Lipinskas, C. Samsonoff, S. Mercer and D. Panicali. 1984. *Construction of live vaccines using genetically engineered poxviruses: biological activity of vaccinia virus recombinants expressing the hepatitis B virus surface antigen and the herpes simplex virus glycoprotein D.* Proc. Nat. Acad. Sci. USA 81:193–197.

Puckett, C. and B. Moss. 1983. *Selective transcription of vaccinia virus genes in template dependent soluble extracts of infected cells*. Cell 35:441–448.

Rigby, P. W. J. 1983. *Cloning vectors derived from animal viruses*. J. Gen. Virol. 64:255–266.

Smith, G. L., M. Mackett and B. Moss. 1983. *Infectious vaccinia virus recombinants that express hepatitis B virus surface antigen*. Nature 302:490–495.

Smith, G. L., G. N. Godson, V. Nussenzweig, R. S. Nussenzweig, J. Barnwell and B. Moss. 1984. *Plasmodium knowlesi sporozoite antigen: expression by infectious recombinant vaccinia virus*. Science 224:397–399.

Smith, G. L., B. R. Murphy and B. Moss. 1983. *Construction and characterization of an infectious vaccinia virus recombinant that expresses the influenza hemagglutinin gene and induces resistance to influenza virus infection in hamsters.* Proc. Nat. Acad. Sci. USA 80: 7155–7159.

Weir, J. P. and B. Moss. 1984. *Regulation of expression and nucleotide sequence of a late vaccinia virus gene*. J. Virol. 51:662–669.

Wiktor, T. J., R. I. McFarlan, K. J. Reagen, B. Dietzschold, P. J. Curtis, N. H. Wunner, M. P. Kieny, R. Lathe, J. P. Lecocq, M. Mackett, B. Moss and H. Koprowski. 1984. *Protection from rabies by a vaccinia virus recombinant containing the rabies virus glycoprotein gene.* Proc. Nat. Acad. Sci. USA 81:7194–7198.

A genetic engineering methodology for insect pest control: female sterilizing genes

JAMES B. COURTRIGHT and A. KRISHNA KUMARAN
Department of Biology, Marquette University, Milwaukee, Wisconsin 53233, U.S.A.

Abstract We propose a novel method to control insect pest populations by integrating into the genome of the target species a gene fusion that serves as a dominant female sterilizing (FS) gene. Based on the developmental regulation, regulatory sequences from vitellogenin and/or chorion protein genes have been identified as suitable for construction of FS gene fusions. These promoters, which respond to tissue- and stage-specific developmental signals, can be linked to a variety of structural gene elements such as those of a scrambled chorion protein, a signal sequence-deleted neutral protease, a phospholipase, an altered tubulin or a toxin. Such a FS gene fusion, because of its tissue and developmental stage specificity, could be expected to be expressed only in adult ovaries with resulting disruption of ovarian functions. In order to introduce the FS gene into the genome of the target species we suggest placing it at a neomycin resistance or other appropriate selectable genetic marker between the termini of a *Drosophila* transposing P-element and transforming the germline cells in embryo. Males carrying the S gene will transmit it to their progeny which would be all female sterile. Theoretical calculations suggest that target insect populations will be reduced substantially in two to four generations if the ratio of FS males to endemic males is maintained 10:1. It is also theoretically possible to construct a FS gene which would undergo replicative transposition and thus maintain its self in the population in a high copy number thus causing continued decline in the number of pests. The current genetic engineering method allows adapting these principles to control a wide variety of insect pest species.

Introduction

The control of insect pest populations is essential to agriculture and over the decades several different methods of control have been developed. The application of a variety of insecticides, development of insect resistant cultivars, pheromone and other insect attractant traps, biological control by parasites and/or predators, utilization of insect growth regulators and release of sterile males are among the most effective in the management of insect pests. While there are certain advantages to each of these methods there are many attendant problems. For example, rapid emergence of resistant strains in treated areas as well as the harmful effects of insecticides on the ecosystems have rendered long-term, continued effective use of pesticides unacceptable (Mani, 1985). Insect growth regulators, which are generally less severe on ecosystems, are not as effective as insecticides in field tests. Hence, integrated pest management by a judicious mixture of the available methods has been extensively used in insect control. Of all the currently available methods of insect control, genetic control by sterile male release has been regarded as the most environmentally safe and economically sound method. These methods involve the production of dominant lethal mutations in the released males with irradiation or alkylating chemosterilants, and sterility caused either through chromosomal or cytoplasmic incompatibility of the released males with the endemic populations. However, the latter method has not been extensively used because of the difficulty in isolating the appropriate strains with chromosomal or cytoplasmic incompatibility and because the irradiated and chemosterilized males are so weakened by the treatment that they have difficulty in competing effectively with endogenous males for mates.

We propose here a new concept in the genetic control of insect populations by means of genetic engineering methodology. With this method dominant female sterilizing (FS) genes will be inserted into the male genome of target species. Release of such nondebilitated males, we predict, would lead to a decrease in insect populations through a reduction of the fecundity of the breeding population. Moreover, male progeny carrying the FS gene would be available to mate with immigrating or surviving fertile females, and thus control of the population size would be maintained for several generations without additional FS male release.

Components of the female sterilizing gene

A consideration of the known molecular features of gene expression suggests that the proposed dominant FS gene should contain a regulatory sequence which responds only to the developmental signals present in female tissues, preferably those of ovaries. The proposed structural gene sequences can be from among a variety of genes which can disrupt cell organization, oocyte architecture or eggshell composition, or alternatively is cytotoxic or lethal to the cells in which it is expressed.

Regulatory sequences. Theoretically, three classes of genes identified in *Drosophila* and other insects can be the source of a regulatory sequence for the FS gene. One class of genes, transformer (*tra*), double sex (*dsx*) and intersex (*isx*) are involved in defining the sexual status of the individual (Baker and Belote, 1983; Belote et al., 1985). Only when the organism is in the "female mode" by the appropriate activity of this class of genes can other female-specific genes function (Belote et al., 1985). Thus, regulatory sequences from these genes seem suitable for construction of FS genes. Since all alleles of these genes are recessive and because the level of their expression is likely to be less than that of such structural genes as vitellogenins and chorion proteins, these promoters are probably unsuitable for FS gene construction. Although a dominant allele of *dsx* has been described (Baker and Belote, 1983) the gene has not been cloned and its size and molecular organization are not known.

The second class of genes, which when disrupted cause female sterility, have been mapped but have not yet been identified (Garabadian, Hung and Wensink, 1985; Komitopoulou et al., 1983). Since disruption of one of the copies of the genes seems to interfere with female fecundity these genes and/or their promoters may prove valuable in constructing FS genes in the future.

The third class of genes, whose developmental expression as well as their molecular features have been studied in detail, include those of vitellogenins and chorion and vitelline membrane proteins (Barnett et al., 1980; Postlethwait and Jowett, 1980; Fargnoli and Waring, 1982; Mindrinos et al., 1985; Griffin-Shea, Thireos and Kafatos, 1982; Margaritis, Kafatos and Petri, 1980). These genes are expressed only in females and only in fat body and ovarian follicle cells, with a well-defined developmental stage specificity (Brennan et al., 1982; Garabadian et al., 1985; Waring and Mahowald, 1979). All alleles of these genes are apparently expressed even when present in heterozygous condition or even in heterologous position in the genome in transformed files (DeCicco and Spradling, 1984; Garabadian et al., 1985). Thus, these genes are suitable donors of promoters for the FS gene fusion.

More importantly, the regulatory sequences from the chorion protein and vitellogenin genes have been identified and were found to maintain sex-, tissue- and stage-specific expression of the linked structural genes. For example, chorion gene promoters linked in vitro with β-galactosidase structural gene from *E. coli* (DeCicco and Spradling, 1984) and inserted into the *Drosophila* genome maintain the original chorion promoter pattern of expression. Similarly, *Drosophila* vitellogenin promoter maintains some of its original tissue specificity in its pattern of expression following insertion at exogenous sites in the genome (Garabadian et al., 1985). This evidence suggests that tissue-specific expression of female reproductive function of genes is determined in part by definite sequences immediately upstream from the start of transcription. Therefore, we propose to use such promoter sequences from chorion proteins or vitellogenins in the FS gene fusion in order to ensure efficient and tissue-specific expression of genes.

Structural gene sequences. The FS gene could disrupt female fecundity without killing the insect by either interfering directly with the organization of the eggshell membranes or indirectly by affecting the function of the ovarian follicle cells that produce the eggshell.

Alternatively, the FS gene may produce a cytotoxic product that may prove lethal to the insect. Since the FS would be expressed only in ovaries, the FS gene could be expected to function only as a female lethal gene.

The simplest and the least deleterious, from the point of potential danger to nontarget organisms, is a scrambled gene of an eggshell protein. A scrambled chorion protein gene in the FS gene construct would be expected to produce the altered protein in sufficient quantity. If the alteration in the protein is of such a nature that it can form complexes with chorion proteins, it will be incorporated into the chorion leading to the formation of nonfunctional aggregates. Thus it would disrupt the functional organization of the chorion, and the resulting eggs would be nonviable because of the altered water-proofing properties. On the other hand, it is conceivable that the altered chorion protein cannot be incorporated into the chorion, and consequently the FS gene would not cause sterility. In this case, other arrangements would need to be tested. In many instances, the FS gene construct may need to contain a species-specific altered chorion gene, thereby requiring a different FS gene construct for each target species. For these reasons it may be economical to choose a structural gene sequence that can be used on almost any insect species.

Genes coding for tubulins and actin or for neutral protease and phospholipase can be expected to be functional in all insect species, and are generally conserved in all organisms (Buckingham and Minty, 1983; Lemischka et al., 1981). Thus a FS gene containing a mutation for these genes could be expected to produce the altered protein in the follicle cells. If the alteration is subtle it would be incorporated into the cytoskeletal elements but produce nonfunctional microtubules (Lloyd, Raff and Raff, 1981). Consequently, the follicle cell organization would be sufficiently disrupted so as to cause follicular dysfunction and female sterility. Again, as in the case of scrambled chorion genes, the mutant gene product may not be incorporated into the cytoskeletal elements and thus escape the deleterious effects of the FS gene. However, the presence of dominant mutations that affect microtubule organization (Kemphues et al., 1979) suggests that this strategy is likely to succeed in causing female sterility.

Genes coding for neutral protease or phospholipase have been cloned and were found to be expressed in bacterial systems. Their structural gene sequences, appropriately altered to delete the signal sequences, would produce mRNAs that would be translated on nonmembrane bound ribosomes. The newly synthesized enzymes could be expected to disrupt cellular organization in the ovarian follicle cells and even cause their death, thus leading to female sterility.

Lastly, if the above three types of structural genes prove ineffective as agents of female sterility, we visualize using the cytotoxin gene from *Bacillus thuringiensis* (Bulla et al., 1980). The *thuringiensis* toxin is being used as an insecticide effective on lepidopteran larvae. Its selective toxicity to lepidopterans is due to the fact that the bacteria produce a protoxin that is activated to the toxic state only following alkaline hydrolysis in the gut of lepidopteran larvae. However, other cytotoxic fragments of the *thuringiensis* protoxin had been identified and even varieties of *thuringiensis* species that are toxic to dipterans had been isolated (Gonzalez and Carlton, 1984). Hence we propose, as a last resort, to use the structural gene sequence of the cytotoxic gene fragment of the *thuringiensis* protoxin gene as the structural gene component of FS gene.

This this cursory survey, it is clear that several relatively innocuous gene products when expressed in insect ovarian tissues can be expected to disrupt their function, thus causing female sterility.

Insertion of FS genes into genome

The first step in the introduction of the FS gene, generated in vitro, into a population is its insertion into the genome of the target insect species. Insertion of foreign genes into the germline of the desired species can be achieved by P-element mediated transformation (Rubin and Spradling, 1982; Spradling and Rubin, 1982). Any sequence of heterologous DNA located between the 31 base pair inverted repeats of the *Drosophila melanogaster* P-element system appears to be integrated into the genome of not only *D. melanogaster* but also for other species of *Drosophila* as well as mammalian cells in vitro (Clough et al., 1985; Brennan et al., 1984). Furthermore, the amount of DNA that can be integrated into the host genome appears to be limited only by the size of the plasmid which can be maintained in the bacterial host. Cosmid vectors carrying up to 40 kb of DNA could be integrated into the genome of recipient flu using the P-element mediated transformation (Haenlin et al., 1985).

In order to integrate the FS gene into the genome of the target insect species the FS gene will be placed between the inverted repeat sequences of an incomplete P-element and injected into the posterior end of an early embryo along with a "wings clipped" P-element (Rubin and Spradling, 1983). The latter provides the transposase, an enzyme necessary for the integration of plasmid DNA containing the inverted repeats of the P-element. Thus, P-element mediated insertion of FS gene into target species genome can be achieved by injection of the DNA into an embryo. This method would be applicable to all insects whose eggshells are relatively soft or could be removed after treatment with mild detergents or oxidants. However, this method may not be applicable to eggs of lepidopterans and other insects with hard eggshells. Genetic transformation of such insects would probably require development of alternative methodology. For example, injection of the FS gene construct into adult ovary or infection with viral vectors are among the alternative methods of insertion of heterologous DNA into the genome (Pennock, Shoemaker and Miller, 1984).

Selection of transformants. The transformation of the target species occurs, like all transformation systems, with low probability. Hence, another important obstacle for successful application of this genetic engineering methodology is detection and selection of the transformed individuals. Nearly all of the transformation studies in *Drosophila* to date have employed recipient strains deficient for either alcohol dehydrogenase or xanthine dehydrogenase, and transformants have been selected on the basis of restoration of the enzyme function. This approach is practical for only those species with strains deficient for a given enzyme. However, the detection of transformants can be more readily accomplished, even in the absence of enzyme-deficient strains, by incorporating within the transformation vector a gene for neomycin phosphotransferase. This confers on the transformants a resistance to kanamycin or its analog, geneticin (G418). Such vectors bearing the inverted repeated P-element segments and neomycin transferase linked to a heat shock gene promoter have been constructed and tested for selection of transformants (Steller and Pirrotta, 1985). By incorporating the FS gene in such a vector the transformants can be selected even in species lacking mutant phenotypic markers. The composition of the FS vector is shown in Fig. 1.

Maintenance and propagation of FS genes

What is clearly desired for the effective use of FS genes is a capability of transmitting these genes to female progeny with a probability approaching 1. This condition can only exist in three cases. First, when the FS gene is present in the X chromosome of males, all resulting female progeny will possess the patroclinous X chromosome. Secondly, when the FS gene is present in a homozygous state on an autosome, all offspring receive either one or the other of a given set of autosomes and all resulting progeny would then have one dose of the FS gene. Thirdly, if the FS gene were to be present in multiple copies in the male genome, it becomes highly probable that any given female would receive at least one FS gene from the distribution and independent assortment of chromosomes.

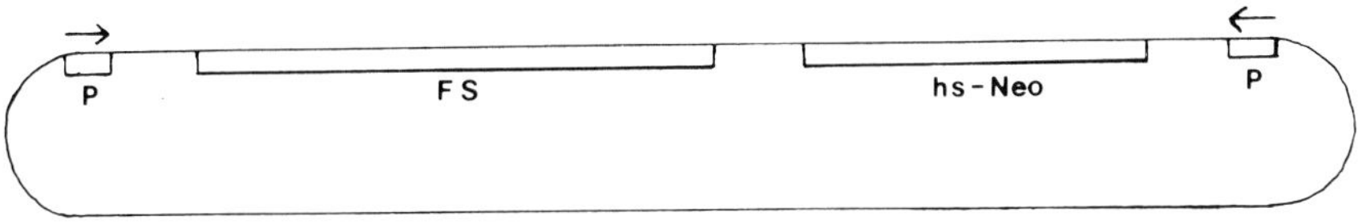

Figure 1. Structure for a FS vector plasmid. This plasmid contains the dominant FS gene, the neomycin selectable marker, and the inverted terminal repeats required for integration into the genome.

Each of these conditions must be qualified. The first condition, that of FS transmission by means of the X chromosome, is clearly an approach which would be effective for at least one generation. The FS males for such a release program could, however, be only maintained in laboratory stocks when they are continually mated to attached X females. This option is available for *Drosophila* and could be extended to other species where the selection of attached X chromosome-bearing females is possible. The second condition, that of autosomal homozygosity, is not, however, obtainable with available genetic systems. The reason for this is that homozygous males can only be obtained by mating with FS-bearing females, and by definition, such FS females would be sterile. This problem may be obviated if there were some means of engineering the FS gene so that it would be conditionally expressed, or selectively inactivated, such as linking the FS gene to a heat shock promoter which would facilitate expression.

The third condition, that of dispersed multiple copy FS genes in the male genome, is obtainable with existing transposable elements, as described below. Importantly, multiple copies of the FS gene at diverse chromosomal sites would result in sperm receiving at least one copy of FS due to simple segregational mechanisms. Moreover, the fact that recombination does not occur in dipteran males, and probably other insects orders as well, means that non-FS chromosomes could not be generated by recombination. Lastly, in species with many chromosomes it is likely that if the copy number is only a fraction of the total number of chromosomes, the probability that a given gamete (sperm) is totally lacking FS genes may be calculated by the Poisson formula.

Another condition which should be considered would be the number of FS genes which would have to be present in the male germline in order to assure that at least one pair of homologs each contained an FS gene. Thus, after segregation, each gamete would contain at least one FS gene. In general, it can be shown that the probability of no pair of chromosomes containing a copy each of the FS gene can be calculated according to the formula:

$$P(0) = \left(1 - \frac{1}{C}\right)^{n(n+1)/2}$$

where n = FS copy number, and C = diploid chromosome number (S. Merrill, personal communication).

Using the Poisson distribution formula for calculating P_0 class and the above expression for calculating P_0 class of FS-bearing homologs of any chromosomal pair, some estimates of the approximate copy numbers may be calculated (Table 1). As seen from the data presented in Table 1, the number of FS genes required to ensure that 99% of the gametes would contain at least one FS gene is approximately the same whether the gene is obtained by distributive or segregational means for organisms with low ploidy numbers. For organisms with a large number of chromosomes, such as lepidopterans (n = 30), an effective transmission of the FS genes could be achieved by the random distribution of FS genes throughout the genome. Interestingly, the minimal number of FS genes required under both modes of calculations is substantially less than the number of P-elements maintained in natural *Drosophila* populations, and which, therefore, should be established in the proposed FS male populations. Thus, resulting females will be genetically sterile as long as an appreciable number of the FS sequences remain functional.

Transposition and copy number. In view of the promising mathematical predictions, it would then be necessary to develop a means of generating high FS copy number solely through the male germline. Interestingly, all evidence to date suggests that this may be accomplished by transposition with the P-type transposase encoded by the P-element (Karess and Rubin, 1984). A complete 2.9 kilobase (kb) P-element, containing two transposase gene, flanked by the P-element 31 basepair (bp) inverted repeats can integrate into the genome of the recipient and has functional transposition activity, as assayed by rearrangements at the *sn* locus (Rubin and Spradling, 1982). Other evidence of P activity is seen in chromosomal rearrangements (Engels and Preston, 1984) or by hybrid dysgenesis and resulting gonadal malformations (Kidwell, 1983). For transposition to occur, it is only necessary (1) that a given sequence contain two 31 bp inverse repeats separated by a minimum of approximately 500 bp DNA and (2) that a source of transposase (P-element) be provided (Rubin, Kidwell and Bingham, 1982; O'Hare and Rubin, 1983; Rubin and Spradling, 1983). The transposase activity provided in *trans* is sufficient for transposition of any P-element containing sequences (Spradling and Rubin, 1982). Hence, foreign DNA

Table 1. The number of FS gene copies required for 0.99 probability of transmission of at least one FS gene to each gamete

Organism	Haploid chromosome number	FS gene copy number required	
		Poisson distribution	Segregation* in homologs
Anopheles (mosquito)	3	7.7	6.6
Ceratitis (Med fly)	6	8.6	9.8
Galleria (wax moth)	30	9.4	22.9

* The number of copies required to have a high probability ($p = 0.99$) of at least one homologous pair of chromosomes, each containing an FS gene.

can be included in a region outside that of the four open reading frames of the P-element (Karess and Rubin, 1984) and the entire transposon used for transformation (Steller and Pirrotta, 1985).

Although the actual mechanism of P-element transposition is unknown, available evidence would suggest that it primarily occurs by means of replicative transposition rather than excision and reinsertion. When intact P-elements are introduced into the nucleus of an M cytotype cell, there is induction of transposition, and the rate of P-element insertion can be as great as 0.8 per chromosome per generation (Bingham, Kidwell and Rubin, 1982). The actual rate of transposition, however, is subject to genes on other chromosomes (Kidwell, 1983). Evidently, such high rates explain the large number of P-elements found in genomes of P-cytotype *Drosophila* strains (Engels and Preston, 1981; Bingham et al., 1982; Searles et al., 1982).

This combined evidence suggests that high FS gene copy number in the male germline may be established by two means. The first and most efficient procedure, which can be tested initially in *Drosophila*, would be to include a FS gene along with an intact P-element and associated inverted repeats, termed *FS neo P* (cf. Fig. 2). This *FS neo P* construct would have several useful features. The presence of a neomycin gene would allow for selection of transformants or segregants in every larval generation, the FS gene would assure sterility in females, and the complete P-element would provide for transposition of the entire sequence to additional chromosomal sites in every generation. Secondly, since the P-element acts in *trans*, it may be equally advantageous to construct double transformants, where one construct contains the FS gene plus the neomycin selectable marker and the other construct contains an intact P-element. The uncertainty in this latter approach would be the rate at which the P-related sequence can separately mobilize the *FS-neo* construct.

On the assumption that the *FS neo P* construct is capable of effecting its own transposition, we envision in developing FS males in several

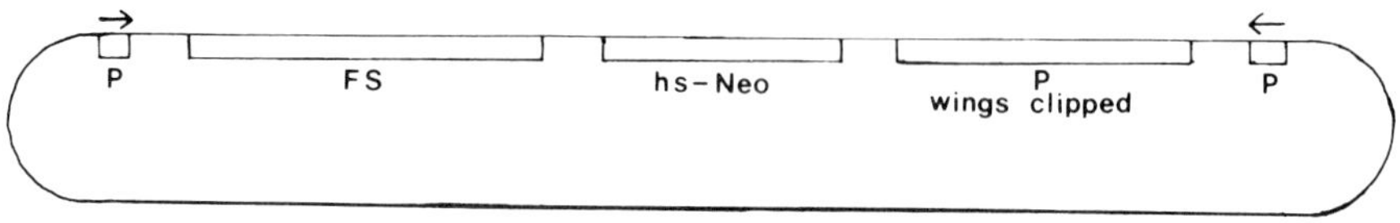

Figure 2. Structure for a transposable FS vector. This plasmid contains the dominant FS gene, the neomycin selectable marker, a functional P transposase, and the inverted P element repeats required for chromosomal integration. The presence of the P transposase within the flanking P element terminal repeats provides for the continued transposition in every generation.

steps. Following transformation, neo-resistant males are selected and mated to M-cytotype females. Since the ooplasm will be M-cytotype, there will be induction of transposase and additional copies of *FS neo P* will be inserted at various other sites in the genome. The resulting progeny will be subjected to additional rounds of selection for neo resistance and continued matings to M-cytotype females, in each instance using only males for subsequent matings. Over several generations of selection and transposition there will be an increase in the number of dispersed FS genes in the male line, perhaps achieving the equilibrium value of 30–50 normallly obtained in nature. The efficiency of this process may be further increased by using attached X females, which, by the simple chromosome mechanics of that configuration, allow retention of the patroclinous *FS neo P* genes on the X chromosome. Thus, prior to release, the male line will have numerous FS genes at sites throughout the genome and these will cause sterility in all female progeny.

Should a similar mechanism for the increase of FS gene constructs be operative in *Ceratitis* or other target species, the release *FS neo P* males will be mated with M cytotype females, thereby causing additional transposition within the male germline in the first generation. As a consequence, the number of *FS neo P* genes should increase by replicative transposition and there should be a partial or complete restoration of the original FS gene number by the completion of spermatogenesis in male progeny. In fact, this rate of transposition may be accelerated by placing the P-element reading frames adjacent to a male-specific promoter which would allow for a higher level of constitutive expression of the P-element functions.

Conclusions and prospects

The proposed genetic engineering method, involving a P-element driven dominant female sterilizing gene, may prove very effective in the management of insect pests. There are many attractive features to the proposed system. For example, it eliminates the need for debilitating radiation or chemotreatments to sterilize males. The male offspring of the sterilizing males would continue to transmit FS genes in the succeeding generations, albeit in lower proportions.

Although there are many attractive features to the P-element driven FS gene mediated insect control, many aspects of this system need to be further explored. For example, it would be helpful to collect data on (1) the rate of P-element transposition under selection pressure and under non-selective conditions, (2) the maximum limits of foreign DNA that can be linked to P-element termini for successful integration, (3) the distribution of transposable elements, analogous or homologous to the *Drosophila* P-elements in other insect species, (4) the variety of insect species in which P-element mediated transformation can occur, (5) the efficiency of the different types of structural gene sequences in disrupting ovarian function, and most importantly, (6) the level of expression of the FS genes, in a model insect such as the fruitfly, and the extent of resulting reduction in fecundity of females.

Acknowledgements The authors thank Dr. Stephen J. Merrill of the Marquette University Mathematics Department for helpful discussions on the quantitative aspects of FS gene mediated insect control. Thanks are also due to Ms. Barbara DeNoyer for typing and to the Marquette University Graduate School and the National Science Foundation for support.

References

Baker, B. S. and J. M. Belote. 1983. *Sex determination and dosage compensation in Drosophila melanogaster*. Ann. Rev. Genet. 17:345–397.

Barnett, T. C. Pachl, J. P. Gergen and P. C. Wensink. 1980. *The isolation and characterization of Drosophila yolk protein genes*. Cell 21:729–738.

Belote, J. M., A. M. Handler, M. F. Wolfner, K. J. Livak and B. S. Baker. 1985. *Sex-specific regulation of yolk protein gene expression in Drosophila*. Cell 40:339–348.

Bingham, P. M., M. G. Kidwell and G. M. Rubin. 1982. *The molecular basis of P-M hybrid dysgenesis: the role of the P element, a P strain specific transposon family*. Cell 29:995–1004.

Brennan, M. D., R. G. Rowan and W. J. Dickinson. 1984. *Introduction of a functional P element into the germline of Drosophila hawaiiensis*. Cell 38:147–151.

Brennan, M. D., A. J. Weiner, T. J. Goralski and A. P. Mahowald. 1982. *The follicle cells are a major site of vitellogenin synthesis in Drosophila melanogaster*. Dev. Biol. 98:225–236.

Buckingham, M. E. and A. J. Minty. 1983. *Contractile protein genes*. Pp. 365–395 in N. McLean, S. P. Gregory, and R. A. Flavell, eds., *Eukaryotic Genes*. Butterworths, London.

Bulla, L. A., D. B. Bechtel, K. J. Kramer, Y. I. Shetna, A. I. Aronson and P. C. Fitz-James. 1980. *Ultrastructure, physiology, and biochemistry of Bacillus thuringiensis*. C.R.C. Cit. Rev. of Microbiol. 8:147–203.

Clough, D. W., H. M. Lepinske, R. L. Davidson and R. V. Storti. 1985. *Drosophila P element-enhanced transfection in mammalian cells*. Mol. Cell. Biol. 5:898–901.

DeCicco, D. V. and A. C. Spradling. 1984. *Localization of a cis acting element responsible for the developmentally regulated amplification of Drosophila chorion genes*. Cell 38:45–54.

Engels, W. R. and C. R. Preston. 1981. *Identifying P factors in Drosophila by means of chromosome breakage hotspots*. Cell 26:421–428.

Engels, W. R. and C. R. Preston. 1984. *Formation of a chromosome rearrangement by P factors in Drosophila*. Genetics 107:657–678.

Fargnoli, J. and G. L. Waring. 1982. *Identification of vitelline membrane proteins in Drosophila melanogaster*. Dev. Biol. 92:306–314.

Garabadian, M. J., M. -C. Hung and P. C. Wensink. 1985. *Independent control elements that determine yolk protein gene expression in alternative Drosophila tissue*. Proc. Nat. Acad. Sci. USA 82:1396–1400.

Gonzalez, J. M. and B. C. Carlton. 1984. *A large transmissible plasmid is required for crystal toxin production in Bacillus thuringiensis variety israelensis*. Plasmid 11:28–38.

Griffin-Shea, R., G. Thireos and F. C. Kafatos. 1982. *Organization of a cluster of four chorion genes in Drosophila and its relationship to developmental expression and amplification*. Dev. Biol. 91:325–336.

Haenlin, M., H. Steller, V. Pirrotta and E. Mohier. 1985. *A 43 kilobase cosmid P transposon rescues the fsD K10 morphogenetic locus and three adjacent Drosophila developmental mutants*. Cell 40:827–837.

Karess, R. E. and G. M. Rubin. 1984. *Analysis of P transposable element functions in Drosophila*. Cell 38:135–146.

Kemphues, T. C., E. C. Raff, R. A. Raff and T. C. Kaufman. 1979. *Mutation in a structural gene for a β-tubulin specific to testis in Drosophila melanogaster*. Proc. Nat. Acad. Sci. USA. 76:3991–3995.

Kidwell, M. G. 1983. *Hybrid dysgenesis in Drosophila melanogaster: factors affecting chromosomal contamination in the P-M system*. Genetics 104:317–341.

Komitopoulou, K., M. Gans, L. H. Margaritis, F. C. Kafatos and M. C. Masson. 1983. *Isolation and characterization of sex-linked female-sterile mutants in Drosophila melanogaster with special attention to eggshell mutants*. Genetics 105:897–920.

Lemischka, I. R., S. Farmer, V. R. Racaniello and P. Sharp. 1981. *Nucleotide sequence and evolution of a mammalian α tubulin mRNA*. J. Mol. Biol. 151:101–120.

Lloyd, J. E., E. C. Raff and R. A. Raff. 1981. *Site and timing of synthesis of tubulin and other proteins during oogenesis in Drosophila melanogaster*. Dev. Biol. 86:272–284.

Mani, G. S. 1985. *Evolution of resistance in the presence of two insecticides*. Genetics 109: 761–783.

Margaritis, L. H., F. C. Kafatos and W. H. Petri. 1980. *The eggshell of Drosophila melanogaster. I. Fine structure of the layer and regions of the wild type eggshell.* J. Cell Sci. 43:1–35.

Mindrinos, M. N., L. J. Scherer, F. J. Garcini, H. Kwan, K. A. Jacobs and W. H. Petri. 1985. *Isolation and chromosomal location of putative vitelline membrane genes in Drosophila melanogaster*. EMBO J. 4:147–153.

O'Hare, K. and G. M. Rubin. 1983. *Structure of P transposable elements of Drosophila melanogaster and their sites of insertion and excision.* Cell 34:25–35.

Pennock, G. D., C. Shoemaker and L. K. Miller. 1984. *Strong and regulated expression of Escherichia coli B galactosidase in insect cells with Baculovirus vector*. Mol. Cell Biol. 4: 399–406.

Postlethwait, J. H. and T. Jowett. 1980. *Genetic analysis of the hormonally regulated yolk peptides in D. melanogaster.* Cell 20:671–678.

Rubin, G. M., M. G. Kidwell and P. M. Bingham. 1982. *The molecular basis of P-M hybrid dygenesis: the nature of induced mutations*. Cell 29:987–994.

Rubin, G. M. and A. C. Spradling. 1982. *Vectors for P element mediated gene transfer in Drosophila*. Nucl. Acids Res. 11:6341–6351.

Rubin, G. M. and A. C. Spradling. 1983. *Vectors for P element mediated gene transfer in Drosophila*. Nucl. Acids Res. 11:6341–6351.

Searles, L. L., R. S. Jokerst, P. M. Bingham, R. A. Voelker and A. L. Greenleaf. 1982. *Molecular cloning of sequences from a Drosophila RNA polymerase II locus by P element transposon tagging*. Cell 31:585–592.

Spradling, A. C. and G. M. Rubin. 1982. *Transposition of cloned P elements into Drosophila germline chromosomes*. Science 218:341–347.

Steller, H. and V. Pirrotta. 1985. *A transposable P vector that confers selectable G418 resistance to Drosophila larvae*. EMBO J. 4:167–171.

Waring, G. L. and A. P. Mahowald. 1979. *Identification and time of synthesis of chorion proteins in Drosophila melanogaster*. Cell 16:599–607.

Part 5

Natural resources management

Biotechnology in pesticide environmental research

JEFFREY S. KARNS, WALTER W. MULBRY and PHILIP C. KEARNEY
USDA, Agricultural Research Service, AEQI Pesticide Degradation Laboratory, Building 050, BARC-West, Beltsville, Maryland 20705, U.S.A.

Abstract Two agricultural problems that show promise for solution by biotechnology are the safe disposal of pesticide wastewaters and the origin of enhanced pesticide metabolism in soils. Both problems involve microbiological processes that may be capable of either genetic manipulation or may be explained by gene regulation. The waste disposal process involves destroying large volumes of wastewater generated from animal dip solutions used in livestock tick control programs. Prior microbial metabolism of the insecticide coumaphos [O,O-diethyl O-(3-chloro-4-methyl-2-oxo-2H-1-benzopyran-7-yl)phosphorothioate] by a *Flavobacterium* (ATCC 27551) which cleaved the phosphorothioate linkage, followed by fragmentation of the resultant chlorinated phenol by UV-ozonation, proved to be a successful strategy for waste disposal. One important pesticide susceptible to enhanced metabolism is the insecticide carbofuran [2,3-dihydro-2,2-dimethyl-7-benzofuranyl methylcarbamate] used for corn rootworm control. We have isolated a bacterium, an *Achromobacater* species, that is capable of rapidly utilizing carbofuran as a sole source of nitrogen and also rapidly hydrolyzing several other N-methyl carbamate insecticides. By isolating pesticide degradative genes through molecular cloning, we hope to manipulate them to cause more rapid degradation of waste pesticides and afford some protection for pesticides in the field.

Introduction

Biotechnology holds great promise for advancement in many areas of agriculture. In our area of expertise, the degradation of pesticides by microorganisms, there are two distinct problems to which biotechnology may offer some degree of solution. The first problem is the safe and inexpensive disposal of aqueous pesticide wastes generated on the farm. The second problem is the appearance of "aggressive" or "problem" soils in which a pesticide is degraded at an exceptionally rapid rate such that the effectiveness of the pesticide is compromised (Felsot, Maddox and Bruce, 1981; Wilson, 1984). Both of these areas are under active investigation in our laboratory.

Biotechnology in the safe disposal of pesticide wastes

Background. Safe, effective, and low-cost waste disposal is one of the serious problems facing the pesticide applicator and(or) farmer. Recent changes in the Resource Conservation and Recovery Act (RCRA) impose strict new regulations on small source generators (which may include farmers) and could carry heavy potential penalties against those deemed to be responsible for pollution. Inadequate disposal of pesticide wastes might be a more likely source of groundwater pollution than is the actual use of these compounds in the field at todays' low usage rates. Unfortunately for the pesticide user, the technology does not yet exist to fully implement the intent of RCRA in protecting groundwater supplies. Current technology usually involves storing the waste material in open pits and allowing the naturally occurring organisms and sunlight to degrade the pesticide components. Meanwhile, evaporation of the water reduces the volume of the waste (Hall, 1984). This concentration of pesticides in a small area may facilitate excessive migration of some chemicals downward into groundwater. One possible solution to this problem is the use of genetically engineered microoganisms that are more efficient at partially or totally destroying pesticides in wastewater solutions. We are currently working on a combined microbial degradation-physical degradation process that holds great promise for solving the waste disposal problem for one important agricultural chemical.

Microbial-physical degradation of coumaphos

The tick eradication program overseen by the Animal and Plant Health Inspection Service (APHIS) of the U.S. Department of Agriculture relies heavily upon cattle dipping operations along the U.S.-Mexican border. The primary insecticide currently used in these operations is coumaphos [O,O-diethyl O-(3-chloro-4-methyl-2-oxo-2H-1-benzopyran-7-yl)phosphorothioate].

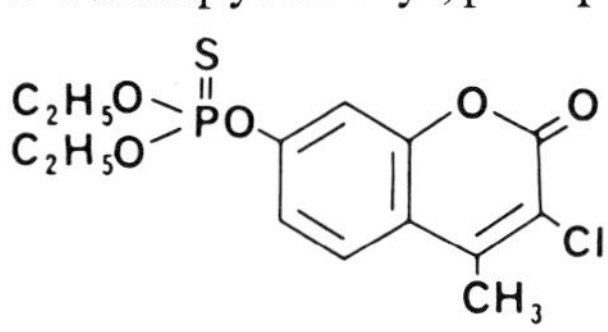

Coumaphos

Cattle are dipped by running them through long, trough-like vats containing aqueous suspensions of coumaphos. There are 42 of these vats in operation on a year-round basis in the APHIS program alone. The average volume of these vats is about 12,500 L (3,300 gallons) and the usual coumaphos concentration is 3000 ppm. Because of fouling by dirt, hair, and bovine fecal matter, these vats are emptied once a year for cleaning. Thus the operation generates over 500,000 L (140,000 gallons) of aqueous wastes containing very high concentrations of coumaphos each year. These liquid wastes currently are put into evaporation ponds. Clearly, a destructive method of disposal would be desirable for the elimination of these wastes.

The half-life of coumaphos incorporated into soil is more than 200 days (Waggoner, unpublished observations). The technique of UV-ozonation, whereby solutions are exposed to oxygen and intensive

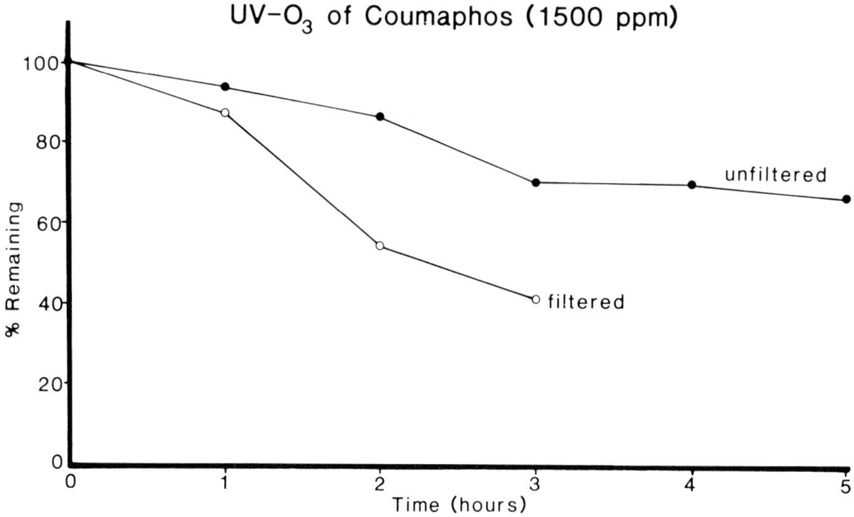

Figure 1. Effect of UV-ozonation treatment on coumaphos residues in filtered and unfiltered dip-vat waste.

ultra-violet radiation as a pretreatment step for soil disposal, accelerated the rate of degradation of waste solutions of the herbicides atrazine, 2,4-dichlorophenoxyacetic acid, and paraquat (Kearney, Plimmer and Li, 1983; Kearney, Zeng and Ruth, 1984). The reaction was very slow on coumaphos in animal dip-vat solutions, requiring over 3 h to degrade 25% of the coumaphos in suspension, with little additional degradation after 5 h of irradiation (Fig. 1). Filtering the dip-vat solution prior to UV-ozonation increased the effectiveness of the process slightly, indicating that large particulate matter in the suspension inhibited the UV-ozonation process (Fig. 1). Even with filtering, the rate of degradation of coumaphos was not rapid enough to make this method practical on a large scale.

Organisms capable of hydrolyzing O,O-diethylphosphorothioate insecticides were previously characterized by several authors (Munnecke and Hsieh, 1974; Nelson, 1982; Siddaramappa, Rajaram and Sethunathan, 1973). Although originally isolated due to their ability to hydrolyze diazinon and parathion, it was subsequently determined that the phosphotriesterase enzyme (parathion hydrolase) of these organisms was able to hydrolyze a wide range of O,O-diethylphosphorothioate compounds. Cell suspensions of the parathion-degrading bacterium

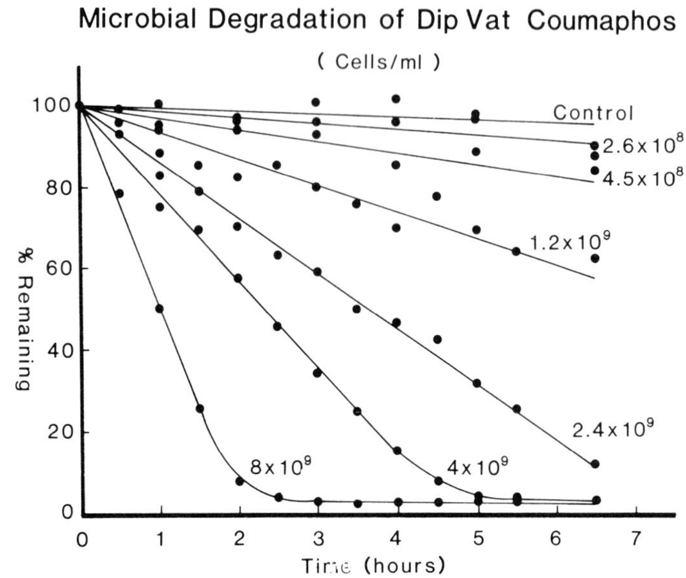

Figure 2. Hydrolysis of dip-vat coumaphos by cell suspension of a parathion-degrading *Flavobacterium* with varying cell densities.

Flavobacterium sp., ATCC 27551, originally characterized by Sethunathan and Yoshida (1972), were capable of completely degrading the 1500 ppm of coumaphos left in spent dip-vat material (Fig. 2). The rate at which coumaphos was degraded was directly dependent upon the cell density, with a very high cell density allowing complete degradation within 2 h. The reaction catalyzed by the phosphotriesterase enzyme produced by *Flavobacterium* sp. is shown in Fig. 3. The only products of this reaction are diethylthiophosphoric acid and a chlorinated, methyl-hydroxy-coumarin, commonly called chlorferon, which represents the intact R group of coumaphos.

Cell-free crude extracts of the *Flavobacterium* sp. added directly to dip-vat solutions displayed very high phosphotriesterase activity (Fig. 4). An enzyme concentration of 600 μg of protein/mL of dip-vat solution catalyzed the complete breakdown of the dip-vat coumaphos within 1 h, while 30 μg of protein/mL of dip-vat solution required 7 h to hydrolyze coumaphos completely to chlorferon. A concentration of 3 μg protein/mL of dip-vat solution catalyzed some degradation of coumaphos, but even after 24 h about 50% of the coumaphos remained. There was little additional degradation of coumaphos, perhaps indicating that the enzyme itself is slowly degraded by the micro-organisms naturally present in the unsterilized dip-vat material.

We have partially purified the phosphotriesterase enzyme from

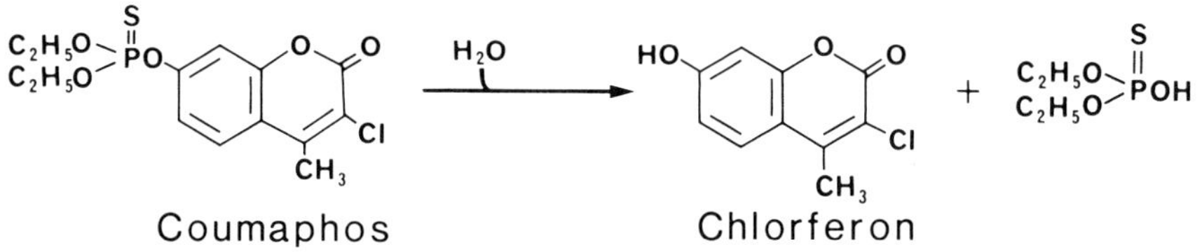

Figure 3. Coumaphos hydrolysis reaction catalyzed by the phosphotriesterase enzyme (parathion hydrolase) of *Flavobacterium*.

Flavobacterium sp. and have found that about 20% of the total enzyme activity is found in the cytoplasmic fraction of the cell, while the remaining 80% is associated with the cell membrane. From the work of Munnecke (1976) and that of McDaniel, Wild and O'Donovan (1985), as well as our own work, we know that this enzyme exhibits a very broad substrate specificity and may be useful for the elimination of other organophosphate insecticides. Among the options that we are considering for the use of this enzyme is to fix the enzyme to a solid support and use it directly in the decontamination process in the form of a cartridge or a fixed-bed reactor.

The chlorferon produced by the enzymatic hydrolysis of coumaphos was much more susceptible to destruction by the UV-ozonation process

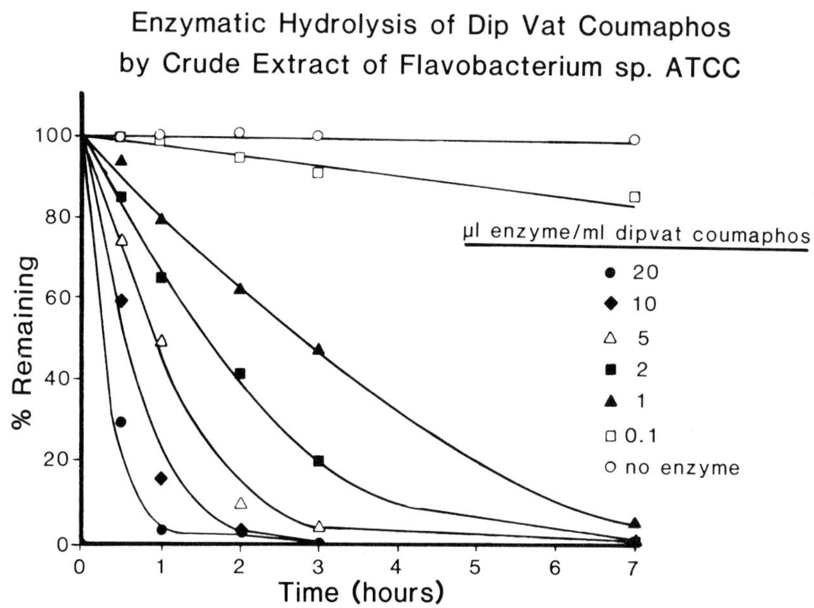

Figure 4. Hydrolysis of dip-vat coumaphos by different concentrations of a crude extract of *Flavobacterium* containing approximately 30 mg/mL protein.

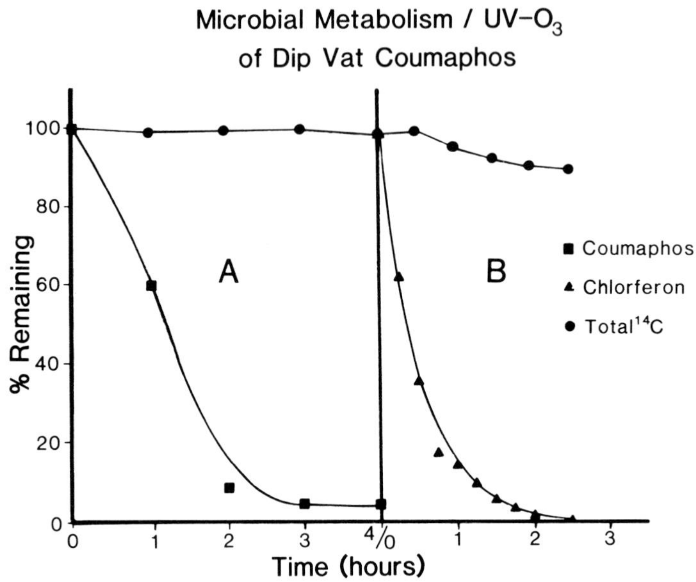

Figure 5. Effect of combining a microbial hydrolysis pretreatment (A) of dip-vat material with UV-ozonation (B) on coumaphos (A) and its primary hydrolysis product chlorferon (B). Total ^{14}C represents the amount of radioactivity from coumaphos labeled with ^{14}C in the benzene portion of the molecule.

than was coumaphos itselt (Fig. 5). This figure represents the complete microbial hydrolysis-UV-ozonation process where 1500 ppm dip-vat coumaphos is degraded by exposure to whole cells of *Flavobacterium* sp. ATCC 27551 for 4 h, followed by complete destruction of the chlorferon product by UV-ozonation for 3 h. Although the chlorferon was no longer detectable after this treatment, most of the carbon derived from the benzene moiety of chlorferon (as measured by ^{14}C label) remained in solution. Therefore, the oxidation of the chlorferon was incomplete and the products of this oxidation were probably other water-soluble organics. However, when this material was applied to topsoil, it was rapidly and completely oxidized to CO_2 and water by indigenous soil bacteria (Fig. 6). The metabolism of dip-vat coumaphos prior to UV-ozonation was required for this complete oxidation since there was little CO_2 release from coumaphos-containing dip-vat material that had been extensively treated in the UV-ozonation unit.

It is also interesting that the UV-ozonation process kills the *Flavobacterium* cells that were added to the dip-vat material for hydrolysis of coumaphos (Fig. 7).

Role of biotechnology in optimizing the coumaphos degradation process

The biological hydrolysis of coumaphos by *Flavobacterium* sp.

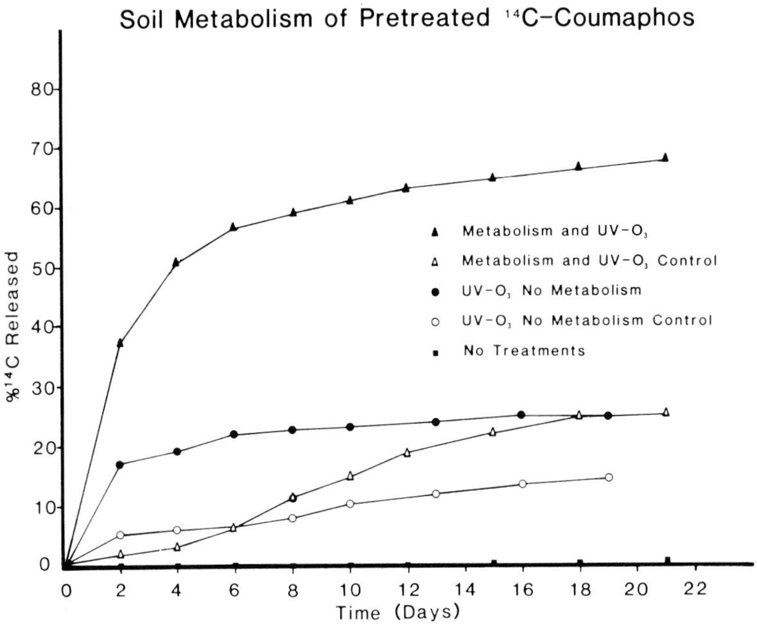

Figure 6. $^{14}CO_2$ release from treated dip-vat material applied to a Drummer soil (Urbana, ILL) that had been activated by incubation with 0.5% glucose and 0.01% yeast extract for 2 weeks prior to the addition of the treated dip-vat materials. Control flasks contained autoclaved soil (1 time for 25 min). No effort was made to keep controls sterile.

ATCC 27551 is very effective in degrading coumaphos to chlorferon. However, at this stage a fairly high cell density is needed to carry out the reaction in the time frame we feel would be required to allow for practical elimination of 12,000 L of waste coumaphos. One objective of our biotechnology program is to use recombinant DNA techniques to clone the gene that encodes the phosphotriesterase enzyme of *Flavobacterium*. Once we have isolated this gene, we will place it into plasmid DNA vectors, which will allow its expression at very high levels in such bacteria as *Escherichia coli* or *Bacillus subtilis*. These bacteria can be integrated into a mass production scheme. In this way we hope to maximize the efficiency and cost effectiveness of the production of the cells and(or) enzyme for use in the biological pretreatment of coumaphos-containing dip-vat material.

A parathion hydrolase enzyme from *Pseudomonas diminuta* was found to be encoded on a 44 Kbp plasmid (Serdar et al., 1982). The gene that encodes this enzyme has been cloned in *E. coli* (McDaniel et al., 1985; Serdar and Gibson, 1985) and is being sequenced. The expression of this cloned gene in *E. coli* is very poor. The poor expression is probably due to the fact that the gene was cloned in close

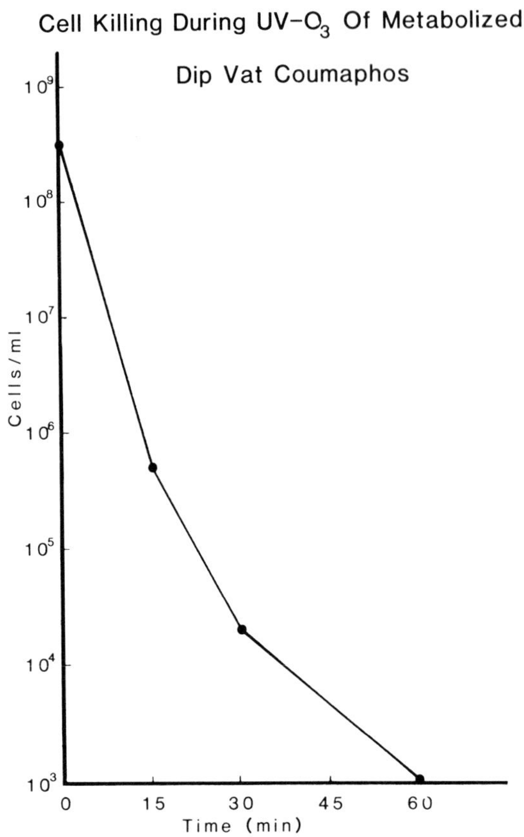

Figure 7. Effect of UV-ozonation process on organisms contained in *Flavobacterium*-treated dip-vat coumaphos.

proximity to the promoter of an antibiotic resistance gene present on the cloning vector (McDaniel, unpublished observations; Serdar and Gibson, 1985). The parathion hydrolase enzyme produced by *Flavobacterium* sp. ATCC 27551 is also plasmid-encoded. The native *Flavobacterium* strain contains several plasmids that appear to be of similar size. In the course of selecting streptomycin-resistant mutants for use in genetic studies, it was noted that about 50% of the streptomycin-resistant *Flavobacterium* colonies were no longer capable of hydrolyzing parathion. Such a high rate of loss of the parathion hydrolysis trait strongly suggests that the loss of this character is due to the curing of a plasmid rather than to the mutation of chromosomal genes. Analysis of intact plasmid DNA from both parathion hydrolase positive and negative strains indicated that curing of plasmids had indeed occurred (Fig. 8). Lanes 1, 3, 7, 8 and 10 contain DNA from streptomycin-resistant colonies that are unable to hydrolyze parathion

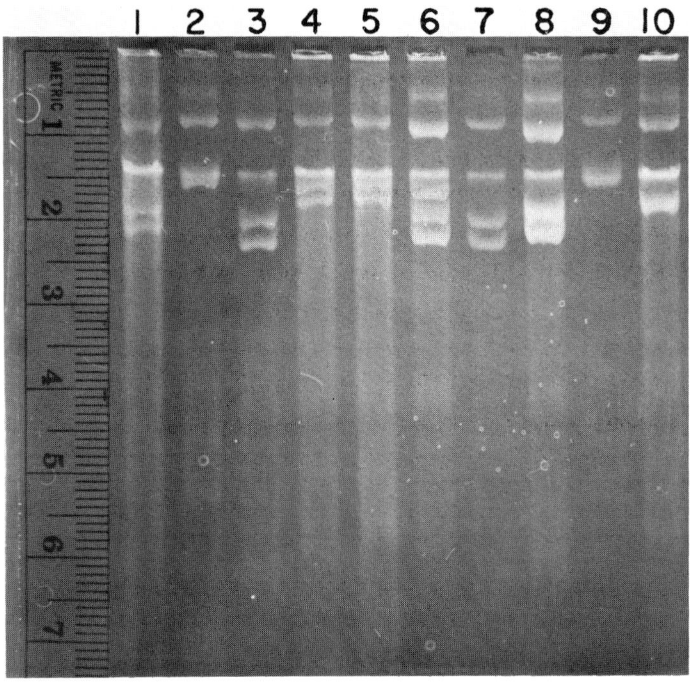

Figure 8. Electrophoresis of intact plasmid DNA from *Flavobacterium* sp. ATCC 27551 parathion hydrolase positive (lanes 2, 4, 5, 6, 9) and parathion hydrolase negative (lanes 1, 3, 7, 8, 10) isolates obtained after treatment with streptomycin on a 0.7% agarose gel containing 0.5 μg/mL ethidium bromide.

(par$^-$) while lanes 2, 4, 5, 6 and 9 contain DNA from streptomycin-resistant isolates that retain parathion hydrolase activity (par$^+$). The two smaller plasmids that migrated at approximately 1.9 and 2.1 cm cannot encode parathion hydrolase since they are present in some par$^-$ strains (lanes 1, 3 and 7) and are absent in some par$^+$ strains (lanes 2, 4, 5 and 9). Only the faint plasmid band that migrated at 1.6 cm is present in all par$^+$ strains (except lane 6, which looks as if it might have incurred a deletion in this plasmid, as it ran at about 1.7 cm) and was absent in all par$^-$ strains. A more noticeable difference in the plasmid DNA content of par$^+$ and par$^-$ strains is seen when the plasmid DNA is cut with the restriction enzyme Eco RI (Fig. 9). The cut DNA from par$^+$ strains in lanes 1, 2 and 3 display a unique Eco RI restriction fragment of about 7 Kb (migration distance 4.1–4.2 cm) that is absent in all par$^-$ strains. We feel that these data convincingly show that the parathion hydrolase of *Flavobacterium* sp. ATCC 27551 is plasmid-encoded. The lack of a transformation system for *Flavobacterium* and the lack of a positive selection procedure for phosphotriesterase activity

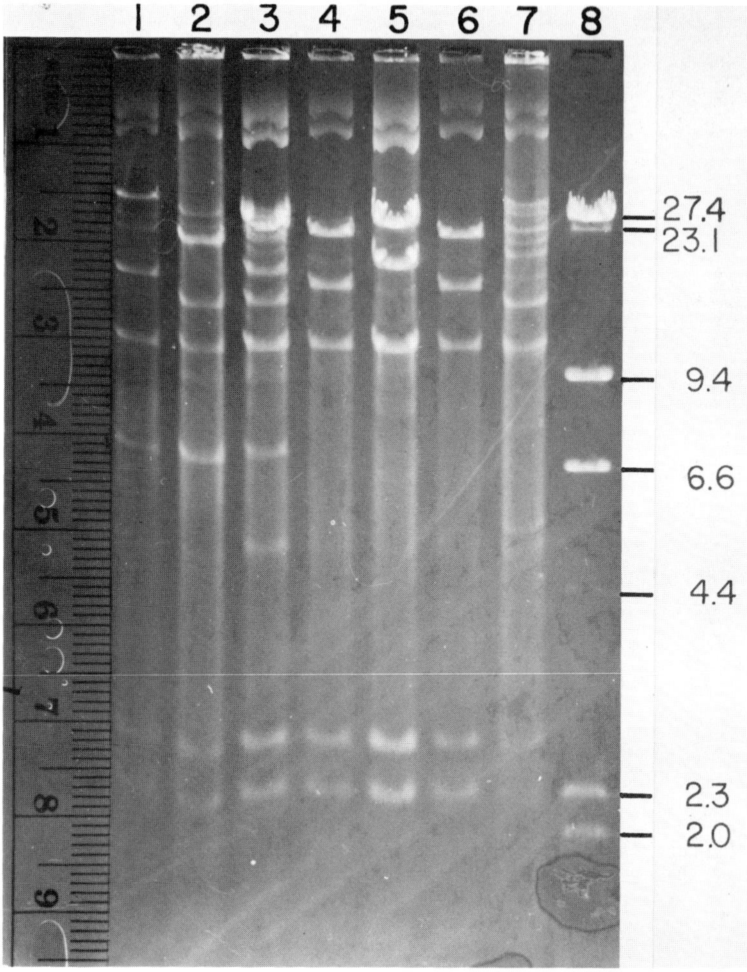

Figure 9. Agarose gel electrophoresis of Eco RI digested plasmid DNA from parathion hydrolase positive (lanes 1, 2, 3) and parathion hydrolase negative (lanes 4–7) isolates of *Flavobacterium* sp. ATCC 27551. Lane 8 contains a molecular weight standard of phage lambda DNA digested with Hind III; the fragment sizes are indicated in kilobase pairs.

have thus far prevented us from performing the necessary confirming experiments. We have cloned both Eco RI and Pst I fragments of the plasmid DNA from this organism into pBR322 and transformed *E. coli* with the resulting chimeric DNA. In both cases, several hundred colonies were screened for the presence of parathion hydrolase but no positive clones were found. Thus, we were probably not as fortunate as Serdar and Gibson (1985) or McDaniel et al. (1985) in inserting the gene close enough to a promoter on the cloning vector. In both cases the restriction fragments are much larger than those generated by digestion of the

P. diminuta plasmid. We are presently working at cloning the fragments of the *Flavobacterium* plasmid DNA directly into vectors with strong concensus promoters to see if proximity to a very strong promoter will result in expression of the parathion hydrolase at detectable levels in *E. coli*. We are also designing new expression vectors that have more potential cloning sites directly downstream from a strong promoter. We are also examining the possibility of using more than one restriction enzyme to digest the *Flavobacterium* plasmid DNA, and using synthetic oligonucleotide linkers to clone smaller fragments into expression vectors.

Investigations into the origin of enhanced pesticide metabolism in soils

Background. There is increasing evidence that a number of biodegradable pesticides currently in use are, at times, failing to control their target pests (Wilson, 1984). The reason for these performance failures is complex, and in some cases, may involve unusual climatic or soil conditions. However, convincing evidence is emerging which suggests that many of these failures are due to the appearance of soil microorganisms that rapidly metabolize certain soil-incorporated pesticides (Harris et al., 1984; Kaufman et al., 1985). Currently about 10 to 15 compounds are suspected of being susceptible to enhanced microbial metabolism in soil. We have chosen to study the degradation of the insecticide carbofuran as a potential model for determining the origin of enhanced metabolism in soil.

Our ultimate objective is to determine the genetic basis for the origin of problem soils. This includes determining whether the degradative enzymes are encoded by plasmid-borne genes and whether these genes can move between members of the soil microbial community through conjugation or other processes.

Microbial degradation of carbofuran

Carbofuran [2,3-dihydro-2,2-dimethyl-7-benzofuranyl methylcarbamate] is a soil-incorporated insecticide and nematicide that is widely used to control corn rootworm in the midwestern United States. Several authors have demonstrated that this compound is subject to enhanced degradation in many of the geographical areas in which it is used (Felsot et al., 1981; Reed, 1983; Venkateswarlu, Siddarame-Gouda, and Sethunathan, 1977; Williams, Pepin and Brown, 1976). We have isolated and described (Karns et al., 1985) a soil bacterium (*Achromobacter* sp.) that rapidly hydrolyzes the carbofuran molecule and uses the *N*-methyl carbamate side chain as a source of cellular nitrogen (Fig. 10). This organism seems to be many times more

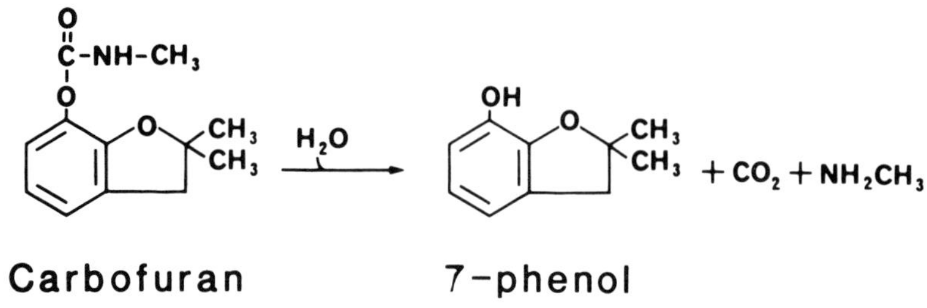

Figure 10. Proposed hydrolysis reaction of carbofuran carried out by carbofuran-degrading *Achromobacter* sp.

effective at degrading carbofuran than any organism previously reported (Fig. 11). The organism can attack a number of other *N*-methylcarbamate insecticides (Table 1), a laboratory result that mimics what some investigators have observed in the field (Harris et al., 1984). At this time we have no direct evidence to suggest that this *Achromobacter* bacterium is involved in the appearance of the problem soils

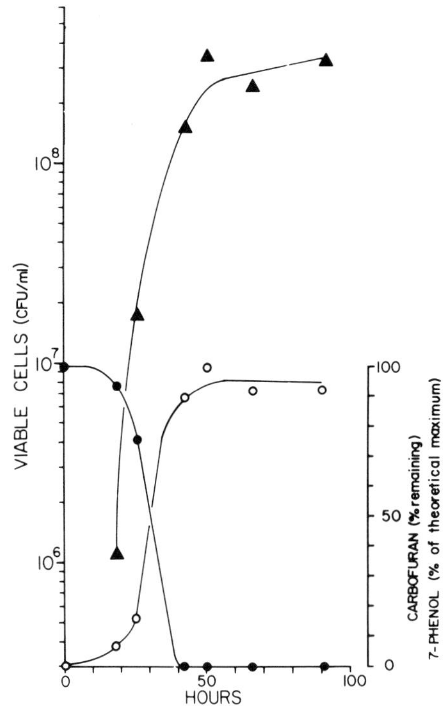

Figure 11. Growth of *Achromobacter* sp. WM111 in nitrogen-free basal salts medium containing glucose as carbon source and carbofuran as sole nitrogen source. (▲) Viable cells; (●) Carbofuran level; (○) 7-Phenol hydrolysis product of carbofuran.

Table 1. Degradation of *N*-methylcarbamate insecticides by resting cell suspensions of carbofuran-degrading *Achromobacter* sp

Compound[a]	% Degradation 1 h	2 h	5 h
Carbofuran	100	100	100
Carbaryl	58	93	100
Baygon	30	70	96
3, 5-Dimethylphenyl-*N*-methylcarbamate	95	96	96
Aldicarb	20	44	80

[a] Pesticides were added to resting cell suspension to a final concentration of 100 μg/mL. Levels were determined by HPLC of culture fluids and are reported as percent of that present at the time of initiation of the experiment.

phenomenon. However, it is our contention that the study of the biochemistry and genetics of carbofuran metabolism in this organism will lead to a better understanding of the means by which new degradative functions evolve in the field.

Role of biotechnology in the problem soils phenomenon

We have a cooperative agreement with the FMC Corporation, the manufacturer of carbofuran, to isolate and purify the enzyme or enzymes responsible for the degradation of carbofuran. Although this project has just started and the results are very preliminary, we have been successful in assaying the enzymatic activity in crude lysates of the *Achromobacter* and have carried out some initial purification steps. Our agreement with FMC is a good example of industry-government cooperation on a problem of mutual interest and high visibility. The FMC Corporation hopes to gain some knowledge that will help protect this compound in the field. We hope to be able to exploit the degradative properties of this enzyme for use in eliminating *N*-methylcarbamate insecticide wastes.

We also are attempting to clone the gene or genes that encode the enzyme(s) required for carbofuran degradation. With a cloned gene we hope to be able to probe the DNA of the *Achromobacter* isolate to determine whether the genetic information for carbofuran degradation is contained on the chromosomal DNA or on the very large plasmid (greater than 100 kilobases in size) we have found in this organism. This information may help us to determine how new degradative functions evolve and how they spread among the microbial populations of soil.

The plasmid from the carbofuran-degrading *Achromobacter* does not run as a distinct band in its native form (lane 4) but when cut with restriction enzymes Eco RI (lane 1), Bam HI (lane 2), or Hind III (lane 3) gives many discrete bands whose cumulative molecular size is greater

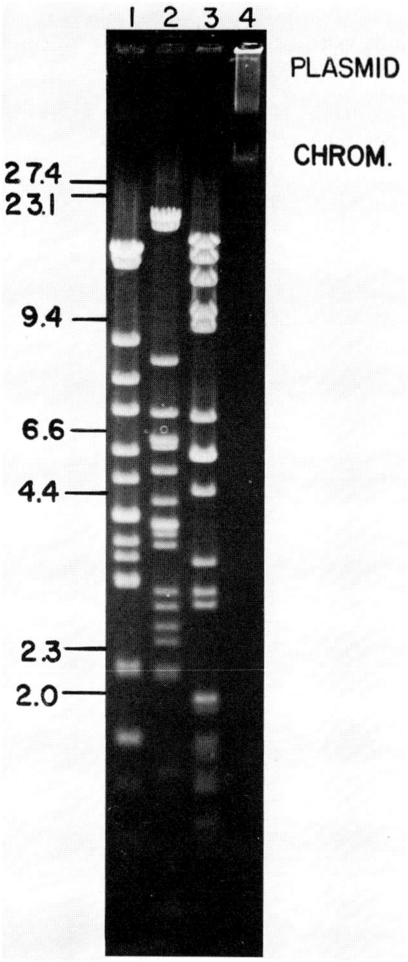

Figure 12. Agarose gel electrophoresis of DNA obtained from the carbofuran degrading *Achromobacter* sp. WM111. Lane 1 – Eco RI digested; Lane 2 – Bam HI digested; Lane 3 – Hind III digested; and Lane 4 – untreated. The numbers to the left indicate the migration distance of the fragments in a Hind III digest of phage lambda included in the same gel and indicate the size of the fragments in kilobase pairs.

than 100 Kb (Fig. 12). We have been unable to link the carbofuran degradation capability with the presence of this plasmid since the conventional techniques for plasmid curing or mutant generation have been unsuccessful with this organism. We have cloned the total DNA (plasmid and chromosome) from the *Achromobacter* into the cosmid vector pCP13 (Friedman et al., 1982). To do this, we partially digested the DNA to be cloned with the restriction enzyme Eco RI and mixed the resulting fragments of linear DNA with Eco RI-digested pCP13. The resulting DNA concatemers were packaged into lambda phage heads using a commercially available packaging kit. *Escherichia coli* was then infected with the lambda phage, and cells containing pCP13 and cloned DNA fragments were selected by growth on plates containing tetracycline (pCP13 encodes tetracycline resistance).

Screening recombinant DNA (r-DNA)-containing organisms for the presence of the carbofuran degradative genes requires a different approach from that described previously for the parathion hydrolase gene. In this case, the r-DNA plasmids contained in the *E. coli* clone bank were transferred into *P. putida* through conjugation, and tetracycline-resistant *P. putida* were plated onto nitrogen-free basal salts medium (NFB) containing glucose and carbofuran. *Pseudomonas putida* is capable of utilizing any methylamine generated by the hydrolysis of carbofuran as a sole nitrogen source; thus *P. putida* containing a r-DNA plasmid encoding the enzyme that hydrolyzes carbofuran should grow on these plates. This experiment is currently in progress. Again, a positive result requires that the gene encoding the carbofuran hydrolase enzyme be expressed in *P. putida*.

Conclusions

Biotechnology offers the potential for very effective and perhaps cost-efficient methods of solving agricultural problems in the area of chemical waste disposal. The combination of biological and physical methods as described for coumaphos disposal may greatly expand the range of chemicals that can be treated. There are many more contributions that biotechnology can make in this area. With such techniques as site-directed mutagenesis, it may be possible to expand the substrate range of existing enzymes so that new classes of agricultural chemicals can be more readily degraded. The genes encoding useful enzymes may be cloned into organisms for which mass production methods already exist, so that large amounts of the enzymes can be produced at low cost. The genes may also be cloned into organisms that are better equipped to endure unique environmental conditions, such as low or high pH, increased salinity, or high temperature. The techniques usually associated with biotechnology will also be useful in developing an understanding of how field conditions, such as those seen in the problem soils phenomenon, are developed. It is clear that there is much work to be done in this area and that the potential rewards are tremendous.

References

Felsot, A., J. V. Maddox and W. Bruce. 1981. *Enhanced microbial degradation of carbofuran in soils with histories of carbofuran use*. Bull. Environ. Contam. Toxicol. 26:781–788.

Friedman, A. M., S. R. Long, S. E. Brown, W. J. Buikema and F. M. Ausubel. 1982. *Construction of a broad host range cosmid cloning vector and its use in the genetic analysis of Rhizobium mutants*. Gene 18:289–296.

Hall, C. V. 1984. *Pesticide waste disposal in agriculture*. in R. G. Krueger, and J. N. Seiber, eds. *Treatment and Disposal of Pesticide Wastes*. ACS Symp. Ser. 259.

Harris, C. R., R. A. Chapman, C. Harris and E. M. Tu. 1984. *Biodegradation of pesticides in soil: Rapid induction of carbamate degrading factors after carbofuran treatment*. J. Environ. Sci. Health B19:1–11.

Karns, J. S., W. W. Mulbry, J. O. Nelson and P. C. Kearney. 1985. *Metabolism of carbofuran by a pure bacterial culture*. Pest. Biochem. Physiol. (In press).

Kaufman, D. D., Y. Katan, D. F. Edwards and E. G. Jordan. 1985. *Microbial adaptation and metabolism of pesticides*. Pp. 437–451 in J. L. Hilton, ed., *Agricultural Chemicals of the Future. Beltsville Symposia in Agricultural Research 8*. Rowman & Allanheld, Totowa, N.J.

Kearney, P. C., J. R. Plimmer and Li, Z-M. 1983. *UV-ozonation and land disposal of aqueous pesticide wastes*. Pp. 397–400 in J. Miyamoto, ed., *Pesticide Chemistry-Human Welfare and the Environment, Vol. 4*. Pergamon Press, Oxford.

Kearney, P. C., Q. Zeng and J. M. Ruth. 1984. A large scale UV-ozonation degradation unit – Field trials on soil pesticide waste disposal. In R. F. Krueger and J. M. Seiber, eds., *Treatment and Disposal of Pesticide Wastes*. ACS Symp. Ser. 259.

McDaniel, C. S., J. R. Wild and G. A. O'Donovan. 1985. *Cloning and characterization of the plasmid-encoded gene for parathion degradation from Pseudomonas diminuta*. J. Bacteriol. (In press).

Munnecke, D. M. 1976. *Enzymatic hydrolysis of organophosphate insecticides, a possible pesticide disposal method*. Appl. Environ. Microbial. 32:7–13.

Munnecke, D. M. and D. P. H. Hsieh. 1974. *Microbial decontamination of parathion and p-nitrophenol in aqueous media*. Appl. Microbiol. 28:212–217.

Nelson, L. M. 1982. *Biologically induced hydrolysis of parathion in soil: Isolation of hydrolyzing bacteria*. Soil Biol. Biochem. 14:219–222.

Reed, D. C. 1983. *Enhanced microbial degradation of carbofuran and fensulfothion after repeated application to acid mineral soil*. Agric. Ecosys. Environ. 10:37–46.

Serdar, C. M. and D. T. Gibson. 1985. *Enzymatic hydrolysis of organophosphates: Cloning and expression of a parathion hydrolase gene from Pseudomonas diminuta*. Biotechnology 3: 567–571.

Serdar, C. M., D. T. Gibson, D. M. Munnecke and J. H. Lancaster. 1982. *Plasmid involvement in parathion hydrolysis by Pseudomonas diminuta*. Appl. Environ. Microbiol. 44:246–249.

Sethunathan, N. and T. Yoshida. 1972. *Conversion of parathion to paranitrophenol by diazinon-degrading Flavobacterium sp*. Proc. Inst. Environ. Sci. 18:255–257. New York.

Siddaramappa, R., K. P. Rajaram and N. Sethunathan. 1973. *Degradation of parathion by bacteria isolated from flooded soil*. Appl. Microbiol. 26:846–849.

Venkateswarlu, K., T. K. Siddarame-Gouda and N. Sethunathan. 1977. *Persistence and biodegradation of carbofuran in flooded soils*. J. Agric. Food Chem. 25:533–536.

Williams, I. H., H. S. Pepin and M. J. Brown. 1976. *Degradation of carbofuran by soil microorganisms*. Bull. Environ. Contam. Toxicol. 15:244–248.

Wilson, R. G. 1984. *Accelerated degradation of thiocarbamate herbicides in soil with prior thiocarbamate herbicide exposure*. Weed Sci. 32:264–268.

Genetic and molecular basis of the microbial degradation of herbicides and pesticides

PAUL H. TOMASEK, BETSY FRANTZ, D. K. CHATTERJEE and A. M. CHAKRABARTY
Department of Microbiology and immunology, University of Illinois at Chicago; Chicago, IL 60680, U.S.A.

Abstract The genes for the degradation of chlorinated compounds normally used as herbicides or pesticides, such as various chlorinated benzoic acids, 2,4-dichlorophenoxyacetic acid, 2,4,5-trichlorophenoxyacetic acid, etc., are usually found to be clustered on various plasmids such as pJP4 and pAC27. The clustering of the degradative genes are allowed their cloning in broad host-range vectors, as well as the sequencing of parts of the gene cluster such as that involved in the degradation of 3-chlorobenzoic acid. The promoters involved in the efficient expression of these genes appear to be substantially different from those in *Escherichia coli*. Plasmid gene sequences of the 2,4,5-trichlorophenoxyacetate-degrading *Pseudomonas cepacia* AC1100 strain demonstrate substantial homology to a cluster of genes present on the 2,4-dichlorophenoxyacetate degradative plasmid pJP4. There is little AC1100 DNA homology to the chlorocatechol genes present on the 3-chlorobenzoate degradative plasmid pAC27. The homology to the 2,4-dichlorophenoxyacetate degradative plasmid pJP4 may therefore be attributed to other genes involved in the degradation of chlorophenoxyacetates. Transposon mutagenesis has shown that at least some of the 2,4,5-trichlorophenoxyacetate genes are chromosomal. A short chromosomal DNA sequence associated with a putative catabolic gene is repeated many times on both the chromosome and plasmid of strain AC1100. Such a repeated sequence may be important in the evolution of the degradative plasmids as well as in genetic rearrangements that often allow efficient degradation of novel substrates.

Introduction

Production and usage of chlorinated herbicides, pesticides and other agrochemicals, while important for agricultural productivity, have also led to massive pollution problems because of the persistence of such chemicals in the environment (Schneider, 1979). The persistence of the chemicals can generally be traced to the presence of chlorine atoms in such molecules in as much as chlorinated compounds, particularly chlorinated aromatics, occur infrequently in nature and native microorganisms demonstrate a limited ability to cleave the carbon-chlorine bonds from highly chlorinated compounds. Yet, attempts to isolate bacterial strains capable of growing with simple chlorinated compounds as their sole sources of carbon and energy have occasionally been successful, particularly for compounds such as 3-chlorobenzoic acid (3Cba), 2,4-dichlorophenoxyacetic acid (2,4-D), etc. This presumably means that microorganisms in nature, in response to the release of large quantities of such compounds, have evolved the capability to biodegrade many of these compounds. Since the synthetic herbicides and pesticides have been manufactured and released in the environment during the last four decades or so, it is likely, but not certain, that the evolution of such biodegradative genes has occurred during the last four decades. A study of the molecular and genetic basis of the biodegradation of chlorinated herbicides and pesticides may therefore provide an insight into the mode of evolution of new biochemical pathways.

Cloning of the genes specifying degradation of 3-chlorobenzoic acid from plasmids pAC27 and pJP4

Considerable effort has been directed towards determining the relatedness and regulation of expression of the 3Cba genes present on two different plasmids, pAC27 and pJP4. The degradative pathways coded by these two plasmids are given in Fig. 1. pJP4 is a P1-incompatibility (inc) plasmid which encodes resistance to mercury, dissimilation of 2,4-D and 3Cba while pAC27 plasmid does not belong to any known inc group and encodes degradation of 3Cba and 4Cba (Don and Pemberton, 1981; Chatterjee and Chakrabarty, 1982). The physical map of pAC27 has been constructed by Ghosal et al. (1985a), while that of pJP4 has been developed by Weightman et al. (1984) and by Don and Pemberton (1985). These maps are shown in Fig. 2. The 3Cba genes of the plasmid pAC27 have been located on a 21 kilobase (Kb) EcoR1 (EcoR1-B) segment cloned into the cosmid pLAFRI in *E. coli.* The 3Cba^{+} character was subsequently selected for by mobilizing the recombinant plasmid, pDC10, from *E. coli* to *P. putida* (Ghosal et al., 1985a). Subcloning of pAC27 BglII fragments in pRK290 as a vector demonstrates that the 4.2 kb BglII fragment will allow growth of *P. putida* with 3Cba. At least three chlorocatechol degradative genes and a regulatory gene are expected to be present on this 4.2 kb fragment. The two enzymes, benzoate oxygenase and dihydroxyhydrobenzoate dehydrogenase catalyzing the conversion of 3Cba to 3-chlorocatechol are believed to be encoded by the host chromosomal genes (Weightman et al., 1984). We have received from Dr. Teruko Nakazawa a mutant defective in benzoate oxygenase. Transfer of pDC25 containing the 4.2 Kb BglII-E fragments into the mutant does not allow growth on either benzoate or 3Cba, suggesting that the chromosomally encoded enzyme is essential for conversion of 3Cba to 3-chlorocatechol. Presence of regulatory gene(s) in this 4.2 kb fragment is indicated by the fact that induction of 3Cba oxidation and release of chloride ions take place only when the pDC25 cells (Fig. 2) are grown in presence of 3Cba. Growth in presence of succinate or benzoate does not elicit either 3Cba oxidation or release of chloride ions.

There appears to be additional regulatory gene(s) not contained on the BglII-E fragment of pAC27. This is suggested by the rapid growth

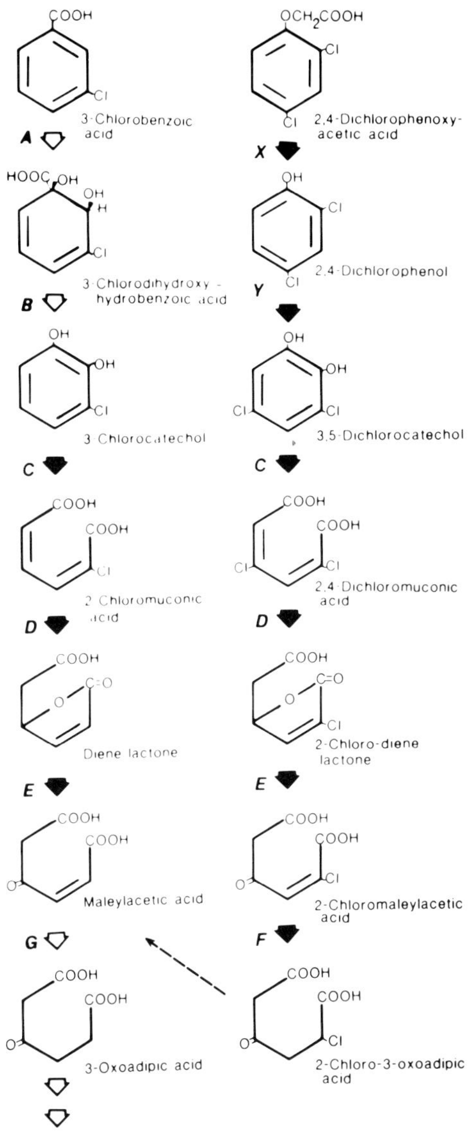

Figure 1. Proposed pathways for the degradation of 3CBA and 2,4-D. Solid arrows indicate plasmid-encoded enzymes while open arrows represent chromosome-encoded enzymes in an appropriate host. The plasmid-encoded pyrocatechase II, and cycloisomerase II have broader substrate specificities than their chromosomal isofunctional counterparts and thus catalyze reactions with chlorinated substrates, Hydrolase II specifically recognizes chlorinated substrates. Dechlorination of chloromuconates is thought to occur spontaneously as the result of lactonization by cycloisomerase II. A-Benzoate Dioxygenase, B-Dehydrogenase, C-Pyrocatechase II, D-Cycloisomerase II, E-Hydrolase II, F-Chloromaleylacetate Reductase, G-Maleylacetate Reductase, X-2,4-D Monooxygenase, Y-2,4-Dichlorophenol Monooxygenase (after K. N. Timmis et al., 1985).

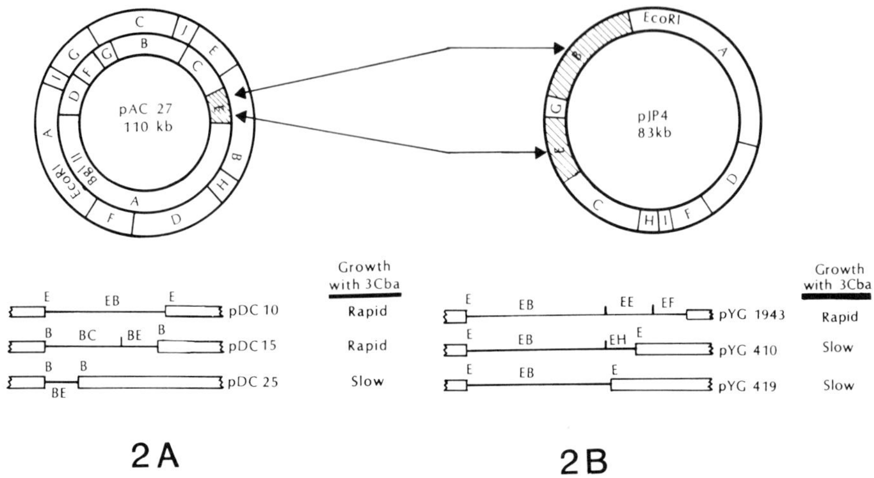

Figure 2. Physical maps of the plasmid pAC27 with EcoR1 and BglII and pJP4 with EcoR1. Regions of homology are indicated by shaded areas. Lines below plasmid maps indicate cloned restriction fragments of the respective plasmids (single line) in broad host range vectors (double lines). The ability of such recombinant plasmids to allow growth of the lost cells with 3Cba is indicated. pDC10, pYG1943, pYG410, and pYG419 are recombinant plasmids with pLAFRI as the vector; pDC15 and pDC25 contain the pRK290 vector (after Ghosal et al., 1985a).

on 3Cba of clone DC15 (containing BglII fragments E + C) compared to the slow growth of clone DC25 (lacking the BglII-C fragment, see Fig. 2). Continued selection on 3Cba plates allowed DC25 to grow more rapidly in 3Cba-containing medium. Electron microscopic contour length determination and restriction analysis of the plasmid DNA from these DC25 colonies show approximately 8–10 fold amplification of the 4.2 kb BglII-E fragment. Amplification may be necessary to compensate for the absence of a positively-acting activator protein (Ghosal et al., 1985a), presumably coded by a gene on the BglII-C fragment.

Cloning of the 3Cba degradative genes from pJP4 was attempted using the cosmid pLAFRI as a vector with EcoR1-treated pJP4 fragments. Two types of 3Cba$^+$ colonies were observed on *en masse* mobilization of the recombinant plasmids from *E. coli*; slow growing colonies that produced brown coloration (clones YG410 and YG419) on 3Cba plates, and normal rapidly growing 3Cba$^+$ colonies (YG1943, see Fig. 2). Isolation of plasmid DNA from the slow growing YG419 showed the presence of only the pJP4 EcoR1-B fragment along with the vector (Ghosal et al., 1985b) while the plasmid DNA from rapid growing 3Cba$^+$ clone YG1943 showed the presence of fragments EB, EE and EF. The alignment of these fragments on the map of pJP4 is shown in Fig. 2. The EcoR1-B fragment from clone YG419 appeared to have a higher intensity than the vector DNA and electron microscopic contour length measurement demonstrated the presence of plasmid species of about 200 kb size. The EcoR1 fragment B in YG419 must therefore have undergone extensive amplification. On the other hand, the plasmid pYG1943 isolated from clone YG1943 demonstrated the presence of a single plasmid species of about 48 kb size, which was non-amplified and showed the presence of fragments, EB, EE and EF of pJP4 on EcoR1 digestion (Ghosal et al., 1985b).

Southern hybridization between various restriction fragments of plasmid pJP4 and pAC27

Since both plasmid pAC27 and pJP4 encode chlorocatechol degradation, it was of interest to determine if they exhibit some degree of homology. In order to localize the area of homology, fragments EB, EE and EF of pJP4 were individually purified, nick-translated and used as probes to check the homology with EcoR1, BglII and HindIII digested pAC27 (Ghosal et al., 1985c). The pJP4 fragment EF did not show any homology while EB and EE showed significant homology with only the 4.2 kb BglII-E fragment of pAC27. Control experiments with EcoR1 fragments EA, EC, ED and EG of pJP4 did not show any hybridization with any of the pAC27 fragments. In a parallel experiment, the same 4.2 kb BglII-E fragment of pAC27 was purified and used as a probe to check for homology with EcoR1-BamH1 digested pJP4. The homology was restricted to a 10 kb region of EB and proximal 3 kb region of EE of pJP4 (Ghosal et al., 1985b). These results in combination with cloning data in which two of the $3Cba^+$ clones from pJP4 library contain only the EB fragment in addition to vector pLAFRI, suggest that all three structural genes needed for chlorocatechol degradation are present on the EB fragment and the regulatory function(s) are presumably provided by the EE fragment. These two homologous segments are continuous in pAC27 as part of the 4.2 kb BglII fragment E whereas they are interrupted by about 2 kb of DNA (fragment EG) in pJP4.

Partial sequencing of the 4.2 kb BglII-E fragment: presence of regulatory signals and open reading frames

We have initiated sequencing of the 4.2 kb BglII-E fragment which contains all the chlorocatechol degradative genes. A restriciton map of this fragment and the sequencing strategy is shown in Fig. 3. The partial sequence obtained so far demonstrates the presence of 3 open reading frames ORF 1, 2, and 3 (Fig. 3) corresponding to polypeptide products of sizes 17, 30 and 11 kilodaltons.

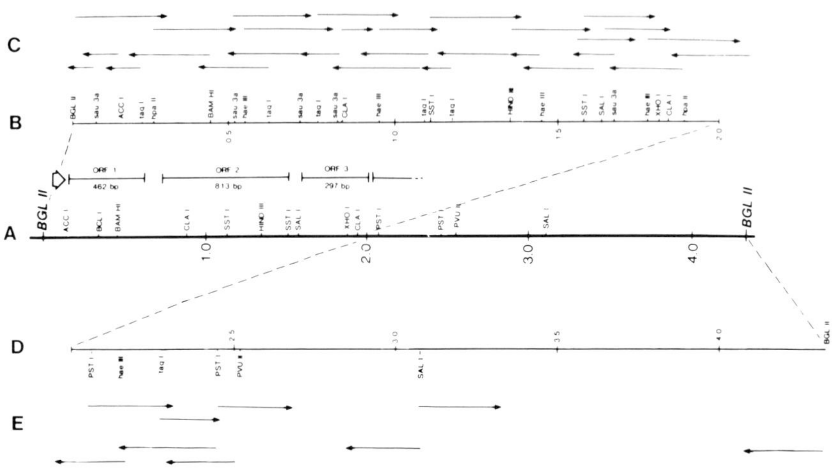

Figure 3. Restriction map and dideoxy DNA sequencing strategy for the 4.2 kb BglII-E fragment of pAC27 which encodes chlorocatechol degrading functions. Line A shows the 4.2 kb BglII fragment with major restriction sites. Line B is an enlargement of the first 2 kb of the BglII fragment showing only those restriction sites used in making M13mp18 and mp19 sequencing clones. Lines at C show the direction of sequencing of the first 2 kb of DNA. Line D is an enlargement of the second 2.2 kb of DNA with restriction sites used for cloning, and lines at E show the direction of sequencing.

An important question in the study of regulation of the 3Cba degradative genes is whether they are clustered as a single or multiple regulons (operons), with operator/promoter sites present on the 4.2 kb BglII-E fragment. It is therefore important to determine the size(s) of the messenger RNA(s) formed in vivo when *P. putida* cells harboring the plasmid pDC25 are grown in presence and in absence of 3Cba. Since the enzymes are known to be inducible, transcripts hybridizable to the 4.2 kb BglII-E fragment are likely to be detected only when the RNA preparation is isolated from 3Cba-grown cells. The nature of transcripts detected by Northern blotting (Alwine, Kemp and Stark, 1977) demonstrated the presence of two bands; control cells grown with glucose-Tc (pDC25$^+$) or glucose alone (pDC25$^-$) showed no apparent RNA species hybridizable to the BglII-E fragment. When RNA was isolated from pDC25$^+$ cells grown in presence of 3Cba and hybridized to the BglII-E fragment, two transcript bands corresponding to 2.3 and 1.7 kb size were visible. Since none of the bands are visible in absence of 3Cba growth, it appears that two major transcripts are formed during expression of the 3Cba genes.

Localization of transcription initiation site by S1 nuclease mapping: detection of a promoter by the use of a promoter probe

Since two transcripts have been detected hybridizable to the 4.2 kb BglII-E fragment, it was of interest to determine if there is a promoter

Table 1. List of plasmids encoding degradation of some halogenated compounds

Plasmid	Degradative pathway	Molecular size (Mdal)	Reference
pU01	Fluoroacetate	43	Kawasaki et al. (1981)
pAC21	p-chlorobiphenyl	65	Kamp and Chakrabarty (1979)
pKF1	p-chlorobiphenyl	53	Furukawa and Chakrabarty (1982)
pJP2	2,4-D	36	Don and Pemberton (1981)
pJP4	2,4-D chlorobenzoate	56	Don and Pemberton (1981)
pWR1 (pB13)	3-chlorobenzoate	72	Reineke and Knackmuss (1979)
pAC25	3-chlorobenzoate	76	Chatterjee et al. (1981)
pAC27	3-chlorobenzoate 4-chlorobenzoate	72	Chatterjee and Chakrabarty (1983)
pAC31	3-chloro-, 4-chloro- and 3,5-dichloro-benzoate	72	Chatterjee and Chakrabarty (1982)
No designation	2,6-dichlorotoluene	63	Vandenbergh et al. (1981)

sequence on the BglII-E fragment and to determine the transcription initiation site. For the latter, we used a modified S1 nuclease protection mapping (Miles and Guest, 1984). For sequencing purposes, we had already cloned the 1.3 kb BglII-HindIII (Fig. 3) fragment into the BamH1-HindIII polylinker sites of M13mp18. Since there is a BamH1 site 421 bases downstream of the BglII site on the BglII-E fragment (Fig. 3) and an EcoR1 site 21 base upstream of this site (on the mp18 polylinker), it was possible to reclone the 442 bp BamH1-EcoR1 fragment into mp18 at the BamH1-EcoR1 sites.

This clone was used to make a single stranded M13 sense strand template from which a labeled antisense strand was generated using the M13 primer extension procedure with radioactive deoxy CTP (Burke, 1984). A 15 base primer was used which hybridized with the sequence 47 bases from the BamH1 site. After the formation of the labeled antisense strand, the double stranded DNA was cut with EcoR1 and run on a 5% strand separating gel. The labeled fragment which consisted of 504 bases (442 bp BamH1-EcoR1 fragment, 47 bases intervening segments + 15 base primer) was eluted from the gel.

The labeled 504 base strand was used in the hybridization of RNA (Ebina and Nakazawa, 1983) isolated from pDC25$^+$ cells grown with either 3Cba or glucose-Tc. No RNA-DNA hybrid was detected when RNA from glucose-Tc grown cells was used. The RNA-DNA hybrid from 3Cba grown cells was treated with S1 nuclease. The double stranded hybrid was precipitated, denatured, and run on an 8% polyacrylamide denaturing (sequencing) gel along with dideoxy reactions using the mp18 442 bp BamH1-EcoR1 fragment as a sequencing ladder. The RNA-DNA hybrid cut with S1 nuclease in this manner showed a reduction in size from the probe fragment of 132 bases, 62 (15 + 47) of which were primer and intervening sequences from the 5′ end, and 70 bases from the 3′ end. Since 21 of the 132 bases were from the mp18 vector, we concluded that the first base of the message originated approximately 49 bases 3′ from the BglII-site.

For determining the presence of any promoter sequence within the BglII-HindIII fragment, a promoter probe (a XhoI fragment containing *xyl*E gene without a promoter) was used. *xyl*E gene, when inserted downstream of a promoter and expressed, codes for catechol 2,3-dioxygenase, which in the presence of catechol would convert this substrate to the intensely yellow compound of α-hydroxymuconic semialdehyde. Spraying the colonies with catechol and screening for yellow colonies thus allow easy recognition of the expression of *xyl*E gene in the cells of a given colony. The cloning of the promoterless *xyl*E gene downstream of the 1.3 kb BglII-HindIII segment of pAc27 allowed production of intense yellow color upon being sprayed with catechol while other transformed colonies remained colorless because of an opposite orientation of this fragment. The presence of the *xyl*E gene downstream of the 3Cba fragment has been confirmed by plasmid DNA isolation and restriction mapping.

Nature of promoter sequences in Pseudomonas species

It is known that *Pseudomonas* genes are poorly expressed in *E. coli* and other bacteria (Nakai et al., 1983), although nothing is known about the nature or sequence specificity of *Pseudomonas* promoters. In a detailed study conducted on the nucleotide sequence surrounding transcription initiation site of *xyl*ABC operon of the TOL plasmid of *P. putida*, Inouye et al. (1984a) failed to find the consensus sequence of *E. coli* promoters in either the −10 region (Prinbnow box) or the −35 region (RNA polymerase recognition site). The *Pseudomonas* promoters could not be identified since no information was available regarding the promoter or other signals necessary for the initiation of transcription in *Pseudomonas*.

Mermod et al. (1984) recently demonstrated that transcription of the TOL plasmid-borne *meta* cleavage operon, which is positively regulated by the *xyl*S gene product, is determined by two overlapping promoters whose activities are coordinately regulated by the *xyl*S product. The isolation and analysis of constitutive mutants enabled identification of newly created *xyl*S product-independent promoters. Comparison of the promoter sequence with those of *E. coli* suggests a tentative consensus sequence as shown below:

	−35		−10	
E. coli	TTGACA	>15–21<	TATAAT	
P. putida	A-AGGC-T	>7–12<	GC$^{T}_{A}$ATA	>3–7<A

It is interesting to note in this context that the data on sequences upstream of the transcription initiation site of *xyl*ABC operon and *xyl*DEGF operon (Inouye et al., 1984a; Inouye et al., 1984b) demonstrate a putative promoter sequence at −40 to −60 region which is somewhat similar to a sequence in the upstream region of the 3Cba gene cluster, as shown below:

```
           -70         -60         -50          -40          -30          -20          -10          +1
xylABC      AAAATCAATAATTTAGATGAAATAAGGGGATCGGTATAAGCAATGGCATGGCGGTTGCTAGCTATACGAGA
xylDEFG    TTGCAAGAAGCGGATACAGGAGTGCAAAAAATGGCTATCTCTAGAAAGGCCTACCCCTTAGGCTTTATGCA
3Cba                                 AGATCTTGAAAGGAGACGAGTCATGGATAAACGAGTTGCCGAGGTCGCAG
```

Genetics of 2,4-D degradation

A number of different 2,4-D degrading bacterial strains have been isolated (Don and Pemberton, 1981; Bollag et al., 1968; Evans et al., 1971). The most extensively studied strain is the *Alcaligenes eutrophus* strain JMP134, carrying the plasmid pJP4. In addition to carrying the pyrocatechase II, cycloisomerase II, and hydrolase II genes (C, D, E; Fig. 1) which allow chlorocatechol metabolism, the pJP4 plasmid also encodes the two initial pathway enzymes, 2,4-D monooxygenase and 2,4-dichlorophenol monooxygenase (X, Y; Fig. 1), as well as 2-chloromaleylacetate reductase (F; Fig. 1). With the possible exception of the latter enzyme, most 2,4-D catabolic enzymes are believed to be encoded by the 15 kb EcoR1-B fragment of pJP4. This fragment has been cloned (Ghosal et al., 1985a), and transfer of the recombinant plasmid to a plasmid-free *A. eutrophus* JMP134 derivative allowed growth on 2,4-D and 3Cba, provided selection pressure was maintained (Timmis et al., 1985; Don et al., 1985). There have been recent attempts to further localize the genes within this cluster. Liu and Chapman (1985) recently localized the pyrocatechase II (dichlorocatechol dioxygenase) gene to a 4.9 kb PstI restriction fragment of pJP4. They also reported that the hydrolase II (chloro-diene lactone hydrolase) and the chloromaleylacetate reductase genes were both present on a 3.4 kb PstI fragment of pJP4. The relationship of these fragments to the established physical map of pJP4, however, remains unknown.

Genetics of the 2,4,5-T degrading strain

Unlike the 3Cba and 2,4-D degrading genes, little is known concerning the organization of the 2,4,5-T degradative genes present in a 2,4,5-T$^+$ strain of *P. cepacia* AC1100 (Kilbane et al., 1982). Determination of the role of plasmid genes encoding 2,4,5-T biodegradative enzymes has been hampered by the presence of several different plasmids and their variable yield from AC1100. Through electron microscopic measurements it has been determined that a 170 kb plasmid (pDG3) comprises about 85% of the total plasmid population. A 40 kb plasmid (pDG4) accounts for 10% of the plasmid population while the remainder is made up of a heterogeneous mixture of plasmids ranging from 3 to 30 kb (Ghosal et al. 1985c).

In order to determine if there is appreciable homology between the 2,4-D and 2,4,5-T degradative genes, the EcoR1-digested pJP4 DNA fragments were hybridized with the chromosomal and plasmid DNA isolated from AC1100. A 4 kb segment of the plasmid DNA from AC1100 showed appreciable homology with the pJP4 EcoR1-B fragment which is known to harbor the chlorocatechol degradative genes, and presumably other 2,4-D degradative genes (Ghosal et al., 1985a). This fragment, however, failed to show appreciable homology with the pAC27 BglII-E fragment, which also encodes chlorocatechol degradation, suggesting that the homology with pJP4 EcoR1-B fragment could be due to other portions of the 2,4-D degradative pathway, or to genes of unknown functions. In order to determine the location of other 2,4,5-T degradative genes, transposon Tn5 insertion mutants, which were defective in 2,4,5-T metabolism, were isolated from AC1100. Several mutant classes were defined based on phenotypic differences. The kanamycin resistance marker of one such transposon insertion mutant was then used to clone chromosomal DNA sequences flanking the site of insertion. Total genomic DNA was partially digested with Sau3A restriction endonuclease. The large (20–40 kb) fragments generated were cloned into the vector pHC79, and the recombinant plasmids introduced into *E. coli*. Kanamycin resistant colonies were used to identify clones containing the Tn5 and the flanking AC1100 chromosomal DNA. Using radiolabeled Tn5 DNA as a probe, two clones which shared a common 6 kb SalI restriction fragment with the original mutant chromosome were identified. This 6 Kb DNA fragment contained approximately half of the Tn5 sequence including the kanamycin resistance gene. The 6 kb SalI fragment was isolated from one of the clones and used to probe native AC1100 chromosomal digests in an effort to localize the putative 2,4,5-T structural gene. As shown in Fig. 4A, the 6 kb probe identified several homologous regions on both the *P. cepacia* AC1100 chromosomal (lanes 1, 4, 7) and

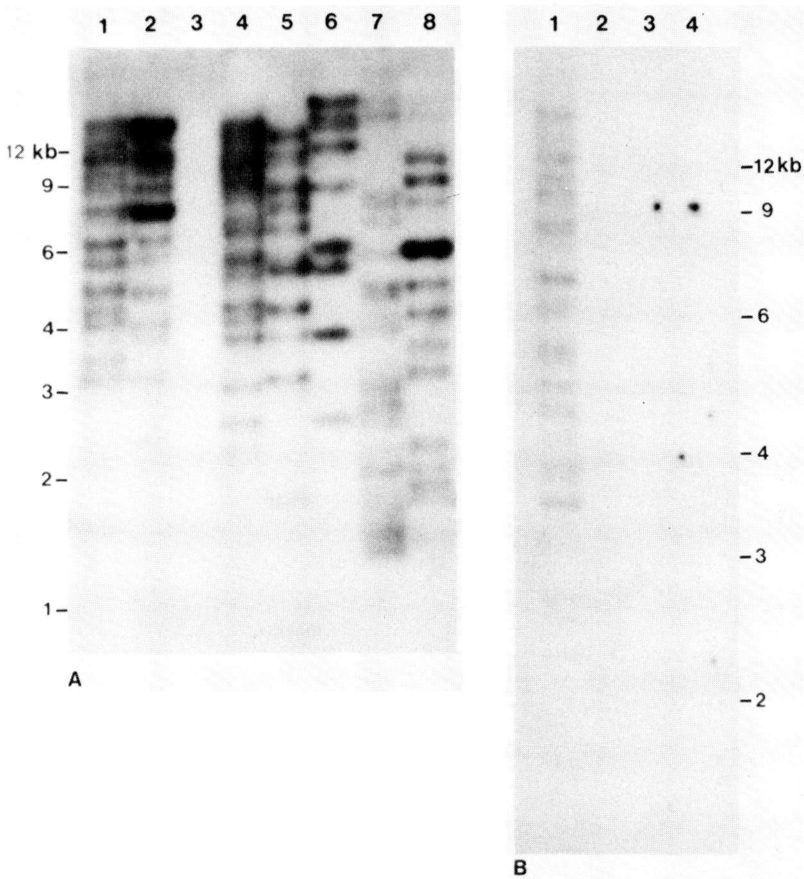

Figure 4. Hybridization of various chromosome and plasmid restriction fragments probed with radiolabeled DNA fragments containing a repeated sequence from *P. cepacia* AC1100 chromosome. *A*. Lanes 1–3: AC1100, PT88, and *P. cepacia* 383 chromosomal DNAs, respectively, digested with EcoR1; Lanes 4–6: AC1100 and PT88 chromosomal DNAs, and AC1100 plasmid DNA, respectively, digested with PvuII; Lanes 7–8: AC1100 and PT88 chromosomal DNAs, respectively, digested with SalI. The hybridization probe was a 6 kb SalI fragment isolated from the Tn5-generated 2,4,5-T mutant PT88. The 6 kb fragment contains part of Tn5 and flanking chromosomal sequences at the site of Tn5 insertion. *B*. Lanes 1–4: *P. cepacia* AC1100, *P. aeruginosa, P. putida*, and *P. mendocina* chromosomal DNAs, respectively digested with EcoR1. The hybridization probe was a 1.4 kb SalI-PvuII fragment from the 6 kb SalI fragment described above.

plasmid (lane 6) digests. It did not hybridize with *P. cepacia* 383 chromosomal DNA (lane 3) included as a control. The highly repeated sequence present on the 6 kb SalI fragment has been further localized to a 1.4 kb SalI-PvuII restriction fragment. When this fragment was used as a probe against *P. cepacia* AC1100, *P. aeruginosa, P. putida*, and *P. mendocina* chromosomal digest only the AC1100 chromosomal and plasmid genes showed homology (Fig. 4B).

Concluding remarks

A genetic and molecular analysis of the plasmids encoding biodegradation of chlorinated herbicides demonstrates not only a functional clustering of the degradative genes, but an interesting mode of expression of such genes through promoter sequences which differ substantially from those in *E. coli*, and which at present remain largely uncharacterized. Two different chlorinated substrates such as 3Cba and 2,4-D are utilized via a common chlorocatechol degradative pathway – which, although specified by two entirely different plasmids, appears to have a common ancestry. That genetic rearrangements play an important role in the ability of the microorganisms to evolve new pathways for the degradation of synthetic compounds is illustrated by amplification and deletion – fusion events which occur when pJP4-harboring cells, which normally utilize 2,4-D, are exposed to strong select ion pressure for the utilization of 3Cba (Ghosal et al., 1985b). It is also interesting that biodegradative genes from two different degradative plasmids (pJP4 and pAC27) are amplified 7 to 8 times when cloned in isolation from the normal regulatory genes and subjected to strong selection pressure (Ghosal et al., 1985b). Such gene amplification is presumably necessary to compensate for the absence of activator proteins that usually control the expression of the genes in a positive way.

Repeated sequences in tandem or inverted orientation are often associated with gene amplification and deletion events (Anderson and Roth, 1977). These sequences may be partially or completely homologous. When the plasmid pJP4 was denatured, rapidly reanneled, and examined by electron microscopy, Ghosal et al. (1985b) observed many small stem-loop structures (indicating inverted repeats) up to 1.8 kb in length. Several such structures were present in the EcoR1 fragments B and E, which contain the structural and putative regulatory gene(s) respectively, for chlorocatechol metabolism.

A highly repeated DNA sequence present in both the plasmid and the chromosome of *P. cepacia* AC1100 has been identified. The repeated element was isolated by its close proximity to a chromosomal Tn5 insertion affecting 2,4,5-T biodegradation. It is likely that this element is associated with a 2,4,5-T biodegradative or regulatory gene. It is tempting to speculate that this or a similar repeated sequence led to the recruitment and rearrangement of various biodegradative genes into an efficiently regulated 2,4,5-T catabolic system. Further efforts are currently underway to establish whether the repeated sequence present in a 1.4 kb segment will enhance genetic recombination given the multiple chromosomal and plasmid sites showing strong homology to the repeat.

Acknowledgements This work was supported in part by cooperative agreement (CR809666) from the U.S. Environmental Protection Agency, and a grant from the National Science Foundation (PCM 81-13558); P. H. T. acknowledges the support of a National Research Award (PHS ES 5264-02) from the National Institute of Environmental Health Sciences.

References

Alwine, J. C., D. J. Kemp and G. R. Stark. 1977. *Method for detection of specific RNAs in agarose gels by transfer to diazobenzyloxymethyl-paper hybridization with DNA probes.* Proc. Nat. Acad. Sci. USA 74:5350–5354.

Anderson, R. Pl and J. R. Roth. 1977. *Tandem genetic duplications in phage and bacteria.* Ann. Rev. Microbiol. 31:473–505.

Bollag, J. -M., G. G. Biggs, J. E. Dawson and M. Alexander. 1968. *2,4-D metabolism: enzymatic degradation of chlorocatechols.* J. Agric. Food Chem. 16:829–833.

Burke, F. B. 1984. *High-sensitivity S1 mapping with single-stranded (32P) DNA probes synthesized from bacteriophage M13mp templates.* Gene 30:63–68.

Chatterjee, D. K. and A. M. Chakrabarty. 1982. *Genetic rearrangements in plasmids specifying total degradation of chlorinated benzoic acids.* Mol. Gen. Genet. 188:279–285.

Chatterjee, D. K. and A. M. Chakrabarty. 1983. *Genetic homology between independently isolated chlorobenzoate-degradative plasmids.* J. Bacteriol. 153:532–534.

Chatterjee, D. K., S. T. Kellogg, S. Hamada and A. M. Chakrabarty. 1981. *Plasmid specifying total degradation of 3-chlorobenzoate by a modified ortho pathway.* J. Bacteriol. 146: 639–646.

Don, R. H. and J. M. Pemberton. 1981. *Properties of six pesticide degradation plasmids isolated from Alcaligenes paradoxus and Alcaligenes eutrophus.* J. Bacteriol. 145:681–686.

Don, R. H. and J. M. Pemberton. 1985. *Genetic and physical map of the 3,4-dichlorophenoxyacetic acid-degradative plasmid pJP4.* J. Bacteriol. 161:466–468.

Don, R. H., A. J. Weightman, H. J. Knackmuss and K. N. Timmis. 1985. *Transposon mutagenesis and cloning analysis of the pathways for degradation of 2,4-dichlorophenoxyacetic acid and 3-chlorobenzoate in Alcaligenes eutrophus JMP134 (pJP4).* J. Bacteriol. 145: 681–685.

Ebina, Y. and A. Nakazawa. 1983. *Cyclic AMP-dependent initiation and rho-dependent termination of colicin E1 gene transcription.* J. Biol. Chem. 258:7072–7078.

Evans, W. C., B. S. W. Smith, N. H. Fernley and J. I. Davies. 1971. *Bacterial metabolism of 2,4-dichlorophenoxyacetate.* Biochem. J. 122:543–551.

Furukawa, K. and A. M. Chakrabarty. 1982. *Involvement of plasmids in total degradation of chlorinated biphenyls.* Appl. Environ. Microbiol. 44:619–626.

Ghosal, D., I. S. You, D. K. Chatterjee and A. M. Chakrabarty. 1985a. *Microbial degradation of halogenated compounds.* Science 228:135–142.

Ghosal, D., I. S. You, D. K. Chatterjee and A. M. Chakrabarty. 1985b. *Genes specifying degradation of 3-chlorobenzoic acid in plasmids pAC27 and pJP4.* Proc. Nat. Acad. Sci. USA 82:1638–1642.

Ghosal, D., I. S. You, D. K. Chatterjee and A. M. Chakrabarty. 1985c. *Plasmids in the degradation of chlorinated aromatic compounds.* Pp. 667–686 in D. Helinski, S. Cohen, D. Clewell, D. Jackson, and A. Hollaender, eds., *Plasmids in Bacteria.* Plenum Press, New York.

Inouye, S., Y. Ebina, A. Nakazawa and T. Nakazawa. 1984a. *Nucleotide sequence surrounding transcription initiation site of xyl ABC operon on TOL plasmid of Pseudomonas putida.* Proc. Nat. Acad. Sci. USA 81:1688–1691.

Inouye, S., A. Nakazawa and T. Nakazawa. 1984b. *Nucleotide sequence of the promoter region of the xyl DEFG operon on TOL plasmid of Pseudomonas putida.* Gene 291:323–330.

Kamp, P. F. and A. M. Chakrabarty. 1979. *Plasmids specifying p-chlorobiphenyl degradation in enteric bacteria.* Pp. 275–285 in K. N. Timmis and A. Puhler, eds., *Plasmids of Medical, Environmental, and Commercial Importance.* Elsevier Biomedical Press, Amsterdam.

Kawasaki, H., N. Tone and K. Tonomura. 1981. *Plasmid determined dehalogenation of haloacetates in Moraxella species*. Agric. Biol. Chem. 45:29–34.

Kilbane, J. J., D. K. Chatterjee, J. S. Karns, S. T. Kellogg and A. M. Chakrabarty. 1982. *Biodegradation of 2,4,5-trichlorophenoxyacetic acid by a pure culture of Pseudomonas cepacia*. Appl. Environ. Microbiol. 44:72–78.

Liu, T. and P. J. Chapman. 1985. *Cloning and expression of 2,4-dichlorophenoxyacetate-degradative genes in Escherichia coli and Alcaligenes eutrophus.* P. H-154 (Abst.). Annual Meeting Amer. Soc. Microbiol.

Mermod, N., P. R. Lehrbach, W. Reineke and K. N. Timmis. 1984. *Transcription of the TOL plasmid toluate catabolic pathway operon of Pseudomonas putida is determined by a pair of co-ordinately and positively regulated overlapping promoters*. EMBO J. 3:2461–2466.

Miles, J. S. and J. R. Guest. 1984. *Nucleotide sequence and transcriptional start point of the phosphomannose isomerase gene (man A) of Escherichia coli*. Gene 32:41–48.

Nakai, C., H. Kagamiyama, M. Nozaki, T. Nakazawa, S. Inouye, Y. Ebina and A. Nakazawa. 1983. *Complete nucleotide sequence of the metapyrocatechase gene on the TOL plasmid of Pseudomonas putida mt-2*. J. Biol. Chem. 258:2923–2928.

Reineke, W. and H. -J. Knackmuss. 1979. *Construction of haloaromatics utilizing bacteria*. Nature 277:385–386.

Schneider, M. J. 1979. *Persistent Poisons, Chemical Pollutants in the Environment*. The New York Academy of Sciences, New York.

Timmis, K. N., P. R. Lehrbach, S. Harayama, R. H. Don, N. Mermod, S. Bas, R. Leppik, A. J. Weightman, W. Reineke and H. -J. Knackmuss. 1985. *Analysis and manipulation of plasmid-encoded pathways for the catobolism of aromatic compounds by soil bacteria.* Pp. 719–739 in D. R. Helinski, S. N. Cohen, D. B. Clowell, D. A. Jackson, and A. Hollaender, eds., *Plasmids in Bacteria*. Plenum Press, New York.

Vandenbergh, P. A., R. H. Olsen and J. F. Colaruotolo. 1981. *Isolation and genetic characterization of bacteria that degrade chloroaromatic compounds*. Appl. Environ. Microbiol. 42: 737–739.

Weightman, A. J., R. H. Don, P. R. Lehrbach and K. N. Timmis. 1984. *The identification and cloning of genes encoding haloaromatic catabolic enzymes and the construction of hybrid pathways for substrate mineralization.* Pp. 47–80 in G. S. Omenn and A. Hollaender, eds., *Genetic Control of Environmental Pollutants.* Plenum Press, New York.

The challenges of conferring herbicide resistance on crops

J. GRESSEL
Department of Plant Genetics, Weizmann Institute of Science Rehovot, IL 76100, Israel

Abstract There is a strong need, with many crops, to obtain resistance to cost-effective herbicides having the best spectrum of weed control. The "promising" results achieved to date using genetic engineering and/or tissue culture techniques have shown the need for carefully choosing economic and scientific criteria in selecting crop, herbicide, and genes for resistance. Too much experimentation has been with a model plant (albeit "weed") tobacco, with major crops not needing herbicide resistance, with expensive new herbicides, or with inappropriate or inadequate genes. Often, as in the case with the maternally inherited, triazine-resistant gene commonly used, the yield and competitiveness of the crop may be considerably reduced due to a genetic load of the defective gene. Results of various groups including our own will be evaluated according to these criteria. We have transferred maternally inherited, triazine resistance from the weed *Solanum nigrum* (nightshade) + *S. tuberosum* (potatoes) by using a protoplast fusion technique in which the nucleus of the donor species was irradiated to prevent division. The regenerated potato plants are resistant to the herbicide but the quality and yield of the potatoes have not yet been evaluated in field trials.

Introduction

The annual loss due to weeds in the United States has been estimated at \$12 billion. To this must be added the \$8 billion already spent on weed control (Shaw, 1983). Herbicides are usually the most cost effective means of controlling weeds, as mechanical cultivation requires considerable fuel, and damages the soil. There are herbicides available capable of killing each major weed, but they also kill crops. The need for supplying selective herbicides for each cropping situation can be expensively met by screening and developing new herbicides or conversely, by making crops resistant to the presently available herbicides. Under rare occasions this can be done by standard genetic techniques (Faulkner, 1982), or, in cases where the inheritance is maternal and chromosome numbers vary, by slightly more complicated breeding techniques (Souza-Machado, 1982). The need for herbicide resistance is not limited to the crops themselves. A case in point is the problem engendered by many pesticides with nitrogen-fixing blue-green algae (cyanobacteria). These are the major source of nitrogen for growing rice in much of the third world, but they are inhibited by many of the pesticides used in rice production (Vaishampayan and Prasad, 1984).

The suggestions for the attempts to use biotechnological techniques (herein defined as more sophisticated than the above) have been around for more than 10 years (Gressel, Zilkah and Ezra, 1978). There are still only a few reports subsequent to the first reports by Chaleff and Parsons (1978) and Radin and Carlson (1978) of regenerating tobacco resistant to picloram or phenmedipham and bentazon, respectively. Almost all these reports are with tobacco, which needs no new herbicides. The number of review articles on obtaining herbicide resistance in crops far surpasses the number of scientific articles describing even partial successes. Some reviews are very optimistic (Netzer, 1984) and some have been more realistic or pessimistic depending on the author's outlook (Gressel, 1985b). We will try to update the latter review in a constructive vein and add more parameters for consideration. This may help in the choices of genes and crops, and to accentuate the pitfalls to help prevent their recurrence. As will be discussed below, there is now sufficient information to allow more sophisticated choices of herbicides and genes for resistance, to accompany the developments in tissue culture and genetic engineering techniques. Unfortunately, the realization of the need for preliminary information, as well as the information itself, is often ignored.

A problem beyond the scope of this review, except to mention it, is the economics of choice of herbicide. The farmer would like crops resistant to the least expensive, broad-range herbicides. This need may be met by some of the private biotechnology companies, but only for those crops where the farmer does not re-use seed – i.e., only for hybrids. The laboratories in, or supported by, chemical companies will be interested in obtaining resistance to their highest priced herbicides or those for which they have an unused capacity for greater production. Clearly, there is no incentive for the commercial sector to develop resistance in each crop to the most cost-effective herbicides. There is thus still a need for having research supported by users' groups and in public laboratories, where it is known that there can be a benefit to society but no great financial benefit to the discoverer in the form of increased sales of herbicides or royalties on seed.

Choice of herbicides and genes

Herbicides: Will weed field resistance appear? Resistance has occurred to many herbicides with many weed species under field conditions (LeBaron and Gressel, 1982; Gressel, 1985c). There is one common denominator for all such cases: the use of a single herbicide with the same crop, year after year without rotation of herbicide. Thus, it would be highly unadvisable to select for resistance to a herbicide which is normally used as part of a rotation. The triazines are the classical case. Triazine resistance appeared throughout the world in mono-culture

Zea mays (corn) but not in the U.S. corn-belt where corn is almost always rotated with other crops such as *Glycine max* (soybeans). Recent history would suggest a disasterous increase in resistant weeds if triazines were used as the sole herbicide with both corn and soybeans. This need not be the case with all herbicides and is a function of the selection pressure of the herbicide itself (Gressel and Segel, 1982).

Fitness of resistance genes. "There is a cost to more genetic selection, the cost is in fitness" (Haldane, 1960). Thus, there is a potential loss of yield and/or competitive ability following selection for a trait such as herbicide resistance. The reasons for loss of fitness with herbicide resistance can be assessed where the biochemical reasons for resistance have been characterized. The potential loss of fitness conferred by a gene should be evaluated early and all alternatives should be explored, especially if different approaches to make a crop resistant to a herbicide are available. The farmer will not buy the resistant crop variety with the lowest fitness. Lost fitness can often be allayed by breeding, but the same breeding material is available to all. Thus, those that use the more fit genes should win.

The histories of more than 40 cases where fitness of herbicide resistance material was measured have been exhaustively summarized (Gressel, 1985b) and the picture looks depressing. The case histories of newly-evolved herbicide resistant weeds and specially developed crops should be viewed with two questions in mind; what is the productivity of the crop (yield) and what is the competitive ability? These are measured in two different ways. The resistant and susceptible strains are grown separately and productivity (yields) is measured. In most cases where fitness has been checked, the material was not isogenic for all other genes except for that conferring herbicide resistance, rendering the comparisons potentially incorrect. Recently, the eighth backcross of triazine resistant *Brassica napus* cv Regent (rape-seed) was checked against the parent variety. This material is nearly isogenic in the nucleus, but the whole plastome was derived from *B. compestris*, the evolutionary source of the resistant gene. When the resistant biotype was grown separately from the parent biotype, the yield was about 30% less. When they were grown mixed together at 5 cm spacing the yield of the resistant biotype was 25% that of the susceptible parent cultivar (Gressel and Ben Sinai, 1985). Similar results were found in Canada, and the official recommendation is to use the resistant biotype only in areas heavily infested with weeds that only triazines can control. Only a small portion of the total area of rape-seed grown is this heavily infested.

There are many cases where there is enough information about resistant genes to ascertain the likelihood of their being unfit, for example, a shift in K_m for the natural substrate as well as other problems in enzymatic activity. Some of these will be reviewed in specific cases for triazine and glyphosphate resistance.

The choice of genetic traits for resistance. The more that is known about the physiology and genetics of resistance, the more can be determined in advance about the possibilities of success in obtaining resistance. This author knows of projects that presumably failed because not enough advanced groundwork was done before they were undertaken. Theoretical examples are discussed below:

A. Herbicide detoxification by genetically engineering glutathione-*S*-transferase (GST) from corn to other crops. Corn has a dominant gene conferring triazine resistance via conjugation with a GST (Shimabakuro et al., 1971). Corn has at least two other GST's responsible for detoxification of thiocarbamates (Lay and Niland, 1984) and chloroacetamides (Mozer, Tiemeier and Jaworski, 1982). What could be easier than transferring a dominant gene of a constitutive protein that represents nearly 1% of cellular protein? During the depression there was a saying "If I had some ham, I'd have a ham sandwich, if I had some bread." The sandwich for detoxifying all of these herbicides is the specific GST, with glutathione as bread. Corn has 10 to 20 times more glutathione than all the rest of a group of species that have been tested (Gressel et al., 1978). If a species does not have enough glutathione, it will not be able to conjugate the herbicide. The amount needed is probably the normal level in corn, as some treatments of corn with herbicide protectants eventually double the level of cellular glutathione. Thus, if a species is low in glutathione, one may well have to engineer in the gene or genes that are needed to elevate glutathione as well as the GST. We learn something important that clearly was not apparent; a gene that is dominant in one species, may be ineffective in another species because one of its substrates is limiting.

B. Resistance to paraquat has evolved in a few weed species and, in two cases, has been correlated with increased levels of the enzyme, superoxide dismutase, which catalyzes a reaction in which oxygen radicals become hydrogen peroxide (Harper and Harvey, 1982). The gene for this enzyme has been cloned from mammalian sources. Why not splice control sequences to cause high levels of production of this enzyme and then engineer plants to become paraquat resistance? There were two reasons that should have suggested potential problems: (a) the product of this superoxide dismutase reaction is hydrogen peroxide

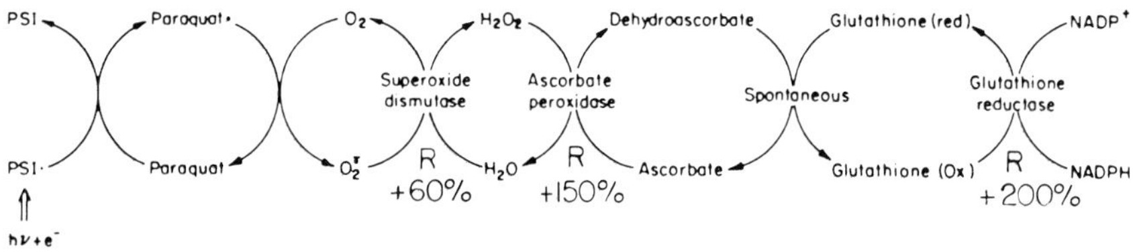

Figure 1. Alleviation of paraquat effects. Paraquat interacts with photosystem I and forms oxygen radicals which can be detoxified by three enzymes as outlined by Halliwell (1981). These three enzymes have greater activity in chloroplasts of paraquat resistant *Conyza bonariensis* than in the wild type, by the amounts shown. Enzyme activity data of Shaaltiel and Gressel (unpublished observations).

which is rather toxic, and (b) paraquat resistance is inherited quantitatively in *Lolium perenne*, the only species studied (Faulkner, 1982), suggesting that more than one gene is involved. Indeed, we have recently found in paraquat resistant *Conyza bonariensis* that there are increased levels of two other enzymes which may be part of a complete oxygen radical detoxification scheme (Fig. 1), and more than one enzyme may be required in other species as well.

The system is further complicated by finding no difference in total cellular superoxide dismutase activity in resistant *Conyza*. The increase was found to be due to a greater activity of one of the isoenzymes in the chloroplast (Shaaltiel and Gressel, unpublished observations). Thus resistance may not be a function of enhanced gene expression – just different compartmentation.

There is reason to justify the concept of herbicide protectants in such cases. Indeed, a paraquat derivation gave a medium of protectant superoxide dismutase activity in plastids (Lewinsohn and Gressel, 1984).

Methods of obtaining resistance

It will clearly be difficult to obtain resistance during normal field treatments with herbicides because it is almost impossible to obtain more than approximately 99% kill even at overkill rates of herbicide application. The vast majority of surviving plants are not resistant, they are escapees which did not get treated with herbicides. Methodology must be designed to transfer resistance from natural sources (weeds which evolved resistance in the field, crops which are normally resistant, and microorganisms which have been selected for resistance) or techniques must be used to select for resistance at the crop cellular level. The methodologies will first be discussed and results will be analyzed with a few particular herbicides.

Herbicide application. If one wishes to work at supraoptimal rates of application, it is necessary to use pure and not technical grades of herbicide. Often, as with diuron, the technical grade has a 1–2% contamination with phytotoxic compounds which may have a different site of action (Blein, Ducruet, and Gauvrit, 1979). If 10^{-3} mM diuron is used to increase selection pressure, instead of 10^{-5} M which is totally toxic, then the contaminant will be well above a toxic level of 10^{-5} M.

Many herbicides, especially those incorporated into the coil, have a high vapor pressure, which causes problems in controlling dose rates when performing selections with callus and suspension cultures. This was elegantly overcome in efforts to obtain resistance to the volatile thiocarbamate, vernolate (Flashman, 1985). A non-commercial, non-volatile but phytotoxic thiocarbamate derivative with the same putative mode of action was used, and the selected plants were tolerant to both compounds.

Interspecific crosses. Two attempts have been made to transfer maternally inherited triazine resistance from wild weed species that evolved resistance, into crops. In the case of transfer from *Brassica campestris* into *B. napus*, transfer was hampered by different chromosome numbers, and a F_2 generation with low viability (Souza-Machado, 1982). Recently, resistance was transferred from wild *Setaria viridus* (green foxtail) into *S. italica* (foxtail millet) in a much more straightforward manner (Darmency and Pernes, 1985). Of the more than forty weed species that have evolved triazine resistance, these are probably the only two sufficiently related to crops to be transferred by genetic hybridization.

Selection in suspension, callus, and protoplast cultures. Herbicides can be applied uniformly to cells in suspension and to calli, vastly increasing the effective selection pressure. These techniques are not as simple as may be construed as the rare resistant cells seem to need viable cells nearby to keep them alive. The minimal viable plating density and plating efficiencies are all problematic to obtain.

A variety of protocols for selecting for herbicide tolerant biotypes through somaclonal variation are described in detail by Hughes (1983). Many herbicides specifically act on the photosystems of photosynthesis which precludes isolating resistant strains in majority of tissue culture systems which are non-photosynthetic. Radin and Carlson (1978) selected for bentazon and phenmedipham resistance in X-ray mutagenized tobacco plants. They excised resistant green islands from herbicide-bleached leaves and then used tissue culture techniques to regenerate tobacco plants for validation of resistance and to determine nuclear inheritance. Their technique cannot be used as such to isolate mutants whose resistance is related to a photosystem II 32,000 M_r protein (Gressel, 1985a). Resistance via this protein is coded for on chloroplast DNA. There is little plastid DNA per cell, and it is in many copies. X-rays are thus more likely to mutagenize nuclear genes than plastid genes. It may be possible to utilize a new modification of this green-island technique developed by Fluhr et al. (1985) which utilized a plastid-DNA specific nutagen N-nitro-N-methyl urea. They have selected for tobacco plants resistant to antibiotics affecting plastids, but not yet to herbicides.

It is not sufficient to assume that if cells are resistant, that plants will be resistant, or that the resistance will be inherited genetically. Resistance can be lost at all levels. Christianson and Deal (1984) regenerated tobacco plants from green islands selected for resistance to norflurazon and other herbicides inhibitory to carotenogenesis. These plants were resistant for over a year, and cuttings that were grown on the herbicide and self pollinated produced offspring that were all sensitive (Christianson, personal communication). This is a poorly understood "epigenetic trap for the unwary biotechnologist."

Laboratory level successes in obtaining resistance with these methods are limited mainly to tobacco. Some success has been achieved in obtaining resistance to imidazolinone herbicides in corn (Anderson, Georgeson, and Hibberd, 1984). Partial progress in obtaining ametryn, dalapon, and mefluidide tolerant somaclonal regenerants from sugarcane is described by Crocomo and Ochoa-Alejo (1983).

Transfer of resistance by protoplast fusion. There is the possibility of transfer of useful genes by protoplast fusion in cases where standard genetic crosses cannot be made. Such fusions between species result in unstable individuals or those that are infertile because unequal numbers of chromosomes remain in the progeny. Infertility is not an impediment with vegetatively-propagated crops. Protoplast fusion techniques are especially useful when the trait being transferred is not inheritied on the nuclear genome. In such cases, the donor-recipient technique of Zelcer, Aviv and Galun (1978), in which the donor has its nucleus inactivated (e.g., by X-irradiation) is of distinct advantage. Immediately after the first report that triazine resistance is maternally inheritied, it was suggested that the transfer be made by the donor-recipient procedure (Gressel et al., 1978). This procedure, when working well in a given species, can have donor plastid or mitochondrial genomes in crop plants, without changes in nuclear composition of the variety. It takes at least five years to obtain the eighth back-cross needed before breeders consider material sufficiently isogenic for release or for yield tests. In this technique, protoplasts of the donors bearing the cytoplasmically-inherited trait are irradiated just prior to fusion. The recipient can often be treated with compounds such as iodacetate which inactivate mitochondria, which will then allow only those cells with the donor cytoplasm and the recipient nucleus to regenerate into a cell. This technique is good with only those herbicides with maternally inherited resistance. The only herbicides known at present are the photosystem II inhibitors of the triazine, phenylurea, and uracil groups. Maternally inherited resistance to the latter two groups has only been shown in algae (Galloway and Mets, 1984). Thus, this technique is probably only usable at present with the triazines, as it is still unknown to what degree of unrelatedness one can have compatibility between nucleus and alien cytoplasm.

After fusion there is a process of segregation of plastomes that is not clearly understood. It should take an immense number of divisions to obtain a few individuals that are completely of one plastid parental type after fusing two cells with 20–100 plastids in each. Instead there is a rapid sorting out, at least in tobacco, of one type or another. Segregation seems to be much slower in other genera, e.g., *Solanum* (Binding et al., 1982). Segregation may still occur in primary regenerated plants, yielding leaves, or parts of leaves that have herbicide resistant or susceptible cells.

Recombinant DNA technology. The basic technologies of gene isolation, cloning into a plasmid with engineered control sequences, from thence to T_1 or other vector, and then into a plant, have been reviewed at length. Indeed, the most commonly used gene for such experimentation confers resistance to a phytotoic xenobiotic, but no farmer could afford to use kanamycin as a herbicide. What seemed to the genetic engineer to be the easiest step in the process, is actually among the hardest – to find a useful gene. Very few genes conferring herbicide resistance are known. The most famous, the plastid psbA gene (coding for a 32,000 M_r plastid protein) conferring triazine resistance. At present there is no insertion vector for plastids. The idea being discussed in the semi-popular press (Netzer, 1984) is to have this spliced to a sequence for a transport peptide similar to that of other plastid proteins synthesized from nuclear coded genes, and then insert it into the nucleus. One would have to do this in a mutant in which the plastid psbA gene is turned off, for reasons of resistance and economy.

A typical gene being considered for genetic engineering is a bacterial mono-oxidase which cleaves the acetic acid side chain from 2,4-D yielding the less phytotoxic 2,4-dichlorophenol (Perkins, Stiff and Lurquin, 1985).

Practical progress with various herbicides

There has been some progress in the past few years in attempts to develop plants resistant to three groups of herbicides. It is interesting to note that these three groups probably have the most known about their modes of action, which may accentuate the need for ground work. This may not necessarily be the case in the future if the genetic engineers manage to insert the pathways leading to the degradation of key herbicides into crops.

s-Triazine. The s-triazine herbicides are among the least expensive herbicides available and have a broad spectrum of activity, killing most dicots and many grasses. Resistance should be conferred only on crops not grown in rotation with corn for reasons explained above, and only in crops where the long residual activity is needed for weed control. These herbicides seem to act specifically on photosystem II of photosynthesis. Corn degrades triazines before they can do lasting damage. The resistant weeds that have thylakoid modifications that prevent herbicide binding and thus herbicide effect. Besides the genetic methodologies for transfer of the plastid-inherited resistance, the only success has been with transfer by protoplast fusion, although micro injection of chloroplasts should work as well. In our initial experiments to determine the practicality of the procedure, we fused triazine sensitive potato with atrazine resistant night shade, without donor irradiation. We received many regenerants with intermediate morphological properties between the two species, all segregating to resistant and susceptible plants (Binding et al., 1982). The only cybrid was a triazine susceptible strain of nightshade. The DNA restriction analysis showed that this was not due to a back mutation, as the susceptible nightshade plastids had the potato DNA restriction patterns (Gressel, Cohen and Binding, 1984). Potato and nightshade are closely related and the plastids of one are compatible with the nucleus of the other. Indeed no difference between the two species was seen with PVU II and only single band displacements were seen with BGL-1, BAM-H1 and XHO-1 (Gressel et al., 1984).

We have recently regenerated triazine resistant potatoes using donor-irradiation of the European commercial potato cultivar "Mirka" (Perl, Aviv and Gressel, unpublished observations). The plants produce tubers, and a large number of plants are being cloned up for yield trials. The problems of productive fitness are scary, and the product may be non-commercial. Still, there may be less unfit genes for triazine resistance in higher plants (Gressel, 1985a) and there are clearly more fit genes available in algae (Galloway and Mets, 1984), including some that have been mapped to transversions at different positions in the psbA genome (Erikson et al., 1985). These could be inserted as newer techniques are developed. It is unknown to what extent it will be possible to transfer plastids by fusion. If intrafamilial transfers are compatible, it would be possible to use fusion techniques to transfer triazine resistance to many major crops (Gressel, Eza and Jain, 1982).

Glyphosate salts. The presently used isopropylamino salt of glyphosate (IPA-glyphosate) is absorbed by leaves and translocated to growing points where it exerts its toxicity. Glyphosates thus kill many perennial, spreading herbaceous weeds with creeping underground organs. Higher levels of glyphosates than other herbicides must be used, suggesting a lower site affinity than for other herbicides. The major site of glyphosate inhibition is enol-pyruvyl-shikimate phosphate (EPSP) synthase (Amrhein et al., 1983). In all cases, glyphosate competitively inhibited only one of the substrates of the enzyme, phosphoenol pyruvate (PEP). Many approaches have been taken to obtain glyphosate resistance with the hopes of eventually obtaining resistant crops. Glyphosate resistant *Salmonella* and *Klebsiella* mutants possess glyphosate resistant EPSP synthase (Comaisen and Stalker, 1983; Schulz, Sost and Amrhein, 1984; respectively). Sost, Schultz and Amrhein (1984) have carefully characterized highly purified enzymes from R and S *Klebsiella*. They find that the K_m of the resistant enzyme for PEP is increased 16-fold and that for shikimate phosphate is increased 4-fold. The "cost" to a plant of having an enzyme with a far lower substrate affinity may be too high. The organism will clearly have to either produce more enzyme (if it was under stringent control and limiting), or grow more slowly. The group at Calgene switched to presumably more fit glyphosate resistant *Salmonella* mutants that grew at the same rate as the wild type. They have not recognized large shifts in K_m (Comai, personal communication). A single amino difference between wild type and resistant mutant has been deduced from sequence analyses (Stalker, Hiatt and Comai, 1985). The resistant gene was expressed in transformed tobacco plants, and bacterial DNA, mRNA, and protein have been detected in tobacco cells. The regenerated plants are tolerant to glyphosate, but not to field levels of the compound. Calgene is employing different promoters with a further modified gene to achieve higher levels of resistance. They have also introduced the gene into tomatoes (Comai, personal communication).

Bacteria that grossly overproduce EPSP synthase, conferring glyphosate resistance have been isolated. Rogers et al., (1983) have engineered multiple copies of the gene for normal EPSP synthase into plasmids and then back into the host bacteria. These bacteria then make nearly 100-fold more enzyme. The K_m and other properties of the enzyme remained unchanged. What will be the cost of such a vast overproduction of a single enzyme, if engineered into crops?

Sequential increases in glyphosate concentrations in plant tissue culture media have been used to increase EPSP synthase levels, presumably by gene duplication. Neither the carrot cultures of Nafziger et al. (1984) nor the *Cordyalis sempervirens* cultures of Amrhein et al. (1983) could be regenerated into plants. Is there something about gene duplications (which are common in drug resistant cancer cells) that prevents regeneration to plant? There are two *Cordyalis* strains one which is resistant to 20 mM glyphosate and which grows rather slowly (i.e., very unfit) and another strain resistant only to 5 mM glyphosate, which seems to grow at the same rate as the susceptible strain (Amrhein, personal communication). A tobacco culture, tolerant to glyphosate and containing a doubled specific activity of enzyme was regenerated into plants slightly more tolerant to glyphosate (Dyer, Leopold and Weller, personal communication). Singer and McDaniel (personal communication) have also regenerated glyphosate resistant tobacco calli into plants which were resistant to levels of glyphosate that were toxic to the wild type. Some of their lines surprisingly have cross tolerance to amitrole, a herbicide with unknown mode of action. They have not delved into the mode of resistance of their material.

Other manners of conferring resistance are possible. Slowly acting "glyphosatases" might confer resistance. Glyphosate is biodegraded in the soil (Rueppel et al., 1977) and a *Pseudomonas* with an enzyme utilizing the phosphonate has been isolated (Shinaburger et al., 1984). The effect of glyphosate on plant cells in culture can be reversed by a variety of organic acids (Killmer, Widholm and Slife, 1981), presumably by increasing PEP levels. This opens avenues for discovery of chemical protectants which will stimulate an increase in PEP levels, and the engineering of cells with glyphosate-inducible PEP synthesis. These solutions seem less costly than having an EPSP synthase with a K_m decreased 16-fold, or having to synthesize a 100-fold more enzyme, which may be the solution that evolved in the somaclonal variants.

Sulfonylurea and imidazolinone inhibitors of ALS. Two groups of herbicides, the sulfonylureas (e.g., chlorsulfuron and sulfometuron-methyl) and the imidazolinones (e.g., imazaquin), which are quite different chemically, specifically inhibit acetolactate synthase (ALS) (Chaleff and Mauvais, 1984; Shaner, Anderson and Stidham, 1984; respectively). Acetolactate synthase is a key enzyme in the pathway of biosynthesis of the three branch-chained amino acids. Using standard cell culture selection techniques, it was possible to obtain tobacco clones regenerating into plants that are resistant to sulfonylureas (Chaleff and Ray, 1984), and corn plants resistant to field levels of imidazolinones (Anderson et al., 1984). The imidazolinone tolerant material has cross tolerance to sulfonylureas (Anderson et al., 1984) which suggests that both diverse chemical groups act at the same or nearby sites. Resistance was dominant in both species. The enzyme ALS has not yet been purified from higher plant sources, which has complicated gene isolation. It has been possible to isolate, clone, and sequence ALS genes from bacteria and yeast. A single amino acid difference between the resistant and susceptible gene for ALS was found in bacteria (Flaco, Dumas and McDewitt, 1985). Both of the herbicide groups have high soil residues giving exceedingly long action. Considering the ease with which dominant resistant mutants were obtained in the laboratory, the cross reactivity between the herbicides, and the very high selection pressures that they exert, it is almost guaranteed that resistant weeds will rapidly evolve to these ALS inhibitors, unless preventative strategies are used (Gressel, 1985c).

Conclusions

Despite the plethora of reviews on the biotechnology of herbicide resistance, there is only one crop variety on the market derived from difficult crosses. This variety of rape-seed is highly unfit. Beyond tobacco, there is only the imidazolinone resistant corn that may be marketable, and that was derived from "simple" tissue culture selections. Genetic engineering has contributed considerable promise, but the groundwork and homework often seem to be lacking. We can summarize with a check list of necessary criteria, each discussed in this chapter: (a) cost-effectiveness of the proposed herbicide; (b) no yield loss expected due to resistance; (c) information on level and mode of action; (d) which of different types of resistance are preferable and/or achievable; (e) will we have the numbers of separately derived resistant individuals for comparative trials; (f) do we need monogenic or multigenic resistance? The need is there for resistance and it is achievable, but it will not be as simple or as rapid as often propounded.

Acknowledgements The author thanks his collaborators for their research efforts quoted here and various colleagues who have provided unpublished results and manuscripts. The author's research on transfer of triazine resistance is supported by the Schmidt Fund for Applied Research, and the research on resistance to oxidant stress by the Israel Academy for Sciences and Humanities. The author holds the Gilbert de Botton Chair of Plant Sciences.

References

Amrhein, N., D. Johanning, J. Schab and A. Schulz. 1983. *Biochemical basis for glyphosate-tolerance in a bacterium and a plant tissue culture*. FEBS Lett. 157:191–196.

Anderson, P. C., M. A. Georgeson and K. A. Hibberd. 1984. *Cell culture of selection of herbicide resistant corn*. Page 56 (Abst.). Crop Sci. Soc. Amer. Meeting.

Binding, H., S. M. Jain, J. Finger, G. Mordhorst, R. Nehls and J. Gressel. 1982. *Somatic hybridization of an atrazine resistant biotype of Solanum nigrum with S. tuberosum: I. Clonal variation in morphology and in atrazine sensitivity*. Theor. Appl. Genet. 63:273–277.

Blein, J. P., J. M. Ducruet and C. Gauvrit. 1979. *Identification et activité biologique d'une impureté du diuron technique*. Weed Res. 19:117–121.

Chaleff, R. S. and C. J. Mauvais. 1984. *Acetolactate synthase is the site of action of two sulfonyl urea-herbicides in higher plants*. Science 224:1443–1445.

Chaleff, R. S. and M. F. Parsons. 1978. *Direct selection in vitro for herbicide resistant mutants of Nicotiana tabacum*. Proc. Nat. Acad. Sci. USA 75:5104–5107.

Chaleff, R. S. and T. B. Ray. 1984. *Herbicide-resistant mutants from tobacco cell cultures*. Science 223:1148–1151.

Christianson, H. L. and L. M. Deal. 1984. *Tobacco plants resistant to inhibitors of phytoene desaturation*. Genetics 107s:20.

Comai, L., L. C. Sen and D. M. Stalker. 1983. *An altered aroA gene product confers resistance to the herbicide glyphosate*. Science 221:370–371.

Crocomo, O. J. and N. Ochoa-Alejo. 1983. *Herbicide tolerance in regenerated plants*. Pp. 770–781 in D. A. Evans, W. R. Sharp, P. V. Ammirato, and Y. Yamada, eds., *Handbook of Plant Cell Culture*, Vol. 1. MacMillan, New York.

Darmency, H. and J. Pernes. 1985. *Use of wild Setaria viridis to improve triazine resistance in cultivated S. italica by hybridization*. Weed. Res. 25: (in press).

Erikson, J. M., M. Rahire, J. D. Rochaix and L. Mets. 1985. *Herbicide resistance and cross-resistance: changes at three distinct sites in the herbicide binding protein*. Science 228: 204–207.

Falco, S. C., K. S. Dumas and R. E. McDevitt. 1985. *Molecular genetic analysis of sulfonylurea herbicide action resistance in yeast*. In L. van Vlotten-Doting et al., eds., *Molecular Form and Function of the Plant Genome*. Plenum Press, New York (in press).

Faulkner, J. 1982. *Breeding herbicide-tolerant crop cultivars by conventional methods*. Pp. 235–256 in H. M. LeBaron and J. Gressel, eds., *Herbicide Resistance in Plants*. John Wiley and Sons, New York.

Flashman, S. M. 1985. *Selection for thiocarbamate herbicide resistance in plant cell cultures: II. A non volatile selective agent for selection of thiocarbamate resistant cells*. Plant Sci. Lett. (in press).

Fluhr, R., D. Aviv, E. Galun and M. Edleman. 1985. *Efficient induction and selection of chloroplast-coded antibiotic-resistant mutants in Nicotiana*. Proc. Nat. Acad. Sci. USA 82: 1485–1489.

Galloway, R. E. and L. J. Mets. 1984. *Atrazine, bromacil, and diuron resistance in Chlamydomonas*. Plant Physiol. 74:469–479.

Gressel, J. 1985a. *Herbicide tolerance and resistance alteration of site of activity*. Pp. 159–189 in S. O. Duke, ed., *Weed Physiology*, Vol. 2. CRC Press, Boca Raton.

Gressel, J. 1985b. *Biotechnologically conferring herbicide resistance in crops: the present realities*. In L. Van Vloten-Dotin et al., eds., *Molecular Form and Function of the Plant Genome*. Plenum Press, New York.

Gressel, J. 1985c. *Strategies for prevention of herbicide resistance in weeds*. In K. J. Brent and A. Atkin, eds., *Rational Pesticide Use*. Cambridge University Press, Cambridge, London.

Gressel, J. and G. Ben-Sinai. 1985. *Low intra-specific competitive fitness in a triazine resistant, nearly nuclear-isogenic line of Brassica napus*. Plant Sci. Lett. 38: (in press).

Gressel, J. and L. A. Segel. 1982. *Interrelating factors controlling the rate of appearance of resistance: the outlook for the future.* Pp. 327–347 in H. LeBaron and J. Gressel, eds., *Herbicide Resistance in Plants*. John Wiley and Sons, New York.

Gressel, J., S. Zilkah and G. Ezra. 1978. *Herbicide action, resistance, and screening in cultures vs. plants.* Pp. 427–436 in T. A. Thorpe, ed., *Frontiers in Plant Tissue Cultures.* University Calgary Press, Calgary.

Gressel, J., G. Ezra and S. M. Jain. 1982. *Genetic and chemical manipulation of crops to confer tolerance to chemicals.* Pp. 79–91 in J. S. McLaren, ed., *Chemical Manipulation of Crop Growth and Development*. Butterworth, London.

Gressel, J., N. Cohen and H. Binding. 1984. *Somatic hybridization of an atrazine resistant biotype of Solanum nigrum with Solanum tuberosom. 2: Segregation of plastomes*. Theor. Appl. Genet. 67:131–134.

Haldane, J. B. S. 1960. *More precise expressions for the cost of natural selection*. J. Genet. 57:351.

Harper, B. M. R. and D. B. Harvey. 1982. *Tolerance of bipyridylium herbicides.* Pp. 215–234 in H. M. LeBaron and J. Gressel, eds., *Herbicide Resistance in Plants*. John Wiley and Sons, New York.

Halliwell, B. 1981. *Toxic effects of oxygen on plant tissues.* Pp. 179–205 in B. Halliwell, ed., *Chloroplast Metabolism.* Clarendon Press, Oxford.

Hughes, K. 1983. *Selection for herbicide resistance.* Pp. 442–460 in D. A. Evans, W. R. Sharp, P. V. Ammirato and Y. Yamada, eds., *Handbook of Plant Cell Culture.* Vol. 1. MacMillan, New York.

Killmer, J., J. Widholm and F. Slife. 1981. *Reversal of glyphosate inhibition of carrot cell culture growth by glycolytic intermediates and organic and amino acids*. Plant Physiol. 68:1299.

Lay, M. M. and A. M. Niland. 1984. *Biochemical responses of inbred and hybrid corn (Zea mays L.) to R-25788 and its distribution with EPTC in corn seedlings*. Page 191 (abst.). Weed Sci. Soc. Amer. Ann. Meeting.

LeBaron, H. and J. Gressel. 1982. *Herbicide Resistance in Plants.* John Wiley and Sons, New York.

Lewinsohn, E. and J. Gressel. 1984. *Benzyl viologen mediated counteraction of diquat and paraquat phytotoxicities*. Plant Physiol. 76:125–130.

Mozer, T. J., D. C. Tiemeier and E. G. Jaworski. 1982. *Purification and characterization of corn glutathione-S-transferase*. Biochemistry 22:1068–1072.

Nafziger, E. D., J. M. Widholm, H. C. Steinruecken and J. L. Killmer. 1984. *Selection and characterization of a carrot cell line tolerant to glyphosate*. Plant Physiol. 76:571–574.

Netzer, W. J. 1984. *Engineering herbicide tolerance: when is it worthwhile?* Biotechnology 2:939–944.

Perkins, E. J., C. M. Stiff and P. F. Lurquin. 1985. *Use of Alcaligenes eutrophus as a source of genes for 2,4-D resistance in plants*. Weed Sci. (in press).

Radin, D. N. and P. S. Carlson. 1978. *Herbicide tolerant tobacco mutant selected in situ and recovered via regeneration from cell culture.* Genet. Res. 32:85–89.

Rogers, S. G., L. A. Brand, S. B. Holder, E. S. Sharp and M. J. Brackin. 1983. *Amplification of the aro A gene from E. coli results in tolerance to other herbicide glyphosate.* Appl. Environ. Microbiol. 46:37–43.

Rueppel, M. L., B. B. Brightwell, J. Schaefer and J. T. Marvel. 1977. *Metabolism and degradation of glyphosate in soil and water*. J. Agric. Food Chem. 25:517–528.

Schultz, A., D. Sost and N. Amrhein. 1984. *Insensitivity of EPSP synthase to glyphosate confers resistance to this herbicide in a strain of Aerobacter aerogenes*. Arch. Microbiol. 137:121–213.

Shaner, D. L., P. C. Anderson and M. A. Stidham. 1984. *Imidazolinones: potent inhibitors of acetohydroxyacid synthase*. Plant Physiol. 76:545–546.

Shaw, W. C. 1983. *The ARS National Weed Research Program.* Pp. 157–174 in J. E. Hilton, ed., *Agricultural Chemicals of the Future, BARC Symposium VIII*. Rowman and Allenheld, Totowa, New Jersey.

Shimabukuro, R. H., D. S. Frear, H. R. Swanson and W. C. Walsh. 1971. *Glutathione conjugation – an enzymatic basis for atrazine resistance in corn*. Plant Physiol. 47:10–14.

Shinaburger, D. L., E. K. Schmitt, H. D. Braymer and A. D. Larson. 1984. *Phosphonate utilization by the glyphosate degrading Pseudomonas sp. Strain PG 1982*. Appl. Environ. Microbiol. 48:1049–1050.

Sost, D., A. Schulz and N. Amrhein. 1984. *Characterization of a glyphosate-insensitive 5-enolpyruvyl-shikimic acid-3-phosphate synthase*. FEBS Lett. 173:238–242.

Souza-Machado, V. 1982. *Inheritance and breeding potential of triazine tolerance and resistance in plants*. Pp. 257–274 in H. M. LeBaron and J. Gressel, eds., *Herbicide Resistance in Plants*. John Wiley and Sons, New York.

Stalker, D. M., W. R. Hiatt and L. Comai. 1985. *A single amino acid substitution in the enzyme EPSP synthase confers resistance to the herbicide glyphosate*. J. Biol. Chem. (in press).

Vaishampayan, A. and A. B. Prasad. 1984. *Interstrain transfer of a pesticide resistant marker in the N_2 fixing cyanobacterium Nostoc muscorum*. Mol. Gen. Genet. 193:195–197.

Zelcer, A., D. Aviv and E. Galun. 1978. *Interspecific transfer of cytoplasmic male sterility by fusion between protoplasts of normal Nicotiana sylvestris and X-irradiated protoplasts of male sterile N. tabacum*. Z. Pflanzenphysiol. 90:397–407.

Biotechnology, "avian society" and "human society"

H. D. DANFORTH* and R. L. ROBLIN†
*U.S. Department of Agriculture, Agricultural Research Service, Animal Parasitology Institute, Beltsville, Maryland 20705, U.S.A.
†GENEX Corporation, 16020 Industrial Drive, Gaithersburg, Maryland 20877, U.S.A.

Abstract The control of avian coccidiosis by use of anticoccidial compounds has been one of the main reasons for the phenomenal growth of the poultry industry in the last 40 years. However, other means of control may be necessary in the future because of loss of efficacy of these compounds and slow development of new anticoccidials. Through use of hybridoma technology and molecular cloning, progress has been made in the study and utilization of the immune response of the bird as a possible means of control. For the first time, an *Escherichia coli*-elicited coccidial protein has conferred partial protection in immunized-challenged birds. The development of such a model has demonstrated that the production of a subunit vaccine against coccidiosis may be possible.

The application of biotechnology to a major disease problem of chickens and the technical possibility of increasing size and weight of animals through genetic engineering illustrate the complexity of interrelationships emerging between biotechnology and society. Controversy has arisen over the transfer of a genetic trait from a human being or other mammalian species into the germ line of unrelated mammalian species. The transfer of animal genes into human germ lines raises ethical questions, and has not been authorized in the United States. Attempts to prohibit the transfer of human or other animal genes into germ lines of animals have been opposed by the NIH Recombinate DNA Advisory Committee. The Committee's opposition was due in part to the importance of such research on the study and treatment of human and animal diseases, and the potential for development of more efficient food sources. The outcome of this debate will obviously have a direct bearing on what will be ethically acceptable in procedures that modify the germ line of domestic animals.

Introduction

The poultry industry in the United States produced approximately 4.1 billion broilers or meat chickens in the last year at an average live weight cost of \$0.29/1b. One of the main reasons for this efficiency has been the control of coccidiosis, a disease caused by infection of intestinal epithelial cells by protozoan parasites, the coccidia. The control has centered on the use of anticoccidial compounds. However, because of loss of parasite drug sensitivity, and the enormous cost involved with clearing such compounds for use in good animals, the possibility exists that other types of control, such as immunity, may need to be developed.

In this paper, the use of hybridoma antibodies developed against the coccidia and the eliciting of coccidial protein in *Escherichia coli* by use of molecular cloning techniques will be described by Dr. Danforth. This description will show how biotechnology can be used in possible vaccine development. Also, a discussion, authored by Dr. Roblin, will

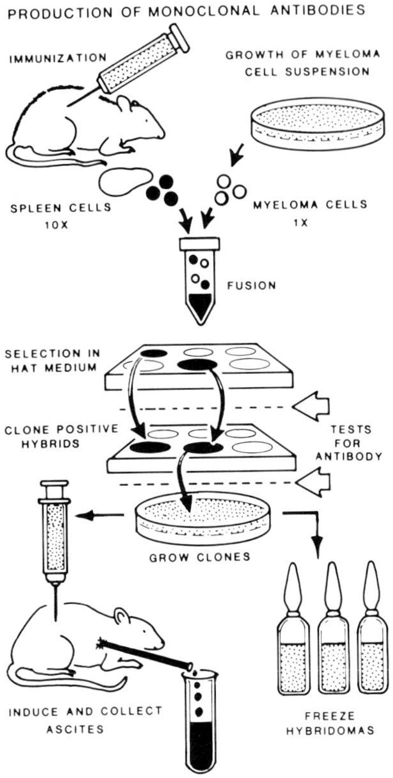

Figure 1. Avian Coccidial Life Cycle. Sporulated oocysts are ingested by bird and excystation of sporozoites occurs. Sporozoites invade intestinal epithelial cells and multiple schizont generations develop intracellularly. Asexual reproduction culminates in production of sexual stages that produce the oocysts that pass out in fecal material. Sporozoite and trophozoite stages, if inhibited by bird immune response, blocks both the pathology associated with the disease and oocyst production. (Used with permission, Animal Nutrition and Health).

be presented on the merging interrelationships between biotechnology and society.

Avian Coccidia: A Background

The coccidia have a complex life cycle that incorporates parasite development both within the animal and on the broiler house floor (Fig. 1). The oocyst stage of the parasite passes out with the fecal material onto the floor of the bird house and undergoes sporulation (becomes infective by forming sporozoites). These infective oocysts are ingested by birds and the initial invasive stage, the sporozoite, subsequently excysts within the intestine. The infection is initiated when the sporozoites penetrate the gut epithelial cells. Once intracellular the parasites undergo a number of asexual generations (schizogony or merogony), producing merozoites which reinvade intestinal cells and ultimately form the sexual stages that produce the oocysts (Fig. 1). These newly-formed occysts pass out in the fecal material and the process of sporulation and infection begins again. Because of this life cycle, the reproductive index of the coccidia is quite high, with a theoretical potential of 100,000 oocysts produced for every oocyst ingested. Therefore, the potential always exists for an explosive infection to occur in a high confinement broiler poultry house.

Control of coccidiosis has been accomplished by the use of anticoccidial compounds that are usually mixed in the feed. These compounds are one of the main contributors to the phenomenal growth of the poultry industry in the last 40 years, and they will continue to be a major defense against coccidia for the foreseeable future. However, even with control, coccidiosis still costs the poultry industry approximately \$300 million a year. Most of this cost (\$200 million) is from decreased feed utilization efficiency, bird weight gain, and skin pigmentation during the infection. Coccidiosis is a subtle disease and may only decrease the feed conversion efficiency by a hundredth of a point. However, if such a decrease is industry-wide, it could amount to a \$20 million loss. The remaining loss to producers (\$100 million) is due to the cost of the anticoccidials put in feed. Mortality contributes to the economic losses only on rare occasions.

Although these losses are presently tolerated by the industry, there is a major problem arising that may eventually demand more effective types of control. Most of the anticoccidials have been on the market for at least 10 years and loss of drug sensitivity by the parasites has become common. Also, because of the enormous cost involved with clearing these compounds for use in food animals, fewer new compounds are being developed and fewer pharmaceutical firms are actively screening for new anticoccidials. If this trend continues, the poultry industry may be faced with a potentially devastating control problem.

Immunity as a possible control

One potential avenue for control of the parasite would be to use the bird's own immune response. Basic research has shown that birds will produce a protective immune response to the pathogenic species of coccidia when infected with viable parasites (Rose, 1982). These infections, however, result in decreased weight gain and feed utilization efficiency. If this means of inducing protective immunity is used in established commercial flocks, careful monitoring of the dosage of coccidial oocysts given to the birds becomes critically important. Also, immunity is species specific so birds must be immunized against all 9 species of coccidia in order to ensure complete protection. Preliminary evidence (Augustine, personal communication) has suggested that the immunological response of birds to the different coccidial species differs and depends in part upon where the parasites invade the intestine. Strain variation adds to the complexity of the immune response. That is, immunizing birds against one strain of a species of coccidia may not protect the birds against infection by a second strain of that species. Strains, as defined in coccidial research, are the same species isolated from different geographical areas of the United States.

Despite these drawbacks, there is a vaccine, Coccivac®, that uses controlled infection for production of flock immunity. As discussed previously, careful management is required to produce a good immune response without clinical effects or weight loss. Also, there is an added disadvantage in that Coccivac contains all 9 pathogenic coccidial species, so new species of coccidia may be introduced into a poultry farm where they had not been present previously. These problems have limited the use of a live vaccine in the poultry industry, but the success of this approach in stimulating immunity has indicated that vaccine control of coccidiosis is feasible.

Use of biotechnology in the study of avian coccidia

Basic research. The two major areas of biotechnology utilized in the study of avian coccidia are hybridoma antibody technique and molecular cloning. Basic information derived from these studies is being applied to the development of a noninfective subunit vaccine against coccidiosis.

The developmental stage that became the object of the studies was the sporozoite, since it initiates the infection in the bird by excysting from the oocysts and penetrating the intestinal epithelial cells (Fig. 1). If this stage could be blocked or inhibited, then all the pathology and physiology associated with the intracellular asexual stages of parasite development would be stopped, and there would be no production of oocysts. Since the immune response in an orally immunized bird is directed against the sporozoite, a knowledge of the antigenic make-up of this stage would also give a greater understanding of which antigens are involved with protection. To acquire this knowledge, hybridoma antibodies (Hab) against numerous sporozoite proteins were produced.

The Balb/CByJ strain of mice were immunized with sporozoites of various coccidia species by tail vein injection. The initial screening of the hybridoma cell lines on air-dried sporozoites of one species of coccidia (*Eimeria tenella*) by use of the immunofluorescent antibody (IFA) technique revealed a number of different IFA patterns and different antibody isotypes (Danforth, 1982). These reactions indicated that the mouse had responded to a number of different antigens either in or on the sporozoite. Similar studies with sporozoites of other species of coccidia also showed a number of different IFA patterns (Danforth and Augustine, 1983), and both species-specific and cross-reactive Hab.

Since *E. tenella* will develop through its complete life cycle in cell cultures (Doran, 1970), it was possible to determine in vitro if any of the Hab had an effect on penetration or intracellular development. The IFA patterns of the sporozoite Hab were a mixed bag. However, only those antibodies that interacted with the surface of the parasite were found to have an inhibitory effect on sporozoite penetration of cells. In contrast, Hab labeling either the surface of the sporozoite or the internal organelles inhibited development to varying degrees (Danforth, 1983). These development-inhibiting antibodies either blocked merogony or prevented the 1st generation merozoites from leaving the mature schizonts and invading new host cells.

One Hab that labeled large internal organelles (refractile bodies) of the sporozoites made it possible to study the dynamics of the immune response in the birds. Through use of IFA labeling techniques and paraffin embedded, sectioned gut tissues (Augustine and Danforth, 1984), the numbers of sporozoites that had penetrated the epithelial cells were easily quantitated. These studies showed that sporozoites of coccidial species that infected the upper intestine were not inhibited from penetrating the epithelial cells of orally-immunized birds. However, sporozoites of species that infected the lower intestinal area (ceca) were partially inhibited from penetration in immune birds. This indicated a difference in the immune response of the birds to the different coccidial species that may be related to the area of the gut in which invasion occurs.

Vaccine development. As a result of these studies, it was possible to utilize Hab that affected parasite invasion and development to isolate and characterize the proteins they recognized. Using the Western blot technique, a number of sporozoite proteins ranging in molecular weight from 7–10 kilodaltons (kd) to >100,000 kd have been identified (Danforth, MacAndrew and Augustine, 1985). With this data, it was possible to identify the type of protein that was needed and, in some cases, to isolate enough of the coccidial protein for amino acid analysis.

Until recently, however, the Animal Parasitology Institute at the Beltsville Agricultural Research Center did not have the molecular cloning that capabilities that are needed to produce the quantities of antigens required for bird immunization studies. Therefore, a memorandum of understanding was established with the Genex Corporation of Rockville, Maryland, whereby the U.S. Department of Agriculture would supply Genex with antibodies and coccidial material, and Genex would use genetic engineering techniques to produce adequate amounts of coccidial protein. Through the unique cooperation of a government agency and a private company it was possible to continue progress in the area of coccidial immunity.

The techniques used at Genex consisted of isolating messenger RNA (mRNA) from sporulating oocysts, producing the complementary DNA (cDNA), and packaging the cDNA into λ gt 11 bacteriophage (Fig. 2) (Danforth et al., 1985). The bacteriophage were plated onto an agar plate and through use of immunoenzymatic techniques with the USDA-supplied antibodies, bacteriophages containing the coccidial cDNA were identified (Fig. 2). The cDNA was removed from the bacteriophage and introduced into a plasmid which was then inserted into *Escherichia coli* bacteria (Fig. 3). The *E. coli* containing the recombinant plasmids were grown on agar plates and again screened with immunoenzymatic techniques. Colonies showing positive antigen or coccidial protein production were isolated, grown in fermentation containers, and lysed to recover the protein (Fig. 3).

One of these partially purifieid proteins, designated 5401, when injected subcutaneously with Freund's complete adjuvant at the base of the neck of the chicken, produced a detectable serum antibody response directed against both the 5401 protein and soluble sporozoite antigen. This antigen at 2400 ng/bird also provided partial protection in chickens challenged with an oral inoculation of coccidial oocysts (unpublished observations). The immunized challenged birds had better weight gains, feed conversions, and lesion scores than unimmunized challenged birds. However, the protection was only partial because these immunized challenged birds did not gain weight or use feed as efficiently as did the uninfected controls.

These results are encouraging. It is possible to stimulate both an antibody response and at least a partially protective response in birds by use of an *E. coli*-elicited coccidial protein. However, the difficult problems of eliciting protective immunity against the coccidia remain to be solved. These include the production of complete immunity, stimulation of cross-protective immunity against the different species of coccidia, and determining the most effective level, methods of delivery, and minimal age of birds for effective vaccination. These problems are compounded by the immunosuppression that occurs in birds during avian viral infections and with the variation in response to vaccines that

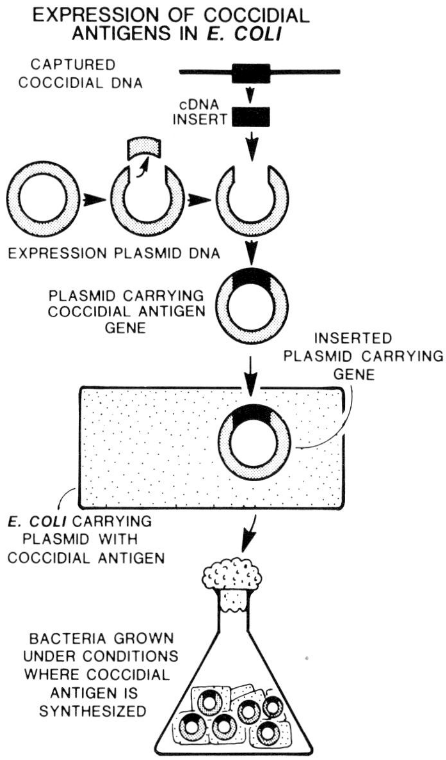

Figure 2. Cloning of Coccidial Antigens. Messenger RNA is isolated and purified from sporulating oocysts and copied into complementary DNA (cDNA). The cDNA is introduced into λ gt 11 bacteriophage that are plated onto agar plates. Proteins released in phage plaques are transferred to introcellulose filter. The filter replica is reacted with antibodies, and by use of immunoenzymatic techniques phage plaques producing coccidial proteins are identified. (Used with permission, Animal Nutrition and Health.)

is due to the genetic make-up of the bird. Collectively, these problems make the task of producing a coccidial vaccine extremely difficult. However, through the use of biotechnology, sufficient progress in the study of the immunity of coccidia has been made so that future control of this disease through vaccination is within the realm of possibility.

Biotechnology and society

The earliest deliberations on ways in which the new discoveries in the area of recombinant DNA research would be dealt with by our society began over a decade ago. Members of the National Academy of Sciences Committee on Recombinant DNA Research wrestled with the complex questions of whether, how and under what circumstances use of these techniques would require new regulations and relationships between the scientific community and the rest of society. The assessment of the net

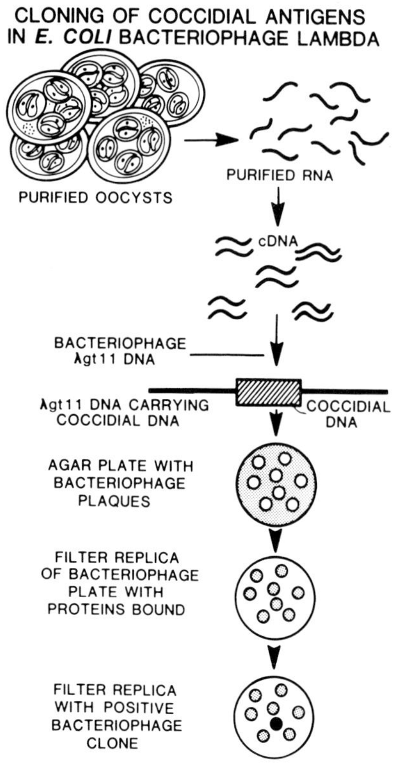

Figure 3. Expression of Coccidial Antigens. The cDNA containing the coccidial antigen gene is isolated from the λgt 11 and introduced into a plasmid. The recombinate plasmid is inserted into *E. coli* bacteria and the transformed bacteria are grown under conditions for coccidial antigen synthesis. (Used with permission, Animal Nutrition and Health.)

costs of setting up the NIH Guidelines and their administration by NIH and the NIH Recombinant DNA Advisory Committee versus the benefits of doing so lies outside the boundaries of what will be discussed in this part of the paper.

The importance of the principle of public discussion of these issues, established early in the course of what came to be called the Recombinant DNA Debate, is however, worth re-emphasizing. In the author's opinion, one of the most positive aspects of consideration of these issues by our society has been the institutionalization of the principle of public discussion and debate in the procedures of the Recombinant DNA Advisory Committee and the administration of the NIH Guidelines. This has helped to ensure that both proponents and opponents of changes in this system carefully articulate their points of view. Our society is strengthened by the way in which this principle leads towards decisions arrived at openly.

In the eleven years that have elapsed since those first deliberations, recombinant DNA research and its applications in what has come to be called biotechnology, have led to some astonishing developments. Microorganisms that synthesize human pharmaceutical proteins such as interferon at more than ten percent of total cell protein have been developed. Recombinant vaccines for improvement of human and animal health have been produced. Genes for herbicide resistance have been isolated and introduced into new species of plants. Biotechnology is clearly going to be able to produce the impacts on the pharmaceutical, chemical, food and agricultural industries that have been forecast over the last ten years.

Several of the examples just cited involved the use of microorganisms to produce substances that they had not been observed to produce in nature. Of developing current interest is the introduction of new genetic material into a wide variety of plants, animals and insects. As a part of this Symposium, reviews of the current status of modification of the germline in animals (Gordon), in plants (Riedel) and in insects (Corces) have been presented. As these talks have made clear, there is an impressive amount of current work being devoted to this area.

In an attempt to define the complex interrelationships between biotechnology and our society, this part of the paper will focus on one application of biotechnology: developing the technical potential to increase the weight and size of animals through genetic engineering. This potential has been clearly demonstrated in mice, and its extension to domestic animals raised for food is now within the realm of technical possibility.

We are now beginning to see a host of new possibilities which involve heritable modification of the genes of plant, animal and insect species. Several examples of this kind of modification have been discussed by speakers at this Symposium. One of the most dramatic examples of this kind of application of genetic engineering to animals appeared on the front cover of the November 18, 1983 issue of SCIENCE. The cover showed two mice, one of normal size and a genetically engineered counterpart that was substantially larger and weighed about 2.3 times as much (Palmiter et al., 1983).

The genetic engineering that produced the larger mice was the result of incorporating the extra copies of the gene for human growth hormone into the germ line of the mice. In these elegant experiments, the extra gene copies were introduced by microinjection of DNA, carrying the human growth hormone gene, into ferilized mouse eggs. Mice developing from such injected eggs produced varying levels of human growth hormone. Interestingly, the largest mice were not those that produced the highest levels of human growth hormone, but those that produced intermediate levels. This is consistent with the possibility of feedback regulation mechanisms in the action of growth promoting hormones.

Of considerable importance in these experiments was the observation that the effects of the cloned human growth hormone gene were heritable. That is, half of the first generation offspring of the genetically-engineered mouse also grew faster and gained more weight than their littermates that did not receive the extra copies of the growth hormone gene. In theory, the extension of these techniques to food animals such as cattle, pigs, sheep and chickens would seem technically feasible. Embryo transplantation is already widely practiced in cattle breeding, and recent experiments suggest that transfection of chicken embryos with viruses carrying foreign genes may be possible (Shuman, personal communication).

The attempted application of these genetic engineering techniques to animals other than mice has provoked some recent controversy. Amendments to the NIH Guidelines seeking to prohibit "any experimentation involving the transfer of a genetic trait from one mammalian species into the germ line of another unrelated mammalian species", or to prohibit "any experimentation involving transfer of a genetic trait from a human being into the germ line of another mammalian species", and to prohibit "any experimentation involving transfer of a genetic trait from any mammalian species into the germ line of a human being" were proposed (Federal Register, 1985). A court suit seeking to prohibit experiments of this type was also filed.

The arguments put forward in support of these prohibitions at the meeting of the NIH Recombinant DNA Advistory Committee meeting on October 29, 1984 were primarily that such experiments violated ethical and moral standards. To the prohibitionists, the cases of human genes transferred into the germ line of animals and animal genes transferred into the germ line of humans were equivalent and both should be prohibited. Many others commentators distinguished these two cases, indicating that transfer of human genes into the germ lines of animals was an important technique for improving understanding of human genetic diseases, and did not raise any ethical problems. Transfer of animal genes into the germ line of humans was viewed as raising ethical questions.

The two types of experiments are also treated very differently by the existing regulatory mechanisms. Transfer of human or animal genes into the germline of other animals requires review and approval of the local Institutional Bio-safety Committee and is already directly covered by

Section III-B of the NIH Guidelines. Transfer of recombinant DNA into humans, in contrast, must be reviewed by the NIH Recombinant DNA Advisory Committee and approved by the National Institutes of Health. In addition, transfer of genes into the human germline would probably have to involve the use of in vitro fertilization. In vitro fertilization research funded by the National Institutes of Health has recently been under a de facto moratorium since such projects required review by an Ethics Advisory Board which had not yet been formed in October 1984.

If the case is narrowed down to the question of the transfer of human or other animal genes into the germ line of animals, what are the major questions? For example, does the purpose of the experiments make a difference? Is transferring the human gene for growth hormone to cattle with the intent to learn more about the fundamental developmental biology of cattle ethically acceptable, while the same experiment with the intent to breed larger cattle for commercial purposes is unacceptable?

If proper attention is given to animal welfare, modifying the germ line of domestic animals raised for food with the intent of improving their properties may become ethically acceptable. By proper attention to animals welfare it is meant that the experiments should be carried out in a way that does not expose the animals to unnecessary pain or suffering. Once the germ line of an animal has been modified by stable incorporation of additional genetic information, the changes will become heritable and will be passed on by conventional breeding methods. Modification of bovine embryos by genetic engineering in vitro before reimplantation is still in its infancy, but may someday be accepted as a standard practice.

The debate will undoubtedly continue for many years and may eventually have to be ruled on by the courts. The NIH Recombinant DNA Advisory Committee has rejected the current proposed prohibitions unanimously. In doing so, it was stated "Both the importance of this class of experiments in current scientific research and the long-term possibilities for treatment of human and animal disease and the development of more efficient food sources make it a moral imperative that we strongly oppose the blanket prohibition of this class of experiments." Hopefully, the outcome of this conflict between biotechnology and society will result in a workable compromise that incorporates the philosophies, needs and concerns of all.

References

Augustine, P. C. and H. D. Danforth. 1984. *Use of monoclonal antibodies to locate Eimeria sporozoites in intestinal tissue*. Proc. Helm. Soc. Wash. 51:361–362.

Danforth, H. D. 1982. *Development of hybridoma-produced antibodies directed against Eimeria tenella and E. mitis*. J. Parasitol. 68:392–397.

Danforth, H. D. 1983. *Development of hybridoma antibodies against merozoites of Eimeria acervulina*. P. 46 (Abst.). Program and Abstracts, 58th Annual Meeting, The American Society of Parasitologists, San Antonio, Texas.

Danforth, H. D. 1983. *Use of monoclonal antibodies directed against Eimeria tenella sporozoites to determine stage specificity and in vitro effect on parasite penetration and development*. Am. J. Vet. Res. 44:1722–1727.

Danforth, H. D. and P. C. Augustine. 1983. *Specificity and crossreactivity of immune serum and hybridoma antibodies to various species of avian coccidia*. Poul. Sci. 62:2145–2151.

Danforth, H. D., S. J. MacAndrew and P. C. Augustine. 1985. *Characterization of avian coccidial antigens recognized by hybridoma antibodies*. P. 1334 (Abst.). Proc. 69th Ann. Meet. Fed. Am. Soc. Exp. Biol., Anaheim, California.

Danforth, H. D., R. McCandliss, M. Libel, P. C. Augustine and M. D. Ruff. 1985. *Development of an avian coccidial antigen by recombinant DNA technology*. Poultry Sci. 64:85.

Doran, D. J. 1970. *Eimeria tenella: From sporozoites to oocysts in cell culture*. Proc. Helm. Soc. Wash. 70:84–92.

Palmiter, R. D., G. Norstedt, R. E. Gelinas, R. E. Hammer and R. L. Brinster. 1984. *Metallothionein-human GH fusion genes stimulate growth of mice*. Science 222:809–814.

Rose, M. E. 1982. *Host immune responses*. Pp. 329–371 in P. L. Long, ed., *The Biology of the Coccidia*. University Park Press, Baltimore, Md.

Address by Frank Press, President National Academy of Sciences to a Symposium on Biotechnology for Solving Agricultural Problems, May 9, 1985

Agricultural Research Center, Beltsville, Maryland, U.S.A.

I must say that I feel a bit abashed after reviewing what you have been talking about in the last several days. What is there left for me to say? I feel a bit like someone, who on marrying for the umpteenth time, said: "I know what I'm supposed to do, but how do I make it interesting." Also, besides being the clean-up hitter for this symposium, I suffer from a second disadvantage. I am a geophysicist, not a biologist. Therefore, it would be both presumptuous and foolish for me to spend a great deal of time talking about genetics, nitrogen fixation, transposons, vaccines and the like. To do so risks the error that the biophysicist, Aaron Katchalski, pointed to in a little story he once told. It goes like this:

> A rich man interested in breeding horses commissioned three experts to find out their best properties. These experts were a veterinarian, an engineer, and theoretical physicist. After a few years, they reported their results. The vet had concluded from a genetic study that brown horses were the fastest. The engineer had found that thin legs were optimum for racing. The theoretical physicist did not give up his quest at the end of the period. Instead he asked for more time to study the question, claiming that the case of the spherical horse was proving very interesting.

So, my task in this talk is to stay clear of spherical horses; to stay within things that I feel comfortable about and that, I hope, you have not endlessly been told before. In brief, I'd like to talk about the special problems which you face in *effectively* applying biotechnology to agriculture and in mounting productive research programs.

There are, of course, no dearth of analyses of what is supposedly wrong with the agricultural research system. As Congressman George Brown once put it, that system is undergoing a "mid-life crisis. It has been criticized, cajoled, praised, and no doubt confused." Many of you, justifiably, feel victimized by the "What have you done for me lately?" syndrome. U.S. agriculture is a spectacular success. U.S. farmers are six times as efficient as they were 50 years ago. Today's farmer produces enough food for 60 people. We grow more food on less land with less people.

Agricultural products are one of our major exports; they have been a major element in blunting some of our trade deficits. Finally, it has become commonplace for Congressmen, Administration officials, and scientific and technological leaders, including myself, to criticize our ability to transfer fundamental knowledge into application. At the same time, the Agricultural Extension Service is cited as possibly our best model for effectively transferring technical knowledge to users.

What, then, is the problem? In part, it is a problem of relativity. Progress in agricultural research has paled before the spectacular advances of the past two decades in the biomedical sciences. Why that happened in an interesting issue all by itself, and I don't propose to describe it at length. Suffice it to say, that several factors were operating. Certainly, one factor was that the very success of agriculture research, relying principally upon classical genetics and other techniques as well as improved fertilizers, irrigation, and the like, cloaked the need for fundamental research. By contrast, the great scourges of cancer, heart disease, the arthritic and metabolic diseases, and multitudes of genetic diseases made their own case for enormous investments in biomedical research. Biomedical research also had a very effective lobby, both at the Bethesda campus and on the Hill. And it also had a small but highly effective and very determined group of private advocates.

Also, scientists tend to be sensible people. They follow market forces. Thus, the spectacular increases in the post-war NIH budget quickly attracted the very best scientists, including those who would have preferred to work in plant biology. As James Bonner, of Caltech and founder of his own plant biology company, once said: "I used to be a plant biologist in the 1960's, but gave it up and worked on rats." The results are quite visible. For example, by mid-1984, the DNA sequences for over 1,100 mammalian genes were known; by contrast, only 86 plant genes had been diciphered by then.

You can add your own illustrations of this vast gap between medical and agricultural research. We have known of that gap for a long time. But, in recent times, it has been sharpened by a number of converging pressures. A major pressure, of course, is that of international competition. That competition is considerable. For example, according to Ralph Hardy, the leading centers for plant science research related to biotechnology are not in the United States. They are in Australia and the United Kingdom.

How well we do in international competition will ultimately depend on the quality of our workers and of our technology. At the same time, as Richard Nelson of Yale has pointed out, technology development is inherently unpredictable, messy, and connected with everything else. That haphazardness confers a large advantage on nations with a broad technology base. Axiomatically, then, a nation with a weak base in plant biology hostages its future. It risks a serious disadvantage in competing in world markets. Certainly, the application of plant biology to biotechnology is obvious. It will help us grow crops more competitively. And, if we do make progress, it will enhance the nation's economic competitiveness.

Unfortunately, we have a growing list of industries that have done badly against competitive pressures from other nations: steel, textiles, consumer electronics, and others. To keep agriculture from going down that sorry road, it is going to have to become even more efficient. And that means exploiting the powers of biotechnology. To take that a bit further, let me quote from a recent op-ed piece written by Ralph Hardy. Dr. Hardy asked his readers to "imagine how beneficial it would be if U.S. farmers could cut in half their bills for pesticides, fertilizers, irrigation equipment, seeds, fuel, and other necessities. This would slash their production costs, raise their profits, and ease the demand for credit in cash-starved farm communities."

Dr. Hardy also pointed to the possibility that "farmers could double their yields of export crops without adding to their production costs. This could substantially reduce the price of U.S. farm products and help win back sales that have been lost as a result of the strong dollar."

Ralph Hardy has it exactly right. Biotechnology can, indeed, contribute to an even more economically-efficient agriculture. It can assure that U.S. agriculture remains globally predominant. And it can assure that its center doesn't move to other countries.

Certainly, the economic returns from investments in agriculture are well documented, beginning in the 1950's with Zvi Griliches' estimation of the economic returns from the development of hybrid corn. Since that time, there have been numerous and very careful studies of economic returns. The outcomes are impressive. Depending on the particular commodity or technology or time period of the study, the returns to society for every additional dollar invested in plant science ranged from 20 to 50 percent annually. That is remarkable.

Of course, progress is not an unalloyed blessing. It often leads to systemic imbalances. And it often raises new and sometimes quite prickly issues. The agricultural sciences are no exception. Let me discuss, briefly, two of those prickly issues. One deals with regulation. And the other deals with the instabilities that begin to pop up as the farmer has to deal with increasingly sophisticated and very costly technologies. The regulatory issues by now are well known to you and have been discussed at this symposium. It is not an issue that can easily be passed by. However, we should note that the issue has shifted now to the application of new genetic engineering techniques and away from the development of the techniques themselves. To date, the federal government has not clearly resolved its policies toward commercializing recombinant DNA and related technologies. The costs of such indecision can be enormous.

However, the government is beginning to move toward shaping a coherent policy on the commercialization of biotechnology. The most recent – and very welcome – step was the issuance in the *Federal Register* of a "Proposal for a coordinated framework for regulation on biotechnology." This proposal, issued for public comment by OSTP, describes the policies of the major regulatory agencies and reviews biotechnology research and products, including USDA. It provides a regulatory matrix outlining the applicable laws, regulations, and guidelines. That proposal is an important first step, although there are a great many quite contentious issues which need to be settled. Among these is the role of the NIH Recombinant DNA Advisory Committee, or RAC. The RAC, as you know, has been spectacularly successful in shepherding research involving recombinant DNA and similar techniques. I believe that the OSTP proposal is a major, but first, step in removing the thorns surrounding the commercialization of biotechnology. While matters will continue to move slowly, while there will continue to be difficulties, and while your patience will be tried, I do feel confident that in time a logical and reasonably stable policy in regard to commercializing biotechnologies will emerge.

Let me now talk briefly about some of the more invisible costs of technological upheavals in agriculture. The nature of those instabilities were described by Bill Brown, of Pioneer Hi-bred International and Chairman of the Board on Agriculture of the National Research Council. Brown commented that, "as we offer the farmer increasingly sophisticated and costly technological packages, we inadvertently exacerbate two related sources of instability in agriculture."

"High yield production systems," he pointed out, "are often more volatile in terms of harvested production and more erratic in terms of profit for the farmer. What policies should we pursue to deal with these growing sources of instability? Who shall pay for the costs associated with them?" As Bill Brown rightly said, "I believe a great deal rides on the answers to these questions."

There is, finally, another pressure related to heightened attention to basic biology and agriculture. Unless this pressure is better understood, it can be seriously damaging. That pressure is one of excessive expectations. We have, of course, seen that already in the health sciences, resulting in the expectations of new cures for devastating diseases and high profits for investors. Certainly, it remains problematic as to where the major impacts of biotechnology on agriculture are likely to be. However, I think we can all agree that many of the applications – especially the more dramatic ones – could take a long time to arrive. Of course, people in agriculture are accustomed to long-term scales. It takes seven to eight years to create and realize the benefits of a new seed variety. However, neither policy makers nor Wall Street investors have the same kind of experience and rarely have that kind of patience. On the other hand, work on some genetically-modulated traits, such as herbicide and pest resistance, seems to be proceeding faster than anyone thought possible three to four years ago. It turned out that these are rather genetically simple characteristics. By contrast, teaching grain crops to fix nitrogen is a much more difficult matter, made so by the genetic complexities and by the need to assure energy sources for the process. It seems likely that, at the moment, the near-term applications of biotechnology to agriculture will be limited. Thus, in animal agriculture, applications are likely to focus on high value animals, such as breeding stock. In plant agriculture, the focus may be on developing new varieties of genetically engineered seeds. (However, I would not be surprised by surprises.)

There is an imperative for your immediate public to understand this. There is a better need for public perception of the differences between expectations and realization. The farmer, quite rightly, looks to total yield, and not what he sees as more nebulous goals, such as improved nutritional qualities of a bean. That bottom-line mentality is understandable; indeed, it is essential to the farmers' economic survival. It is made even more so by the economic disasters that are now visiting on many of the smaller farms.

We have, of course, had some spectacular examples of what can happen when molecular biology is blended with classical techniques. One example is a new vaccine for foot-and-mouth disease. Earlier vaccines made from killed virus risked inadvertently spreading this disease. The new and safer vaccine emerged from a blending of two streams of knowledge. One stream was long-term support for basic research on foot-and-mouth disease by the U.S. Department of Agriculture. The second was the basic research supported by the NIH that led to recombinant DNA technology. That technology was in turn used to produce a new vaccine. As a result, quoting *Science* magazine, "prospects seem good that within this decade foot-and-mouth disease will come under more effective control worldwide."

Even within the context of hard realities, the fact remains that there are vast opportunities in the offing for agriculture, opportunities that we must explore if we are going to meet future needs, both in the nation and internationally. The issue is how to do it effectively. A bit more exactly, the issue is how to strengthen an already robust system, without damaging it. Thus, even the famous – or perhaps to some of you, the infamous – Winrock report commented that, quote, "the continued success of the agricultural research enterprise will depend on its ability to retain the strongest elements of a system in existence for over 120 years."

I think that can be done; indeed, it is being done, thanks to imaginative and effective leadership. However, while a beginning has been made, much of the journey remains to be taken. Let me, then, in finishing up, suggest some points to emphasize, in the difficult but necessary task of wedding frontiers of biological science to improved opportunities for agriculture. Some of these suggestions are taken from the recent publication by the NRC Board on Agriculture entitled *New Directions for Biosciences Research in Agriculture.* I've also taken some suggestions from more recent discussions by the Board of issues in agricultural research.

First, the agricultural research system must move to further enhance the climate for fundamental research. A mechanism for periodic outside review should be established. Such a mechanism, constituting a periodic visiting committee to the Agricultural Research Service, can be a valuable conduit for two-way communications between intramural research programs and the wider scientific community. You already have a model for this kind of group in the advisory committee to the ARS Plum Island Animal Disease Center in New York. And, as you know, scientific advisory boards exist for the National Bureau of Standards, the Environmental Protection Agency, the National Institutes of Health, NASA, the National Science Foundation, and even the Department of Defense.

A second approach to enhancing the climate for basic research is to increase the ecology of understanding between ARS leadership and its research laboratories. One possible approach is to periodically rotate laboratory chiefs to the national program staff. Related to that is the absolute need to ensure some sort of stability for programs. And not only ensure it, but make sure that the scientific staff feels it can trust those assurances.

To continue, ARS must work even harder to enhance cross disciplinary research. As you well know, cross disciplinary research is becoming ever more important in almost all areas of science. Each of those areas struggle in their own fashion to create ways to do cross-disciplinary research, both effectively and without compromising disciplinary standards. The ARS certainly does not lack for opportunities for cross-disciplinary research. For example, there is, at this point, no multidisciplinary program in insect neurobiology. Perhaps a program can be established that brings together the relevant disciplines for this field – chemistry, endocrinology, neurobiology, biochemistry, developmental biology, and others.

In forming such multi-disciplinary programs, ARS ought to at least look for a possible model to the NSF Engineering Research Centers, the awards for which were just recently announced. This program, which is expected to grow to about $100 million in about 5 years, is intended to facilitate multidisciplinary research on campuses, engage students in such research, and, most importantly, involve industry in a working partnership at the research level.

Is such a model apt for agricultural research? I don't know. But it's certainly worth investigating. Further, such models may be one way of dealing with a recurring criticism of the large numbers and often seemingly duplicative work of the 147 ARS research centers. While geographical distribution is important, especially in dealing with regional problems, the case has to be made that critical mass is not lost and economic and scientific resources are applied effectively. I should add, however, that the intramural Centers do include among their advantages a greater ease in fostering the sort of team research discussed above. You're already seeing that with, for example, the future plant gene expression center in Albany, California.

The ARS has been energetic in creating a post-doctoral program. However, it remains a fairly small program, and needs to be expanded. As the report of the Board on Agriculture commented, "this type of program is virtually the best single mechanism for bringing new techniques, new capabilities, and new ideas into the ARS."

Let me address money matters directly. As you well know, modern-day science, including modern-day biology, is intensive in both special materials, such as restriction enzymes and tissue culture supplies, and also in equipment, such as protein synthesizers. Such capabilities must be available to researchers if they are to put their ideas to the test. The point is that research budgeting must allow not only for adequate salaries but also for the purchase and acquisition of these necessary tools. Further, given the extreme rapidity of research advances, the budgets must be highly flexible, to allow quick changes in research directions.

To continue: something has to be done about the inadequate level of research grants which the USDA provides through its competitive grants program. Thus, a three to five year grant from NIH provides about $105,000 per year in direct costs. The grants from the NSF and the Department of Energy average $70,000 per year for a three year grant, including both direct and indirect costs. The average grant awarded by the USDA is only for two years and is funded at $35,000 per year. That $35,000 per year includes both direct and indirect costs. And indirect costs can often represent about 40% of the total grant.

There is simply no way that a researcher can mount an effective research program with about $20,000 a year in direct funds. That level of grants does not allow for effective disciplinary research and certainly precludes any meaningful effort in interdisciplinary work.

The USDA has lagged in its emphasis on training of young scientists. By contrast, the training grants in the biomedical sciences have been wonderfully effective in catalyzing and sustaining research excellence in selected areas of science. In particular, the USDA could use the training grant mechanism to provide integrated training in basic biology, including molecular biology and agricultural science. Such training grants should be complemented by career development awards, offering an outstanding individual a research and teaching position for 5 years at an attractive level of support. These awards could be funded through the USDA or, paralleling the Presidential Young Investigators Awards, could be supported by matching funds from industry or private foundations. Whether career development or training awards, they could provide the basis for training outstanding people to do research that combines basic and applied agricultural biology with molecular techniques.

Let me make a final suggestion. This is based upon work done by my institution. Periodically, we have prepared surveys of disciplinary areas of science. Physics. Chemistry. Astronomy. Mathematics. These are enormous but fruitful efforts. They lay out priorities for a field. They suggest funding levels. They suggest weaknesses in resources. They suggest unusual opportunities and likely future directions. They analyze personnel needs. They often become the skeletal structure on which research agencies hang their annual budgets. While they are effective, their obvious weakness is that they are periodic. What is really needed is a continuing surveillance mechanism – to spot problems and opportunities quickly and to move to deal with them quickly. Therefore, my institution has established a basic sciences fund to provide just such continuing oversight mechanisms of the institutional and resource needs of a field. This fund will provide that for chemistry, physics, and other of the basic physical and biological sciences. We are now seeking to create a comparable mechanism for the agricultural sciences. We believe that such a mechanism, once in place, will become an essential pillar to the efforts already under way to assure that the wonderful agricultural enterprise which the United States has created continues to remain so.

Thank you.

Subject Index